Composites Science and Technology

Series Editor

Mohammad Jawaid, Laboratory of Biocomposite Technology, Universiti Putra Malaysia, INTROP, Serdang, Malaysia

This book series publishes cutting edge research monographs comprehensively covering topics in the field of composite science and technology. The books in this series are edited or authored by top researchers and professional across the globe. The series aims at publishing state-of-the-art research and development in areas including, but not limited to:

- Conventional Composites from natural and synthetic fibers
- Advanced Composites from natural and synthetic fibers
- Chemistry and biology of Composites and Biocomposites
- Fatigue damage modelling of Composites and Biocomposites
- Failure Analysis of Composites and Biocomposites
- Structural Health Monitoring of Composites and Biocomposites
- Durability of Composites and Biocomposites
- Biodegradability of Composites and Biocomposites
- Thermal properties of Composites and Biocomposites
- Flammability of Composites and Biocomposites
- Tribology of Composites and Biocomposites
- Applications of Composites and Biocomposites

Review Process

The proposal for each volume is reviewed by the main editor and/or the advisory board. The chapters in each volume are individually reviewed single blind by expert reviewers (at least two reviews per chapter) and the main editor.

Ethics Statement for this series can be found in the Springer standard guidelines here - https://www.springer.com/us/authors-editors/journal-author/journal-author-helpdesk/before-you-start/before-you-start/1330#c14214

Satish Babu Boppana · C. G. Ramachandra ·
K. Palani Kumar · S. Ramesh
Editors

Structural Composite Materials

Fabrication, Properties, Applications and Challenges

Springer

Editors
Satish Babu Boppana
Department of Mechanical Engineering
School of Engineering
Presidency University
Bengaluru, India

C. G. Ramachandra
Department of Mechanical Engineering
School of Engineering
Presidency University
Bengaluru, India

K. Palani Kumar
Department of Mechanical Engineering
Sri Sairam Institute of Technology
Chennai, India

S. Ramesh
Department of Mechanical Engineering
Jerusalem College of Engineering
Chennai, India

ISSN 2662-1819 ISSN 2662-1827 (electronic)
Composites Science and Technology
ISBN 978-981-99-5981-5 ISBN 978-981-99-5982-2 (eBook)
https://doi.org/10.1007/978-981-99-5982-2

This Springer imprint is published by the registered company Springer Nature Singapore Pte Ltd.
The registered company address is: 152 Beach Road, #21-01/04 Gateway East, Singapore 189721, Singapore

Contents

Introduction to the World of Composites

Composites Overview

Mithun Vinayaka Kulkarni and Satish Babu Boppana

Abstract From providing shelter for early civilisations to paving the way for future developments, composites have been particularly important to the humanity throughout the history. Most notably, composites are advantageous due to their resistance to corrosion, design versatility, longevity, lightweight, and strength. Products made from composites are not limited to a particular field but finds their application spanning across the construction sector, sports, medicine, space applications and many more. It seems unlikely that projects like rocket ships would have been implemented without the use of composite materials. New materials, applications, and methods make this sector fascinating to work in. Using hybrid virgin and recycled fibres speeds up and automates production. Composite materials are rising at 5% annually worldwide. Carbon fibre demand is expanding at around 12% per year. Nowadays, many composites are manufactured for purposes other than just improving the materials' mechanical qualities, such as strength. Composites are also engineered to be excellent heat conductors or insulators, and/or to have magnetic characteristics; these features are highly precise and specialised, but they are also incredibly significant and valuable. Composites find their use even in electrical and electronic devices, such as transistors, photovoltaic cells, various sensors and detectors, semiconductor diodes, and lasers, as well as in the production of anti-corrosive and anti-static surface coatings. Other applications of these composites include lasers and sensors. This chapter is a collection of topics such as importance and history of composites, types and manufacturing process materials used for composite construction, applications and case studies.

Keywords Composites · Carbon fibres · Corrosion resistance · Coatings · Case studies composites · Carbon fibres · Corrosion resistance · Coatings · Case studies

M. V. Kulkarni (✉)
Engineering Department, Mechanical Engineering Section, University of Technology and Applied Sciences—Salalah, Salalah, Oman
e-mail: Mithun.Kulkarni@utas.edu.om

S. B. Boppana
Department of Mechanical Engineering, School of Engineering, Presidency University, Bangalore, India

S. B. Boppana et al. (eds.), *Structural Composite Materials*, Composites Science and Technology, https://doi.org/10.1007/978-981-99-5982-2_1

1 Introduction

Composite materials are solid materials made by mixing two or more unique substances, each having its own properties, to create a new substance with superior properties for a specific use. A structural material (like plastic) with an imbedded fibre element is called a "composite" (like silicon carbide). By embedding fibres of one material in a host matrix of another, composites get their amazing properties [1, 2]. Most traditional composites are constructed in thin layers, which are reinforced by long fibres put down in a single direction, similar to how plywood is constructed. These materials only have stronger properties in the fibre direction. The fibres are woven into a three-dimensional structure where they are laid along three mutually perpendicular axes to create composites that are durable in all directions. Evidence related to the use of composites by the early humans has been recovered at various sites around the globe. It is understood that they used wattle and daub, to build walls of their houses. The technique involved building a woven mesh made of strips of hardwood called wattle (treated as reinforcing material), coated with a mixture prepared from animal manure, moist sand, clay, etc., which is sticky and could be easily bonded with the reinforcements [3]. This building material has been supplanted in more recent decades by concrete, a composite made of cement and reinforcements like gravel (aggregates) [2, 4]. Composite bows, which were at the time extraordinarily effective, were first developed by the Mongols in 1200 AD. The bows were mainly made from materials such as wood and animal remains. The Egyptians began crafting plaster-soaked papyrus or linen into death masks [5]. Canoe makers in the late 1800s began experimenting with paper laminates made from various materials. Shellac was used in an attempt to adhere sheets manufactured from wood pulp. The years 1870–1890, is being considered as an era of research towards the development of synthetic or man-made resins. The molecules of these polymer resins are crosslinked to change their state from liquid to solid. The 1930s were pivotal years in the development of resins for the composites industry. Carleton Ellis patented unsaturated polyester in 1936 (these resins had the ability to harden rapidly that replaced previous resin options in the composites industry). By the end of the 1930s, some performant resins like epoxy resins, were commercially available.

Bakelite, one of the earliest synthetic resins, was developed in 1907 by Belgian-born American scientist Leo Hendrik Baekeland. Baekeland discovered that by mixing the resin with cellulose, it became less brittle and more durable. In 1917, Bakelite was first put to commercial use when it was used to manufacture gearshift knobs for Rolls Royce vehicles. During the 1920s and 1930s, new and improved resins were developed. Polymer resins were further developed by American Cyanamid and DuPont, two American chemical corporations, in the early 1930s. Both firms originally developed their own unique formulae for polyester resin throughout the course of their testing. The Owens-Illinois Glass Company pioneered a method in the late 1930s for pulling glass into thin strands or fibres, which were used in clothing's and apparels, and paved the path for developing light weight and durable glass fibre reinforced polyester resin materials. The first fibreglass and polyester resin dinghy (small boats that could be used in rescue mission or for racings and recreational purposes) was created in 1942 by Ray Greene.

With World War II, development and use of composites took a new level. The war forced composite innovators to create new combinations of materials. These materials were introduced in the areas of aerospace, construction, and transportation. Also, at this time fibre reinforced composites development occurred. The two most popular forms of fibre reinforced plastics (FRP) are glass reinforced plastic (GRP) and carbon fibre reinforced composite (CFRC). As indicated by the names, carbon and glass were the fibre materials (or inclusions) used in the fabrication of CFRC and GRP materials, respectively. More than seven million pounds of fibreglass were employed by 1945, almost all in the defence industry. The public quickly learnt of the advantages of FRP composites, particularly their resistance against corrosion. For instance, the oil sector has been one of fibreglass pipe's most widespread applications since its introduction in 1948. Filament winding method provided the foundation for large-scale rocket engines, which were critical to the advancement of space exploration during the 1960s and beyond. During these years, other fabrication techniques such as pultrusion and vacuum bag moulding were also invented and used for large-scale production of composites.

Carbon fibre composites weren't commercially accessible until many years after the initial carbon fibre patent in 1961. The utilisation of carbon fibre contributed to the development of a wide variety of applications across various industries. A chemist working for DuPont named Stephanie Kwolek came up with the idea for Kevlar in 1966. Kevlar is simply a tough and durable fibre that finds its use in the development of superior quality composites as in the development of body armour (resistant to both bullets and blades). Thus, new and better resins increase composite demand. This was especially true for composites that were intended for use in environments with higher temperature ranges or that were subjected to corrosion. The automotive industry overtook the marine market in the 1970s and is the largest composites market today. In the late 1970s and early 1980s, European and Asian infrastructure projects utilised composites in building their first highway bridge and bridge deck with composite reinforcing tendons which was first of its kind in the history of composites used in infrastructure projects. Other early applications of these composites in infrastructure were in the United States. In the 1990s, Scotland erected its first pedestrian bridge completely made out of composites in Aberfeldy. FRP bridge deck was constructed in West Virginia's McKinleyville. A large number of pedestrian bridges made of FRP composites have been placed in secluded areas of state and national parks in the United States that are inaccessible to heavy construction equipment. These bridges have also been used to span highways and railroads.

Composites are being specified by an increasing number of industrial designers and engineers for use in various sectors. Structures made of reinforced concrete or masonry can be strengthened or seismically upgraded using FRP composites systems. It wasn't until the early 2000s that nanotechnology first made its way into consumer goods and thus certain applications started using carbon nanotubes as reinforcements in PMCs, which enhanced the properties of materials. The year 2010 brought a dramatic change in the area of manufacturing by introducing additive manufacturing

(AM). Businesses and customers could interact and corelate their ideas and concepts with each other using the CAD-based prototypes, which were then printed using AM technology. Several firms that specialise in composites have recently entered the area of 3-D printing using reinforced fibres. Reinforcing plastics in three-dimensional printing techniques often involves the use of discontinuous strands of carbon fibre or fibreglass. This practise is used in all market sectors, including the automotive, aerospace, tooling, medical, and infrastructure industries. The reinforcements in the shortest possible of time enhance the composites strength using minimal material. Additionally, they may be planned and prototyped from a single desktop. In 2014, MarkForged made history by unveiling the world's first 3-D printer made of carbon fibre.

The composites business is constantly growing and developing. The maritime, automotive, and aerospace industries have all been significantly impacted by the increased usage of FRP composites. Also, a large number of unique applications in infrastructure and chemical processing have witnessed remarkable transformations. The Institute for Advanced Composites Manufacturing Innovation was announced in 2015 by the United States Department of Energy as a public–private collaboration with a budget of 259 million dollars. The Institute will concentrate on lowering the cost of producing advanced composites and reducing the amount of energy required for their production, as well as making them recyclable. The resins and fibres developed would be able to assist in the development of an even wider range of applications for composites. As the need for stronger composites, lighter, and more environmentally friendly continues to rise, bio-based composites will undoubtedly be used as alternative materials in the years to come.

2 Composite Manufacturing Processes

The production of composites may be accomplished in one of the three separate ways: (1) open moulding, (2) closed moulding, and (3) cast polymer moulding. Within each of these types, there are several processing techniques, and each one has its own set of advantages.

- Open moulding involves exposing the raw ingredients (which might include resins and fibre reinforcements) to air while they cure or become more solid. Open moulding uses various methods, such as manual lay-up, spray-up, casting, and filament winding.
- Closed moulding refers to the process in which the raw ingredients (fibres and resin) are allowed to cure inside a mould with two sides or in vacuum. Because closed-moulding techniques are often automated and require specialised equipment. They are most commonly used in large factories that manufacture enormous quantities of material (up to 500,000 components annually). Closed moulding uses various methods, such as vacuum bag moulding, centrifugal casting, continuous lamination, and resin transfer moulding.

- Cast polymers are unlike any other kind of material used in the composites industry, since in most cases, they exclude reinforcements in the form of fibres and are engineered to cater for the applications from the strength perspective only. The use of cast polymer moulding allows for the production of pieces of any size or form. Gel coated cultured stone moulding and solid surface moulding are the two types of cast polymer moulding.

3 Materials

Composites consists of a resin matrix that binds the fibre reinforcements. Additionally, they may consist of core materials, fillers, additives, and surface finishes, all of which contribute to the provision of one-of-a-kind performance characteristics.

3.1 Resins

Resins are used in composites to (1) act as a stress transfer agent between the reinforcing fibres, (2) act as bonding agent to hold the fibres intact, and (3) to act like a shield that protects the fibres from mechanical and environmental fibres.

Reinforced polymer composites typically employ either thermoplastic or thermoset resins in their construction. Figure 1 depicts the use of various resins in the composite construction.

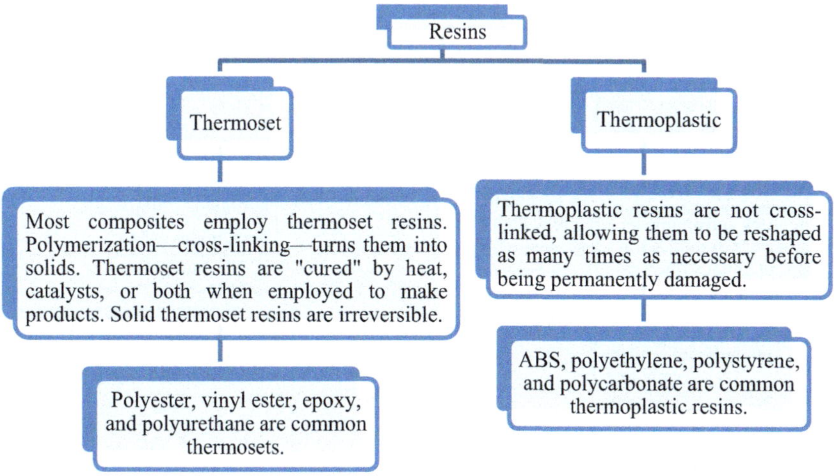

Fig. 1 Types of resins

3.2 Reinforcements

Reinforcements may be orientated to match end-product loads and user design criteria. Most of the commercial reinforcements are manmade. By varying the quantity of reinforcement, product performance, weight reduction, and fabrication related expenses of the composite can be optimised. Due to the low cost and high strength-to-weight ratio, glass fibres make up over 90% of reinforced plastics. Composites as we know are categorised into (1) Metal matrix composites (MMCs), (2) Polymer matrix composites (PMCs), and (3) Ceramic matrix composites (CMCs). Figure 2 depicts the various reinforcements used in MMCs, PMCs and CMCs.

- Composite materials that combine metal with another material are known as "metal matrix composites" (MMCs). Examples of MMCs include Aluminium MMC, Magnesium MMC, Titanium MMC and other MMCs [6, 7].
- Composites made of a polymer (resin) matrix and a fibrous reinforcing phase that is spread throughout the matrix are known as "polymer matrix composites" (PMCs). Examples of PMCs include GRP and CFRP, Kevlar (aramid) fibre reinforced polymers [8].
- When both the reinforcement (refractory fibres) and the matrix material are ceramics, we obtain a composite material known as a ceramic matrix composite (CMC). Sometimes the same ceramic is used for both portions, and sometimes

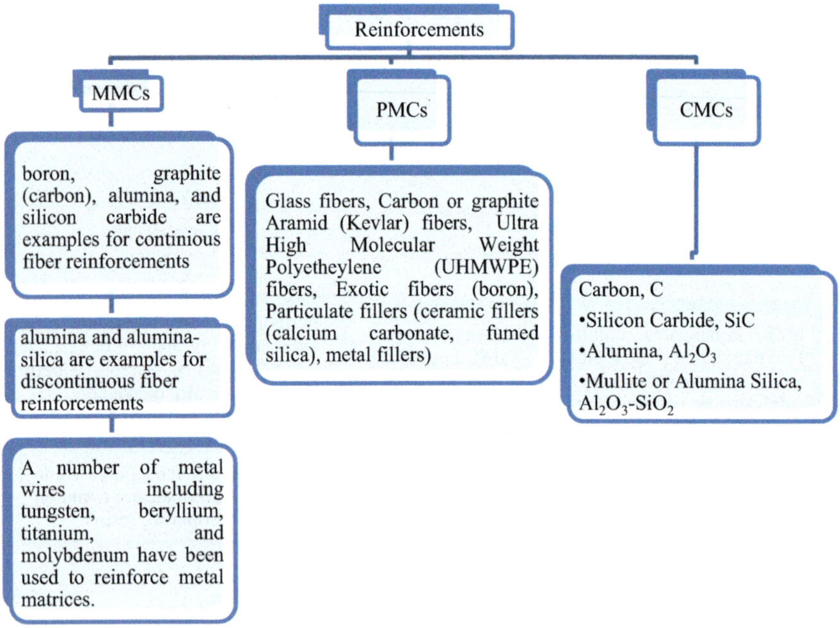

Fig. 2 Reinforcement types [7, 10]

supplementary fibres are added for reinforcement. For this reason, CMCs are classified as a subset of both composites and ceramics [9].

3.2.1 Reinforcement Forms

Reinforcements come in different shapes and sizes to accommodate several manufacturing methods and the needs of the finished products they are used in. Reinforcing materials may be provided in various forms, such as roving, milling fibre, chopped and thermoformable mats. Reinforcement materials may be manufactured using specialised techniques and shaped into the required form, depending on the requirements of the final product and the manufacturing technique. The different reinforcement forms include multi and single-end rovings, 3D fabrics, stitched, mats, woven, unidirectional, and prepeg, etc. [11].

3.3 *Additives and Fillers*

To increase the utility of polymers, improve their processability, or lengthen the product's durability, a broad range of additives and modifier chemicals may be used. In comparison to resins, reinforcements, and fillers, additives are often used in much smaller quantities by weight; yet the tasks that they perform are quite important. Although the cost of the basic material system is often raised by the addition of additives and modifiers, the overall cost and performance of the system is always improved by these components.

3.3.1 Additives

Additives are used to change and improve the characteristics of resins that form a part of the polymer matrix. Some of the important additives include thixotropes, release agents, conductive additives, colour and pigments, fire retardants and suppressants, UV Inhibitors, etc.

3.3.2 Initiators, Promoters and Inhibitors

Catalysts or initiators are the most crucial components in the polyester production. To get a heat-cured finish, benzoyl peroxide is often used with the resin, whereas methylethylketone peroxide (MEKP) is used for room temperature curing methods. Peroxides change into a reactive state (exhibiting free radicals) when heated and or mixed with a promoter like cobalt naphthenate, that causes the resin to cross-link and become solid. To slow down the process, additives such as tertiary butyl catechol (abbreviated as TBC) are often used. Accelerators such as DMA (dimethyl aniline) and others can speed up the curing process.

3.3.3 Fillers

Fillers not only reduce the composite's cost, but they also impart performance advantages to the composites that the reinforcement and resin elements by themselves would not be able to accomplish. Extenders and fillers are two common terms that are used interchangeably. The key elements can be broken down into three categories: fillers, resins, and reinforcements. Fillers can improve mechanical properties such as fire and smoke resistance in composite laminates by decreasing the organic content of the laminates. Because filled resins contract less than empty resins do, the dimensional control of items moulded using filled resins may be enhanced using these resins. By using fillers in the right way, one may increase several key features of a material, including stiffness, resistance to temperature, weathering and water absorption. A growing number of composites are also using inorganic fillers. These fillers are added anywhere between 40 and 65% of the total weight of the composite. Some of the commonly used inorganic fillers include: calcium carbonate, kaolin, alumina trihydrate and calcium sulphate.

3.4 Core

Core materials are used to make stiffer and lightweight products that improve properties such as fire resistance, thermal conductivity, and sound insulation of the sandwich composites. Sandwich composites represents a class of composites that consists of core materials embedded in between the layers of laminates and these laminates are usually known as upper and lower skin. The upper and lower skins are made up of several layers of resins and reinforcements (for example: epoxy and glass fibre (GF) sheets as shown in Fig. 3) and the thickness of the skin depends usually on the nature of application, required flexibility and stiffness without having to increase the weight of the composite.

Over the last 45 years, composites have relied on bonded sandwich structures. It would have been impossible to even imagine stiff, yet light in weight composites, which are durable even in harsh environments by the industries. Sandwich composites, over the period of time, have been used in marine applications (example: fishing boats and rescue operation boats), cars, panels used in the construction sector, blades of the wind turbine, etc. If cores and skin are carefully selected, an increase in weight by 3% may boost the stiffness by 3.5 and 7 times, respectively. Carbon and glass are the most frequent face sheets in composites. To improve mechanical qualities, certain core materials may be formed like waffles or corrugations. The core materials vary in quality and cost. Some of the commonly used core materials are polyurethane (PU),

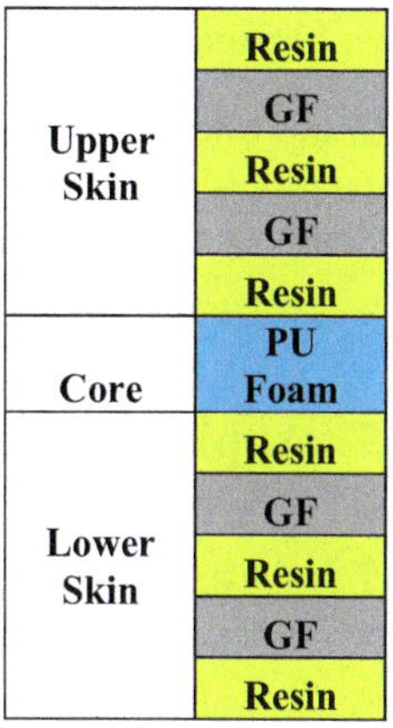

Fig. 3 An example of sandwich composite laminate

syntactic, thermoplastic and PVC foams, apart from these balsa and fibre-reinforced cores are also used.

4 Benefits of Composites

Composite materials comes with several advantages such as strength, low weight, resistance to corrosion, design flexibility, and long-term durability.

- **Strength**

Composites are among the most durable materials currently available. When the material's density is considered, composites supply superior strength to the vast majority of construction materials. It's no wonder that they're used in almost everything from aeroplanes to autos and in applications that are defined based on the strength perspective. There are four important properties that influence the structural design, and these are: specific, tensile, shear and compressive strength.

- **Lightweight**

Composites are much lighter in weight than most metals and woods. But why is it preferable to be lighter? Lighter vehicles and planes use less fuel due to the lower overall weight and since lighter things are simpler to carry and install, anything from utility poles to bridge decks may receive help from this.

- **Resistant**

Composites may be used in places where traditional materials would deteriorate quickly due to exposure to harsh weather and chemicals. Composites have been assessed in environments such as sea water, harmful chemicals, extreme temperatures, etc.

- **Flexibility**

Composites' design versatility stems from the vast variety of possible material combinations. Materials may be modified to meet the precise requirements of any given use. Composites also have the advantage of being simple to form into intricate designs.

5 Applications of Composite Materials

5.1 Aerospace

Top original equipment manufacturers (OEMs) in the aviation industry, like Airbus and Boeing, have shown that composite materials could be used on a large scale. The National Aeronautics and Space Administration (NASA) is always on the lookout for innovative methods and space-related solutions from composite producers for use in rockets and other spacecraft. Thermoset-based composites are being requested for use in an increasing number of bulkhead, fuselage, and wing applications across all three categories of aircraft. Aerofoil surfaces, antenna structures, compressor blades, engine bay doors, fan blades, flywheels, transmission structures in helicopters, jet engines, radar, rocket engines, solar reflectors, satellite structures, turbine blades, turbine shafts, rotor shafts in helicopters, wing box structures, etc., are just some of the many other areas where composites find use. Composites are used in diverse types of businesses nowadays [12, 13]. Mentioned below are some examples of military and civilian planes used by different countries that use a lot of composites in their airframes: [14].

- Fighter aircraft
  - U.S.: AV-8B, F16, F14, F18, YF23, F22, JSF, UCAV
  - Europe: Harrier GR7, Gripen JAS39, Mirage 2000, Rafael, Eurofighter, Lavi, EADS Mako
  - Russia: MIG 29, Su series
- Bomber: B2
- Transport
  - U.S.: KC135, C17, 777, 767, MD11
  - Europe: A320, A340, A380, Tu204, ATR42, Falcon 900, A300-600
- General Aviation: Piaggio, Starship, Premier 1, Cirrus SR 20 and SR 22
- Rotary Aircraft: V22, Eurocopter, Comanche, RAH66, BA609, EH101, Super Lynx 300, S92 [15].

Among the key benefits of employing composites in aeronautical applications include: (1) weight reduction of somewhere between 20 and 50 percent. (2) Structures made of a single shell may have increased strength while maintaining a reduced overall weight. (3) A remarkably high resistance to impact. Armor shields made of Kevlar have, for example, significantly cut down on the amount of damage caused by accidents to the engine pylons that carry fuel lines and engine controls. (4) A high resistance to thermal instability; (5) resistance to fatigue and corrosion. (6) Structural components that are constructed of composite materials are straightforward to put together [16].

5.2 *Space Application*

The usage of composites in space applications has been increasing steadily over the past few decades. Composites are employed in various spacecraft systems, including those that transport humans, satellites and payload supports, and launch vehicles. Because of their low weight and resistance towards the extreme temperatures in the space, composites are a need for spacecraft. They're also seeing the increasing use of rockets for various missions. Composites are frequently used to enhance the structural integrity of solid rocket motors and pressure tanks used to store fuel and gas. As early as the Apollo programme, composites were used for ablative and other high-temperature components, including rocket motor nozzles and re-entry heat shields.

Composite spaceships use high-modulus carbon fibre reinforced laminates. Human-rated crew capsules use composite panels for re-entry thermal shielding. Temperature capability and low thermal expansion may minimise the vehicle's bulk. Satellites and payload support structures use carbon fibre laminates. Bus frames often use aluminium honeycomb sandwich panels coated in carbon fibre or aluminium. Optical benches and other precision constructions are always made of carbon fibre laminates with low moisture absorption resins, usually cyanate ester. These materials help maintain dimensions in space's vacuum and extreme temperatures. RF reflectors and solar array substrates employ stiff, dimensionally stable high modulus carbon fibre laminates.

The fabrication of both the disposable payload fairings and the core booster's interstage uses composite materials. To manufacture the essential components with fewer joints and to maximise the benefits of such structures, larger composite structures require the development of high-quality Out-of-Autoclave composite systems. This was necessary because of the increased size of the composite structures.

In several space applications, filament wrapped composite structures are used. Solid rocket motors, which are frequently employed in spacecraft's upper stages, are made of filament wound—high strength carbon fibre. Pressurised tanks are normally made of very thin wall metallic liners that are then covered in carbon fibre and epoxy resin. These tanks are used for liquid hydrazine fuel and various gases.

The hottest parts of rocket nozzles, such as the throats and exit cones, use special feverish temperature composites. Re-entry vehicle heat shields are also made of similar composite materials. They can be divided into two main groups: ablative and ceramic matrix composites. Carbon or ceramic fibres are reinforced with carbon or a ceramic matrix in ceramic matrix composites. Of these materials, carbon–carbon is the most prevalent. The Space Shuttle's nose and wing leading edge were covered in carbon–carbon panels to protect it from the re-entry temperatures, which reached 2300 °F. Ablative composites are often silica or carbon-fibre reinforced phenolic materials that change their state to absorb heat [8].

5.3 Automotive

The automotive sector has been using composites for quite some time. The most significant advantage of using composite materials is the decrease in overall weight. A lighter lorry or car will take less gasoline to go ahead, making it more fuel-efficient. Composites not only enable innovative vehicle designs but also aid in making cars lighter and more fuel-efficient. Estimates suggest that the weight of the vehicle accounts for 75% of the total amount of gasoline used. Lightweight composite materials, commonly known as fibre reinforced plastics (FRP), provide possibilities for cutting vehicle weight. These materials improve fuel efficiency and minimise emissions of harmful pollutants, making them an attractive choice for lighter vehicles. Apart from these, composite materials are also employed in engines, pistons, cylinders, connecting rods, bodies, and other components. In comparison to steel, composites offer several advantages to the automotive and transportation industries, such as: (1) a weight reduction of 20–40%; (2) greater design freedom in the form of deep drawn panels; (3) a reduction in tooling costs of 40–60%; (4) a shortening of assembly times and costs through part consolidation; (5) resistance to corrosion, scratches, dents, reduced noise vibration harshness (NVH), and higher damping; and (6) the ability to accommodate new materials and processes [17].

5.4 Biomedical Composites

These composites have found application in healthcare. Their medical applications extend from the safe handling of biological samples to the accurate diagnosis and treatment of various illnesses and injuries. If they are manufactured correctly, composite biomaterials can serve as effective replacements for organs compromised by trauma or disease. The ability to produce composite biomaterials with precise predetermined physical, chemical, and mechanical qualities for novel applications is one of the key factors contributing to their current dominance [2]. Biomedical composites are used in several medical applications, including pacemakers, surgical and diagnostic tools, implants, surgical and diagnostic devices, electrodes for the collection or transmission of electrical or optical signals used in diagnosis or treatment, exercise equipment for people with disabilities using wheelchairs, pharmaceutical packaging, and instrumentation for the chemical and physical analysis of medical conditions. Particularly in orthopaedics, they are employed in bone cement, bone transplants, hip replacements, fixation plates, and bone resections. Dental implants can be used to replace missing teeth. They were first put into use in the middle of the 1960s and have since gained considerable attention as an alternative to removable dentures. Dental implants may be used to support crowns, that are used to replace single missing teeth, bridges, which are used to repair many missing teeth, and dentures for patients who have lost all their teeth (toothless). A dental implant is simply a titanium and zirconia screw or cylinder that ranges in length from 4 to 16 mm. It is placed into a bone socket made in the jaw and serves as a substitute root for a tooth lost [18, 19].

6 Case Studies

6.1 Boeing 787 Dreamliner

The Boeing 787 Dreamliner is a twin-engine, long-range aircraft with a wide-body cabin and a lightweight composite frame that accounts for 80% of its capacity [20]. Additionally, the aircraft is composed of ten percent steel, fifteen percent titanium, twenty percent aluminium, and five percent other elements. Because of the significant weight reductions that may be accomplished with the help of this structure, Boeing can realise significant cost savings as a result. The overall weight is reduced by an average of twenty percent despite the fact that composites make up the bulk of the construction [21]. Structures made of composite materials are capable of conforming to any form and thus taking advantage of this, Boeing constructed the fuselage of 787 using a number of separate barrel parts. Instead of producing numerous tiny aluminium panels that would need to be assembled. Analysing every part of the airframe to find the appropriate material for a certain purpose requires considering the operational environment and loads a component endures during its lifetime. Aluminium handles compression well but tension poorly. Composites manage tension better than compression. Thus, Boeing claimed that the increased usage of such materials, especially in the tension-loaded fuselage, significantly reduces fatigue maintenance compared to an aluminium structure. The Boeing 787 became the first commercial aeroplane to have most its airframe constructed out of carbon fibre reinforced polymer (CFRP). This material was used in the empennage, fuselage, wings, doors, and in most other essential components [22]. Each 787 has roughly about 35 tonnes of CFRP, which was manufactured using 23 tonnes of pure carbon fibre [23]. The fact that CFRP materials have a greater strength-to-weight ratio than typical aluminium structural materials adds greatly to the weight reductions of the 787. Furthermore, the fatigue resistance of CFRP materials is superior [24]. In the annals of aviation history, the first CFRP primary structure was installed in a Boeing commercial aircraft in 1984 on the horizontal tail of the Boeing 737 Classic, and in the middle of the 1990s on both the vertical and horizontal tail (empennage) of the Boeing 777 [25]. While researching the potential Sonic Cruiser in the early 2000s,

Boeing constructed and evaluated the first carbon fibre-reinforced plastic (CFRP) fuselage section for a commercial aircraft. This was a 20-foot (6.1 m) long anechoic chamber, which was eventually used for the Dreamliner [26].

Swift Maintenance

Boeing has created a new line of maintenance repair capabilities that can be applied to an aircraft in less than an hour. This is in contrast to the traditional bonded repair, which may require an aeroplane to be grounded for twenty-four hours or more. This swift method provides the opportunity for interim repairs and a quick turnaround, while an aluminium aircraft with such small damage may have been forced to make an emergency landing. That is an interesting take on the situation. The fuselage is built up of tubular sections, which are then connected to one another during the final assembly process. It is estimated that the use of composite materials may save 50,000 rivets per aircraft. As a possible point of failure, each rivet site would have needed to be checked for maintenance every flight [20, 27].

Composites use in the Engines

Both the General Electric (GEnx-1B) and Rolls Royce (Trent 1000) engines, available for the Dreamliner make substantial use of composite materials. There is a clear opportunity for composites in the nacelles, which include the intake and fan cowls. However, composites are employed in so many parts of GE engines that they are even used in the fan blades. Since the days of the Rolls-Royce RB211, there have been significant advancements made to the blade technology. In 1971, when the company's Hyfil carbon fibre fan blades failed in bird attack testing, it caused the company to go bankrupt. Since 1995, General Electric has been at the forefront of innovation in the field of titanium-tipped composite fan blade technology. Composites are used for the first five stages of the low-pressure turbine in the Dreamliner power plant. This turbine has a total of seven stages [20].

6.2 Biomaterial for Tissue Engineering

The use of biomaterials as scaffolds infused with live cells to repair, preserve, or improve deformed or missing anatomical structures dates back to the 1980s. In 1988, a National Science Foundation workshop formally coined the phrase "tissue engineering" [28]. Generally, tissues can be classified as either "hard" (such as bone and teeth) or "soft" (such as skin, blood vessels, cartilage, and ligaments). Hard tissues, as the name implies, are often more rigid and robust than their softer counterparts. Hard tissue applications typically use metals or ceramics due to their structural or mechanical compatibility with tissues, while soft tissue applications typically use polymers due to their biocompatibility [29]. Synthetic polymers, such as polycaprolactone (PCL), polyglycolic acid (PGA), polylactic acid (PLA), and their copolymer poly-lactic-co-glycolic acid (PLGA), as well as poly(ethers), such as polyethylene glycol (PEG), polyvinyl alcohol (PVA), and polyurethane (PU), have all been proposed to develop scaffolds for soft tissue regeneration [30–32]. These polymers can be easily manufactured with customised structural conformation and geometry, and their biodegradation profile can be controlled by changing their chemical composition. They also show physical, chemical, and mechanical properties comparable to those of biological tissues. However, they often lack cell adhesion sites, therefore chemical alterations are needed to improve cellular attachment. Furthermore, note that PEG is not readily produced: the ethoxylation process requires complex reactors and safety precautions due to the thermal instability and high reactivity of ethylene oxide [33].

The goal of tissue engineering is to repair injured tissue by growing new tissue from scratch using a pre-existing structure, or "scaffold," made of biomaterials [33]. The goal of this strategy is to use biomaterial scaffolds to repair the previously injured tissue. Natural extracellular matrix (ECM)-based biocompatible scaffolds are materials that promote tissue integration [34]. Unique to both cells and tissues, these scaffolds are a vital part of cell and tissue growth and maintenance. Human cells can be attracted to and organised efficiently by polymer composites with dimensions smaller than those of the cells [34, 35]. Cell migration, proliferation, and engraftment are all aided by the distinctive shape of the polymer composites. Organ and tissue regeneration relies on advances in polymer composites that promote cell growth and adhesion [36]. Thus, polymer composites, are attractive models for emulating natural morphology [35]. Thin films of porous biocompatible protein polymers were manufactured by Buchko et al. [33] and used in a variety of implanted devices as an example. Additionally, Huang et al. [31] synthesised and electro spun elastin-mimetic peptide polymers. They concluded that, with careful choice of protein polymer type, it is possible to fabricate FRP composites that helps in the creation of artificial organs with superior clinical performances.

6.3 Sandwich Composites in Boat Building

Sandwich structures are a class of composite materials that are made from two or more varied materials used to produce products with enhanced physical and chemical properties in terms of specific strength, specific modulus, corrosion resistance, stiffness, wear resistance, and fatigue life [35–42]. Because of these qualities, sandwich structures are used in many different fields, including athletics, aircraft, infrastructure, the maritime industry, the automobile industry, and commercial organisations. The literature indicates the use of low-density core material such as rigid polyurethane foam (RPUFs) as a matrix and as well as a reinforcing material. RPUFs are currently used in several lightweight industrial and engineering applications. Their excellent dimensional stability, low thermal conductivity, low apparent density, and lower moisture permeability make them an excellent choice as a core material in sandwich construction panels, and shipbuilding (used for making ribs for the hull) [40–42]. They are ideal in the construction of fish feeder hulls because of their increased load-bearing capacity and thermal insulation, decreased water absorption, and decreased fuel consumption, expense, and pollution. Conventional boats are constructed from two long, thin pieces of wood that run the length of the vessel [43]. To build the hull, they are nailed and cemented together. Ribs or perpendicular slabs of wood are used in the construction of an inner skeleton that keeps the two sides of the boat together and provides strength. With the advent of composite materials, boat builders began to experiment with diverse types of fibres (both synthetic and natural) in the construction of boat hulls. With regard to the use of glass fibre in boat hulls, exhaustive work has been conducted by Mohan and Araya [44]. A detailed study on the design and fabrication of FRP boats was undertaken by them. The work also highlighted points related to the reduction in leakage and biofouling of hull surface due to marine growth, and critical design parameters in the construction of small-sized pleasure boats were noted. Tanaka et al. [45] used bamboo fibre-reinforced plastics (BFRP) in the construction of pleasure boats. They noted that moisture absorption and BFRP strength were the critical factors during the fabrication process. Further observations revealed that the delamination was caused due to moisture absorption between the BFRP and the gel coat. Also, the Non-woven bamboo fibre mats were less effective in providing the reinforcing effect compared to the short fibres. Kasda et al. [46] worked on the design of a fish feeder barge boat, focusing mainly on the development of a low-cost remote control barge boat. Barge boats are a kind of cargo-carrying vessels used in both canals and rivers for carrying goods or passengers [47]. Rubino et al. [48] reported that with the use of FRP a drastic weight reduction can be achieved which results in larger cargo capacity, enhanced buoyancy and stability, reduction in cost and fuel savings. The sandwich structures especially in marine-based applications use aramid, carbon and glass FRP skins along with polystyrene and polyvinyl chloride foams. The use of these fibres results in the development of boats that have outstanding mechanical properties and dimensions compared to conventional boat-building materials. Lee et al. [49] conducted research on the structural design and analysis of a composite boat hull built using Resin Infusion Simulation. Aramid fibre resin was used in the research work. The results of the study concluded that RTM (Resin Transfer Moulding) method was a more capable method compared to the other boat hull manufacturing process.

7 Conclusions

Composites excel because of their strength, rigidity, and low weight. Manufacturers may create qualities that are a perfect match for the needs of a given construction by adjusting the ratio of reinforcement to the matrix material. Aviation in the modern era, both in the military and in civilian life, is a great illustration of this. Without composites, the effectiveness would be drastically reduced. In fact, the aviation industry's need for lightweight but durable materials has been a driving factor in the creation of composites. The wings, tails, propellers, and rotor blades, as well as a good portion of the interior structure and fittings, are now often built from sophisticated composites. Composites are less prone to break under stress than metals like aluminium, which is important to know while thinking about aircraft. A minor break in metal may spread rapidly and cause catastrophic damage (especially in the case of aircraft). Fibres in a composite material serve to prevent the spread of any existing cracks and distribute the resulting stress more evenly throughout the material. Composites that are fabricated perfectly are corrosion- and heat-resistant also and thus, they are perfect for use in items that will be subjected to harsh conditions, such as ships, chemical processing machinery, and even spacecraft. Composites' additional value lies in its adaptability in many applications. Moulding composites into intricate forms is a significant benefit when making surfboards or boat hulls. The creation of composite materials derived from recycled sources, such as discarded food and construction materials, or even plastic beverage containers, is also a major area of focus at the moment. The expense is often the main drawback of composites. The raw ingredients for composites are pricey, despite the fact that composites frequently result in more efficient production methods. While composites will never completely replace more conventional materials like steel, they do serve a useful purpose in many situations. There's also the certainty that, as technology advances, new applications will emerge, yet the full potential of composites is to be realised.

References

1. Composite material (2018) Britannica, the editors of encyclopaedia [Online]. Available: https://www.britannica.com/technology/composite-material
2. Egbo MK (2021) A fundamental review on composite materials and some of their applications in biomedical engineering. J King Saud Univ Eng Sci 33(8):557–568. https://doi.org/10.1016/j.jksues.2020.07.007
3. Wattle and daub. Wikipedia, the free encyclopedia, [Online]. Available: https://en.wikipedia.org/wiki/Wattle_and_daub
4. Salit MS, Jawaid M, Bin Yusoff N, Hoque ME (2015) Manufacturing of natural fibre reinforced polymer composites. Manuf Nat Fibre Reinf Polym Compos 1–383. https://doi.org/10.1007/978-3-319-07944-8
5. What is a Composite Material? (A Definitive Guide), TWI Ltd. https://www.twi-global.com/technical-knowledge/faqs/what-is-a-composite-material. Accessed 15 Jul 2022
6. Gotro J (2016) Polymer composites part 3: common reinforcements used in composites. Polym Innov Blog. https://polymerinnovationblog.com/polymer-composites-part-3-common-reinforcements-used-composites/. Accessed 15 Dec 2022
7. Metal-Matrix Composites (2002) Machine design. https://www.machinedesign.com/materials/article/21812641/metalmatrix-composites. Accessed 15 Dec 2022
8. Composites 101 (2022) Composites lab. https://compositeslab.com/composites-101/. Accessed 19 Nov 2022
9. Ngo T-D (2020) Introduction to composite materials. In: Composite and nanocomposite materials—from knowledge to industrial applications. Intechopen, pp 1–27
10. Quilter A (2004) Composites in aerospace applications. Information Handling Services, Inc. (IHS). http://www.aviationpros.com/article/10386441/composites-in-aerospace-applications
11. Mangalgiri PD (1999) Composite materials for aerospace applications. Bull Mater Sci 22(3):657–664. https://doi.org/10.1088/1755-1315/166/1/012006
12. Thomas GP (2013) Composites used in the aerospace industry. Azo Mater 1–5
13. Composites Applications for Space, SAMPE @ 75—Composites applications for space article

14. Friedrich K, Almajid AA (2013) Manufacturing aspects of advanced polymer composites for automotive applications. Appl Compos Mater 20(2):107–128. https://doi.org/10.1007/s10443-012-9258-7
15. Dental implants explained (2022) European Federation of Periodontology. https://www.efp.org/for-patients/dental-implants/dental-implants-explained/. Accessed 27 Nov 2022
16. Boeing 787 Dreamliner (2022) WikiMili. https://wikimili.com/en/Boeing_787_Dreamliner. Accessed 17 Dec 2022
17. Marsh G (2014) Composites flying high (Part 1). https://web.archive.org/web/20150916181258/, http://www.materialstoday.com/composite-applications/features/composites-flying-high-part-1/. Accessed 17 Dec 2022
18. Singh S (2022) Why The Boeing 787 & airbus A350 are built with composite materials. https://simpleflying.com/787-a350-composite/. Accessed 17 Dec 2022
19. Fiber C (2010) Strategic Business Expansion of Toray Industries, Inc.
20. Wallace J (2013) How the 787 'dream' was born. Seattlepi. https://www.seattlepi.com/business/article/How-the-787-Dream-was-born-1242144.php. Accessed 17 Dec 2022
21. William MUK, Roeseler G, Sarh B (2017) Composite structures: the first 100 years [Online]. Available: http://www.iccm-central.org/Proceedings/ICCM16proceedings/contents/pdf/MonA/MoAM1-01sp_roeselerw228184p.pdf
22. Boeing Testing Sample Sonic Cruiser Fuselage (2002) Boeing. https://web.archive.org/web/20081205020633/, http://www.boeing.com/news/releases/2002/photorelease/q3/pr_020724h2.html. Accessed 18 Dec 2022
23. Johnson T (2019) Boeing's 787 dreamliner: how composites and carbon fiber are used. Thought Co. https://www.thoughtco.com/boeings-787-dreamliner-820385?print. Accessed 18 Dec 2022
24. Griffith LG, Naughton G (2002) Tissue engineering—current challenges and expanding opportunities. Science (80–) 295(5557). https://doi.org/10.1126/science.1069210
25. Ramakrishna S, Mayer J, Wintermantel E, Leong KW (2001) Biomedical applications of polymer-composite materials: a review. Compos Sci Technol 61(9):1189–1224. https://doi.org/10.1016/S0266-3538(00)00241-4
26. Reddy MSB, Ponnamma D, Choudhary R, Sadasivuni KK (2021) A comparative review of natural and synthetic biopolymer composite scaffolds. Polymers (Basel) 13(7). https://doi.org/10.3390/polym13071105
27. Place ES, George JH, Williams CK, Stevens MM (2009) Synthetic polymer scaffolds for tissue engineering. Chem Soc Rev 38(4):1139–1151. https://doi.org/10.1039/b811392k
28. Liu S, Qin S, He M, Zhou D, Qin Q, Wang H (2020) Current applications of poly(lactic acid) composites in tissue engineering and drug delivery. Compos Part B Eng 199:108238. https://doi.org/10.1016/J.COMPOSITESB.2020.108238
29. Ernesto Salzano MDS, Santacesariab E (2007) The evaluation of risks of ethoxylation reactors. Process Saf Prog 26(4):304–311. https://doi.org/10.1002/prs
30. Laurencin CT, Ambrosio AMA, Borden MD, Cooper JA (1999) Tissue engineering: orthopedic applications. Annu Rev Biomed Eng 1:19–46. https://doi.org/10.1146/annurev.bioeng.1.1.19
31. Huang ZM, Zhang YZ, Kotaki M, Ramakrishna S (2003) A review on polymer nanofibers by electrospinning and their applications in nanocomposites. Compos Sci Technol 63(15):2223–2253. https://doi.org/10.1016/S0266-3538(03)00178-7
32. Zagho MM, Hussein EA, Elzatahry AA (2018) Recent overviews in functional polymer composites for biomedical applications. Polymers (Basel) 10(7). https://doi.org/10.3390/polym10070739
33. Buchko CJ, Chen LC, Shen Y, Martin DC (1999) Processing and microstructural characterization of porous biocompatible protein polymer thin films. Polymer (Guildf) 40(26):7397–7407. https://doi.org/10.1016/S0032-3861(98)00866-0
34. Huang L, Andrew McMillan R, Apkarian RP, Pourdeyhimi B, Conticello VP, Chaikof EL (2000) Generation of synthetic elastin-mimetic small diameter fibers and fiber networks. Macromolecules 33(8):2989–2997. https://doi.org/10.1021/ma991858f

35. Nagavally RR (2016) Composite materials—history, types, fabrication techniques, advantages, and applications. Int J Mech Prod Eng 2:25–30
36. Baley C et al (2019) Specific features of flax fibres used to manufacture composite materials. Int J Mater Form 12(6):1023–1052. https://doi.org/10.1007/s12289-018-1455-y
37. Chuang W et al (2017) Dispersion of carbon fibers and conductivity of carbon fiber-reinforced cement-based composites. Ceram Int 43(17):15122–15132. https://doi.org/10.1016/j.ceramint.2017.08.041
38. Glassfiber reinforced plastics (GRP) (2022) Amiblu. https://www.amiblu.com/why-grp/. Accessed 05 Jun 2022
39. Shokrieh MM, Rezvani S, Mosalmani R (2017) Mechanical behavior of polyester polymer concrete under low strain rate loading conditions. Polym Test 63:596–604. https://doi.org/10.1016/j.polymertesting.2017.09.015
40. Członka S, Strakowska A, Strzelec K, Adamus-Włodarczyk A, Kairyte A, Vaitkus S (2019) Composites of rigid polyurethane foams reinforced with POSS. Polymers (Basel) 11(2):1–19. https://doi.org/10.3390/polym11020336
41. Kumar M, Kaur R (2017) Glass fiber reinforced rigid polyurethane foam: synthesis and characterization. E-Polymers 17(6):517–521. https://doi.org/10.1515/epoly-2017-0072
42. Reignier J, Alcouffe P, Méchin F, Fenouillot F (2019) The morphology of rigid polyurethane foam matrix and its evolution with time during foaming—new insight by cryogenic scanning electron microscopy. J Colloid Interface Sci 552:153–165. https://doi.org/10.1016/j.jcis.2019.05.032
43. Sunardi, Sukandar EY, Sulkhani, Rahman MA (2019) Repair technique for wooden fishing boats using fiberglass. IOP Conf Ser Earth Environ Sci 370(1). https://doi.org/10.1088/1755-1315/370/1/012081
44. Mohan T, Araya Z (2018) Design and fabrication of fiber reinforced boat in Eritrea 9
45. Tanaka R, Okubo K, Fujii T, Ono M, Sakurai A (2007) Development of a pleasure boat using bamboo fiber reinforced plastics. ICCM Int Conf Compos Mater 1–6
46. Kasda, Kosasih DP, Nugraha HD, Rachman M (2021) Low-cost remote control barge boat to feeder fish. J Mech Eng Res Dev 44(2):112–121
47. Shipping Terms (2020) Cogo Freight Pvt. Ltd. https://www.cogoport.com/shipping-terms/barge-169. Accessed 06 Jun 2022
48. Rubino F, Nisticò A, Tucci F, Carlone P (2020) Marine application of fiber reinforced composites: a review. J Mar Sci Eng 8(1). https://doi.org/10.3390/JMSE8010026
49. Lee H, Jung K, Park H (2021) Study on structural design and analysis of composite boat hull manufactured by resin infusion simulation. Materials 14(20). https://doi.org/10.3390/ma14205918

Structural Composites

Composites in Structural Applications

Satish Babu Boppana and N. Gopalakrishnan

Abstract Composite materials have been used extensively in today's life owing to its many advantages and applications. Mechanical properties related to composites have been improved by choosing right reinforcement and matrix combination. Volume and weight are two important parameters considered while choosing composites for specific applications. Alongside if they are to be used for structural applications, their mechanical properties also play a key role for consideration of composites. Composite materials are often used in automotive, construction and aerospace industries. Nowadays, polymers have been used extensively in preparing such composites. Some of the applications would include their use in aircraft components, nuclear reactors and civil structures. The present paper deals with such structural applications of composites.

Keywords Structural composites · Applications · Natural fibers · Carbon fibers

1 Introduction

A structural composite usually relates to the combination of both homogeneous and composite materials. The geometric design is also found to be the deciding factor for such composites besides the composition of materials. It is generally multi-layered that often relates to low density with applications involving high tensile and torsional strengths. Such composites are often known for structural integrity.

Laminar type of composites and sandwich panels are amongst the prevalent kinds of structural composites.

S. B. Boppana (✉)
Department of Mechanical Engineering, School of Engineering, Presidency University, Bangalore, India
e-mail: satishbabu3@yahoo.co.in

N. Gopalakrishnan
Department of Civil Engineering, School of Engineering, Presidency University, Bangalore, India
e-mail: gopalakrishnan@presidencyuniversity.in

S. B. Boppana et al. (eds.), *Structural Composite Materials*, Composites Science and Technology, https://doi.org/10.1007/978-981-99-5982-2_2

A laminar composite has 2 dimensional sheets which possess high strength as in case of wood. Stacking of layers is done and finally cemented to get orientation of the high-strength direction varied with every successive layer.

If the applications need improved stiffness and strength, relatively light weight beams or panels need to be designed. If two external sheets are separated and bonded by adhesive to a core that is thick, we often term them as sandwich panels. Al alloys, Ti and steel are some of the materials used for preparing the sheets. Certain foams, honeycombs and wood are often used as core that has minimum weight.

The applications of such panels include walls of buildings, floors and roofs and in aviation sector involving fuselage and wings.

Fiber-reinforced composites are being used as structural members over the past few decades since they have good stiffness and comparatively lesser densities. If the said composite is to be an ideal one, then it must be properly organised in terms of laminate, ply and structural component. Such composites would be able to be designed for specific applications with the desired mechanical properties. Hence, they are used widely in construction, generation of energy, sports and transportation.

The need is to develop FRC that have properties as that of metals while they lack fire retardancy and high electrical conductivity. Some of these have been overcome by adding reinforcements in nano scale or by supplementing with additional layers while incorporating desirable properties.

Polymer matrix based composites are considered to be of prime importance in the current scenario owing to their high specific tensile properties which can be tailored easily as per the needs. Hence conventional materials like metals are often replaced by comparatively lighter weight composites. Usually in most of the cases, the matrix would be a polymer based on hydro carbons while the reinforcing phase could be aramid, carbon and glass fibres. Presently, natural fibers have been used often to overcome disadvantages related to using synthetic fibers. The applications of such composites are in diverse fields like equipment related to sports and aerospace components.

2 Various Types of Composites Used in Structural Applications

2.1 Carbon/Glass Fibre Reinforced Composites for Structural Applications

Nowadays FRP based laminates are used in aircrafts in the form of housing of turbine and wings owing to its strength and resistance against corrosion when compared with materials made up of metals. "Dreamliner 787" was considered to be fuel-efficient airliner owing to the use of composite materials during its manufacturing [1]. Challenges were also found in preparing the energy efficient aircraft consisting of composite materials. One such example is in the case of carbon fiber based polymer laminates that has lesser conductivity features. Lightning strikes [2] also damage the structure due to the said feature. To overcome this, the composite synthesized using C fiber is bonded to Cu mesh to synthesize a conductive external-layer. The bond usually adheres metal with the surface of CFRP using a layer of resin [3–5].

In aerospace applications, the structural components are being made conductive through various modification phenomenon. Priority is given between structural performance and conductivity. Carbon fiber based composites have offered good mechanical properties like compressive and tensile strength but often tend to fail through interlaminar fracture. Synthesis of such materials depend on conditions like environmental factors, ratio of resin and reinforcement and finally method used for producing such composites.

The use of glass reinforced concrete in several construction projects including tall structures across the UK, Europe, Middle East and South East Asia has been documented by the GRCA International, a trade association which aims at promoting Glassfibre Reinforced Concrete [6]. The 62 storeyed high-rise 1000 Museum residential complex in Miami where glass fibre reinforced panels are used as the structural exoskeleton is an excellent example of Glass fibre composites being used as a structural material in buildings.

Traditional E-Glass Fibre or borosilicate glass fibre were found to degrade in highly alkaline environment. Considering the alkaline environment in concrete whose pH ranges under usual circumstances between 12.5 to 13, Alkali resistant Glass Fibres popularly known as AR Glass fibres are used in concrete. The Glass fibres are made alkali resistant by treating them with zirconium dioxide. If alkali resistant glass fibres are not available, then the alkalinity of the concrete has to be reduced by addition of pozzolanic materials like silica fume or GGBS to produce a durable glass fibre reinforced concrete [7].

Glass fibre reinforced concrete (GFRC) owing to their enhanced strength and durability coupled with light weight have found applications as pipes, permanent formwork and parapet panels.

Restoration of historical structures is another application where GFRC find wide application [8]. Polymer in the form of white latex acrylic emulsion is added to GFRC helps improve the workability as well as long term mechanical properties while eliminating the need for wet curing [9].

Concrete reinforced with Carbon fibres are gaining popularity not only due to the enhanced strength and durability of these composites but also due to the reducing cost of carbon fibres. Another important aspect of carbon fibres is its ability to bring in substantial improvement in strength and durability that it brings in concrete even when added at volume fractions as low as 0.2%. The chemical stability of carbon fibre in the alkaline concrete environment [10]. The major impediment to the development and use of this material is the challenge of ensuring adequate bond between the fibres and concrete. To improve the bond between the carbon fibres and concrete matrix, several methods of treating the surface of carbon fibres have been found to be effective like heating the fibres to the use of sodium hydroxide solution. Adding admixtures of the likes of silica fume and latex have been found to not only improve the bond but also the fibre dispersion in concrete with silica fume being more effective in fibre dispersion particularly when used in conjunction with a small dosage of methylcellulose. Carbon fibre reinforced concrete has also found application in smart structures owing to the sensing abilities of carbon fibres like strain sensing and damage sensing ability [11]. The benefits of adding Carbon fibres along with steel fibres, steel rebars as well as carbon reinforced polymers in concrete is well documented in literature and is found to improve the tensile strength as well as fracture toughness. Carbon nano fibre reinforced concrete is also gaining popularity in recent times with their enhanced strength and deformation properties both in static and dynamic loading conditions as compared to conventional carbon fibre reinforced concrete [12].

2.2 Fibre Reinforced Polymer Composite for Repair, Rehabilitation and Retrofit

In order to reinforce beam and column joints in construction industry, fiber reinforced composite was used. Usually, the reinforcement is done by wrapping the FRP on an entire surface or at a particular section. A jacket made of steel was initially used to wrap the column and beam joints. The wrapping methods commonly followed were by using carbon anchor and hand layup technique. Reinforced concrete structures wrapped with FRP would improve the structural behaviour [13]. Carbon fiber based composites are preferred as prestressing tendons owing to its tensile behaviour and resistance for creep whereas, glass based reinforced polymer is generally utilised like non-prestressed bar in concrete structures owing to cost factor [14–16].

The beam deflection decreases while improving load bearing capacity when FRP based laminates are used to reinforce concrete. The cracks that formed were less severe and widely spaced. If FRP was added in vertical layers, the load carrying capacity would increase while reducing deflection.

The length of the FRP directly impacts the performance of retrofitted beams. If the FRP stretches out, the ultimate load would also increase. FRP is responsible for load–deflection and absorbing energy for cracking stages of beam reinforced with concrete [13].

In FRP based composites, the volume fraction of reinforcement can be more than 25%. The matrix of the composite can take up compressive and tensile loads satisfactorily. The resins also provide adequate toughness compared to concrete. Bridge deck made up of concrete can be substituted with composite material made up of FRP. The weight reduction seems to be of prime importance in such cases.

Fiber reinforced composites with polymer matrix have been used in aggressive environments due to their excellent resistance to corrosion, high strength and low weights. The composites have been utilised as rods for reinforcing internally in some concrete structures. Highway bridge superstructures are also being constructed using fiber reinforced polymer matrix composites. While laying foundation, composite piles are being used. Space frames are being manufactured to act as primary structures bearing load. Some of the other applications of using the said composite is in making chimneys and pipelines.

Fiberglass polymer rebars which are lightweight in comparison to conventional steel rebars and are also corrosion free have found extensive application in repair and rehabilitation works where its ultra-high tensile strength is also an advantage. Life cycle analysis on fiberglass polymer rebars and conventional steel rebars used for a particular application reveals that the environmental impact and impact on human health are substantially lower for fiberglass polymer rebars. It is to be noted that steel rebar requirement is about 8 times the fiberglass polymer rebars for the same application [17].

Carbon reinforced polymer [CFRP] sheets externally bonded to concrete corbels for rehabilitation or retrofitting applications have been found to almost double the original strength when placed in either horizontal or diagonal configuration in both undamaged condition as well as upto 60% damage levels [18]. Similar results have been reported in finite element investigations [19] of Steel Plate shear wall with CFRP sheets with an increase in the energy absorption capacity and lateral load bearing capacity. The noteworthy outcome observed in the study was that an increase in the number of CFRP sheet layers prevented the diagonal buckling failure of the shear wall under lateral/ seismic loading and only separation of the sheets or fracture of the sheet was observed at the same load which was causing failure in the shear walls without CFRP.

Use of Composites for Structural strengthening has led to a revolution in the construction. Several rehabilitation and restoration projects have been successfully completed in short duration of time without major alteration to the existing structural system thanks to the near surface mount techniques achieved through CFRP wrapping. The fibre wrapping technique is achieved through fabrics made up of unidirectional stitched or woven fibres of glass and carbon impregnated with resins. The fibre wrapping techniques have the advantage of achieving the strengthening required without altering the carpet area or the head room available which was a major drawback in conventional repair methods like plate mounting and RCC jacketing. Epoxy and resin/polymer-based Injection grouts have been successfully used for repair applications in civil engineering structures. Post-tensioned CFRP plates are another example of successful use of CFRP for strengthening of several building projects [20]. FRP composites using the external wet lay-up technique have been successfully adopted in marine and offshore structures for repair, rehabilitation and retrofit of wooden piles, corroded and cracked RC as well as steel components in bridges. Addition of Steel bars and steel collar with shear studs in between the original column and the CFRP layer enhanced the structural integrity [21].

A trenchless method for rehabilitation of steel pipelines onshore and offshore has been successfully developed using liners made up of carbon fibre and epoxy combination. This method is being applied from the past decade or so for industrial, water and sewage pipelines. In addition to Carbon fibre reinforced polymers, aramid fibre reinforced polymers have also been used due to their excellent properties in creep as well as fatigue. The peeling away of the fibre reinforced polymer [FRP] plates from the steel surface has been a challenge in addition to the higher cost of the FRP Plates [22]. FRP prestressing strands and reinforcements have been used for construction of new bridges with examples of CFRP being used in bridges in Canada documented as early as 1993. There is limited knowledge and experience on the long-term behaviour of the bridges reinforced with CFRP which has made the rate of penetration of this technology in the market rather slow [23]. A case study of structural monitoring of CFRP reinforced Beddington Trail bridge in Canada 6 years after its completion revealed that the performance of the CFRP reinforcements were satisfactory and in line with its design [24].

ACI and AASHTO standards/guide are available for design and use of CFRP strands for prestressing. Creep rupture and relaxation stresses should be accounted for while prestressing the strands. Traditional steel anchors can be used for GFRP strands hence effectively using traditional prestressing concepts [25].

FRPs in general are believed to be durable as they are highly resistant to electrochemical corrosion but the durability of FRP deteriorates under harsh environmental conditions and severe mechanical effects [22].

2.3 Green Composites Being Used as Structural Composites

It is found that conventional structural materials are being replaced by light weight composites. The applications of the composite materials are often found to be critical and require to safeguard properties during their use. Hence, they are generally made by utilizing conventional materials that are generally non-degradable. Disposing the composites would always pose a problem. Resins and fibers cannot be separated for using it further. However, natural fiber based reinforced green composites could also be developed to have good mechanical properties. They might replace few of the conventional composites in some secondary structural applications. But liquid crystalline cellulose based green composites can be utilised in primary structural applications. Hybrid green resins can be utilised to increase the properties of green composites. The properties of such composites can also be tailored. Some of the plant based cellulosic fibers are reinforced with cement to prevent cracks. Short length natural fibers have been considered to improve strength and decrease the formation of cracks in cement [26].

Structures involved in critical applications necessitate to maintain properties. Majority of the composites make use of non-biodegradable composites, however, disposing them is a cause of concern. Moreover, petroleum which happens to be a major source while preparing the composites is not a sustainable raw material. Presently there is an increasing need to manufacture materials that are sustainable and could be reused and pose no hazards while disposing. Polymeric based materials need to be developed in such a way that they have minimum interference with green house effect. Manufacturing green polymers based on plants is the current trend being seen to lessen the impact on environment. Resins from soy-protein, linseed oil and natural rubber have been used in preparing composites.

The seats of race car models developed recently are incorporated with carbon fibers. This will enhance the protection to the driver in case of an accident. The fibers would tend to absorb the impact by a cushioning effect [27].

Some of the composites prepared from natural fibers could be used in automotive and construction field as they were comparable with composites reinforced with aramid and E-glass fibers.

As per statistics, North America remains the major consumers of natural fibre composites followed by Europe and the use of natural fibre composites it is expected to see a steady growth rate of about 9.59%. Natural fibre reinforced thermoplastics are the most popular natural or green composites. The strength and other mechanical properties of natural composites is improved by ensuring uniform dispersion of fibres as well as improving/enhancing the interface bonding. Natural nano fibre-based composites have also received significant attention as a new age material to the extent of the adoption of 3D printing in recent times particularly using the fused deposition modelling technique. The tensile properties have received more attention in research while scant attention has been paid to performance of natural composites under impact which in general is found to be low [28]. One such study which focussed on the impact strength of a hybrid composite of Kenaf and Indian almond leaf fibre layers sandwiched by layers of epoxy resin in between them. The hybrid composite of consisting of two layers of kenaf fibre with a layer of Indian almond leaf fibre in between found to have about 40% better impact strength than natural Kenaf fibres while the tensile strength as well as flexural strength was found to reduce marginally by about 6% and 10% respectively. The moisture absorption increased by 1.4 times for the hybrid composite as compared to natural kenaf fibres and also showed higher biodegradability as compared to natural fibres with an additional weight loss of 10%. These hybrid fibre composites owing to higher impact strength can be used in damping applications [29]. Another similar study on kenaf fibres, focussed on a composite of kenaf and jute fibre layers sandwiched between one another with layers of epoxy resin in between. The study shows that the composite with two layers Kenaf fibres between a layer of jute fibre performed better than the composite with two layers of jute in between a layer of kenaf with an improvement in tensile and flexural strength of about 6% and 32% respectively [30]. It is noteworthy that both the studies on Kenaf fibre-based hybrid composites discussed above used the hand lay-up technique with the epoxy resin being made up of a combination of epoxy and a hardener.

The use of natural fibre reinforcement in hybrid polyester composites has also found to give promising results showing their potential for applicability in many structural applications. The use of banana fibre, coconut shell fibre and coir fibre as well as other natural material like cellulose filler and nano clay for improving the mechanical properties of polyester composites have been evaluated in various researches. The issue of high-water absorption and low dimension stability which are considered as major drawbacks of using natural fibres is overcome by chemically treating the fibres with hydroxides of Calcium, Sodium, potassium and lithium which help in removing cellulose content as well as in making the fibres less hydrophilic. A combination of synthetic and natural fibres induced with inorganic fillers reinforcement to polymers have shown promising results in terms of improving the mechanical characteristics [31].

Natural Composites made up of a combination of bamboo and wood in the form of cross laminates with layers of bamboo and wood (timber) have been found to exhibit excellent shear strength as well as high bending stiffness making them suitable for application as load bearing members in multi storeyed structures. The combination of bamboo and wood as layered cross laminates is shown to have a 6–24% increased bending strength as compared to conventional timber cross laminates [32].

The application of a natural self-adhesive geotextile by using basal fibre in road pavements has been found to be effective in reducing the possibility of crack development as well as control the crack development. The finite element modelling predicts that the pavement life with the addition of these fibres increases by about 1.75 times [33]. Jute fibres addition in the subgrade of road pavement has also shown to improve the California bearing ratio by about 5–6% while also improving the stability due to its water retention. The biodegradability of jute fibres makes it a safe option for subgrade of the pavement. The maintenance cost of flexible pavement which in normal circumstance is very high and is a major factor that dissuades the adoption of bitumen based flexible pavement has been found to reduce with the usage of Jute fibre-based geotextile which are bituminized. The paving fabric made up of Grey Jute fibre (Jute fibre containing Jute waste) was impregnated with different bituminous mixes and the mix with polymer modified bitumen mix of penetration grade 40 was found to produce a pavement with 3 times higher fatigue life as compared to conventional bituminous pavement. Field trials under simulation of actual traffic on a stretch of pavement have reported no signs of visible distress including absence of any cracks or potholes even after a year [34].

Green composite mortars have been found to be effective in restoration works of heritage structures where the original lime/sand mortar has been replaced by a composite made up of a combination of lime, silica fume and biopolymers like whey protein from milk and egg albumen. The compressive strength of these modified mortars with silica fume and whey protein derived from milk unfit for human consumption is found to be 3.5 times the strength of conventional lime mortar even at milk protein replacement level of about 6%. The diametric tensile strength of these modified mortars is also found to be about 1.8 times the diametric tensile strength of conventional lime mortars where silica fume plays a major role in the strength improvement. These sustainable alternatives produced by addition of waste material can be been used in the restoration of cultural heritage structures with an improved mechanical strength primarily owing to the increased carbonation level which is attributed to the increase in void ratio/porosity [35].

Several Green concrete mixes have been developed in the recent past. Green concrete with nano waste materials have been developed including lightweight green concrete. A mixture of Ceramic waster powder and nano calcium carbonate has been used to produce green concrete with aeration (light-weight) being achieved through the addition of aluminium powder. The compressive and split tensile strength were found to improve with the addition of nano calcium carbonate and ceramic waste powder although the workability takes a hit with increasing percentage of nano calcium carbonate [36]. A novel green concrete produced with cement being replaced by silica fume and GGBS both of which are industrial waste products along with fine aggregate being replaced by a combination of Scrap Tyre rubber powder and Bambara nut shell (groundnut shell) ash is found to improve workability, strength and durability [37].

2.4 Hybrid Composites for Structural Applications

In advanced applications, the composites that are used are often fiber reinforced. During the recent years, hybrid composites are also being used in various engineering fields. The reason being, its cost effectiveness during its maintenance and operation [38]. Since hybrid composites are having good specific strength and greater resistance against fatigue loads, they are commonly used as frames for automotives and coatings related to fuselage in aircrafts. Researchers tried to use natural fibers along with glass fibers to manufacture brake levers. Glass fiber and carbon based composites have been checked for use in bumpers in automobiles [39–42].

When natural and synthetic fibers are combined and reinforced, the resulting hybrid composite tend to improve mechanical properties. This will also be helpful in reducing cost incurred in synthesizing composites. Such composites would always reduce the usage of synthetic fibers which are non-renewable. The impact on environment would also be lesser due to hybridisation. Rigidity would increase when cotton and glass fiber were used in preparing the composite. Flax and carbon fibers were reinforced to prepare composites and bicycles were manufactured with such composites. Flax fiber is known to have good damping properties while carbon fiber is always responsible for improving mechanical properties.

In the year 1941, hemp and flax fibers were placed in a matrix made of resin and the composite was used while manufacturing a Ford car. The impact strength in such vehicle was considered to be ten times more than that of steel. Another example of using hybrid composite consisting of flax/carbon fibers is construction of a small boat.

The addition of graphene oxide in extremely small quantities in the range of 0.05% by weight of cement has been found to improve the compressive strength as well as flexural strength of cement mortar in addition to making the mortar more durable by reducing the pores and water absorption. It is to be noted that the workability is substantially reduced even at such small dosage of graphene oxide [43].

2.5 SiC-Matrix Composites for Thermo-Structural Applications

If SiC/SiC and SiC/C composites are properly synthesized such that matrix and fiber are properly bonded, they will be regarded as tough ceramics. Some of the applications include nuclear reactors, engines and braking systems.

The hot structures of aircraft were thought of being synthesized by SiC based composites. The structures would be under thermal shocks and cyclic loading. Presently they are thought of manufacturing using carbon fibers. Some coating would also be used to reduce the rate of recession during oxidation. In case of meteorites or any such foreign materials striking the satellites, such kind of composites would be beneficial [44].

Due to its higher thermal stability, hardness and favourable corrosion resistance, SiC is used in many structural applications. But monolithic silicon carbide is not an effective material considering its brittle nature. Hence the same was reinforced with SiC fibers to form a composite that exhibit high fracture energy. They are used in turbines and spacecrafts. Sometimes BN coating is given on silicon carbide based fibers to improve interface between the matrix and the fiber. SiC_f/SiC composites are prepared by stacking sheets followed by hot pressing. Electrochemical process can also be used to fabricate the said composites.

Another unique application of SiC is its addition in concrete used of rigid pavements. Owing to its excellent thermal conductivity, SiC is added to concrete for rigid pavements in cold countries where de-icing of pavement is a major challenge. A combination of magnetite and SiC is found to have double the de-icing efficiency as compared to conventional concrete. SiC in particular is found to enhance the heat transfer capacity of concrete and this is found to have application particularly when the microwave heating technique is used for de-icing pavements [45].

Gamma ray shielding effect is an important parameter to be considered for concrete used in nuclear power plants. Nano particles of Cerium oxide and zirconium oxides are used in concrete to improve the mechanical properties of concrete at high temperatures and at the same time give an enhanced shielding effect against gamma rays. A nanocomposite which is synthesized by a combination of Cerium and Zirconium oxide when added at low concentrations of about 0.25% of OPC in concrete is found to reduce the porosity substantially, increase the bulk density and compressive strength thereby having a pronounced improvement in the shielding effect of concrete [46].

2.6 Composites in Structural Automotive Applications

Sheet molding compounds are used in automotive industry which consists of materials like chopped glass fibers. Some of the applications include doors and panels used for opening of grills.

Leaf springs made up of composites can be found to successfully take up fatigue. The composites could be reinforced with glass fiber. In order to manufacture seats and backup beams for bumpers, thermoplastic stamping is used. Polypropylene which is reinforced with glass fiber tend to offer good properties compared to polypropylene which is filled with wood.

Tail gates and hoods are produced by preparing sheet molding compound and then subjecting to compression molding. The other method of producing components in automobiles is by using high-speed resin transfer molding technique. A preform would be kept in a mold cavity and then a resin is sent in such a way that the reinforcement is intact [47].

Mercedes-Benz had used epoxy based matrix and jute for its panels of door. Audi had panels of door synthesized from polyurethane reinforced with sisal and flax. Toyota also used cover made up of green based composite for its spare tyre [48].

2.7 Energy Storage in Structural Composites

Researchers worked on the synthesis of composites which could store energy. Thin CNT fibers and polymer electrolyte interleaves were used amidst carbon fibers and then filled by epoxy. The structure would act as a composite and a capacitor. The flexural modulus and flexural strength were found to be 60 GPa and 153 MPa respectively [49].

Similarly, two carbon fiber based woven electrodes were separated using a layer of woven glass fiber, they were then added into an electrolyte made up of polymer. The resulting product was a multifunctional composite that would play a structural role as well as store electrical energy [50].

Researchers prepared structural dielectric capacitors which had considerable improvements in the electrical and mechanical properties [51]. The capacitive function was also optimum during loading. Nowadays with the advancements in manufacturing multifunctional composites, higher energy can be easily stored in capacitors and the same could be designed as structural elements. This would also help to save weight in automotives.

In a related research [52], laminated glass and epoxy prepegs were incorporated with metal-based electrode films. Tests were conducted to find improvements in mechanical properties along with high voltage performance tests. Some void content was found to increase when volume of fiber fraction increased, even then the performance of the capacitor still offered better results. The resulting capacitor would decrease weight when compared with existing ones.

3 Conclusion

The use of composites for structural applications depends on the optimum combination of matrix and reinforcement and the way in which it is synthesized with reference to processing conditions. The complexity of the behaviour of composite is a matter of concern for using them as structural materials especially in aerospace industry. The material needs to be reliable, and safety of the aerospace structure is always a priority. The use of composite materials in structural applications leads to saving weight. The technologies associated with synthesizing composites has given a ray of hope to develop complex and delicate structural components. Newer methods like resin transfer moulding have substituted autoclave moulding of prepegs. Using some of the advanced techniques like structural health monitoring will enable to overcome some of the limitations the composites possess due to their complexity. The next generation composites would be highly optimized and smart.

References

1. Das S, Yokozeki T (2021) A brief review of modified conductive carbon/glass fibre reinforced composites for structural applications: lightning strike protection, electromagnetic shielding, and strain sensing. Compos Part C Open Access 5:100162
2. Chakravarthi DK, Khabashesku VN, Vaidyanathan R, Blaine J, Yarlagadda S, Roseman D et al (2011) Carbon fiber–bismaleimide composites filled with nickel-coated single-walled carbon nanotubes for lightning-strike protection. Adv Func Mater 21(13):2527–2533
3. Rajesh P, Sirois F, Therriault D (2018) Damage response of composites coated with conducting materials subjected to emulated lightning strikes. Mater Des 139:45–55. https://doi.org/10.1016/j.matdes.2017.10.017
4. Kumar V, Yokozeki T, Okada T, Hirano Y, Goto T, Takahashi T et al (2018) Effect of through-thickness electrical conductivity of CFRPs on lightning strike damages. Compos Part A: Appl Sci Manuf 114:429–438. https://doi.org/10.1016/j.compositesa.2018.09.007
5. Gagné M, Therriault D (2014) Lightning strike protection of composites. Prog Aerosp Sci 64:1–16. https://doi.org/10.1016/j.paerosci.2013.07.002
6. https://www.grca.online/case-studies
7. Bartos PJM (2017) Glassfibre reinforced concrete: a review. IOP Conf Series Mater Sci Eng 246:012002
8. https://www.grca.online/uses-of-grc/grc-in-building/renovation
9. GRCA International technical notes covering aspects of glass-fibre reinforced concrete (GRC) technology. https://www.grca.online/contentfiles/notes/9.pdf
10. Muthukumarana TV (2023) A review on the variation of mechanical properties of carbon fibre-reinforced concrete. Constr Build Mater 366
11. Chung DDL (2000) Cement reinforced with short carbon fibres: a multifunctional material. Compos Part B Eng 31:511–526
12. Chen T et al (2023) Comparative study on the static and dynamic mechanical properties and micro-mechanism of carbon fiber and carbon nanofiber reinforced concrete. Case Stud Constr Mater 18(2023):e01827
13. Solahuddin BA, Yahaya FM (2023) A state-of-the-art review on experimental investigation and finite element analysis on structural behaviour of fibre reinforced polymer reinforced concrete beam. Heliyon

14. Peng F, Xue W (2018) Design approach for flexural capacity of concrete T-beams with bonded prestressed and nonprestressed FRP reinforcements. Compos Struct 204:333–341. https://doi.org/10.1016/j.compstruct.2018.07.091
15. Dolan CW, Swanson D (2002) Development of flexural capacity of a FRP prestressed beam with vertically distributed tendons. Compos Part B Eng 33(1):1–6. https://doi.org/10.1016/S1359-8368(01)00053-1
16. Grace N, Ushijima K, Baah P, Bebawy M (2013) Flexural behavior of a carbon fiber-reinforced polymer prestressed decked bulb T-beam bridge system. J Compos Constr 17(4):497–506. https://doi.org/10.1061/(ASCE)CC.1943-5614.0000345
17. Işildar GY et al (2020) A comparison LCA of the common steel rebars and FRP. J Build Pathol Rehabil 5(8)
18. Shakir QM et al (2023) Rehabilitation of deteriorated reinforced self-consolidating concrete brackets and corbels using CFRP composites: diagnosis and treatment. J Build Pathol Rehabil 8(16)
19. Ebadi-Jamkhaneh M et al (2023) Numerical finite element investigation of thin steel shear walls retrofitted with CFRP layers under reversed cyclic loading. J Build Pathol Rehabil 7(62)
20. Refurbishment Structural Strengthening with Sika Systems for Owners of Buildings and Civil Engineering Structures. https://www.sika.com/content/dam/dms/corporate/b/glo-structural-strengthening-for-owners.pdf
21. Mohammed AA et al (2020) State-of-the-art of prefabricated FRP composite jackets for structural repair. Eng Sci Technol 23:1244–1258
22. Hollaway LC (2014) Using fibre-reinforced polymer (FRP) composites to rehabilitate differing types of metallic infrastructure. Rehabilitation of metallic civil infrastructure using fiber reinforced polymer (FRP) composites—types properties and testing methods. Woodhead Publishing
23. Belarbi A, Dawood M (2016) Sustainability of fiber-reinforced polymers (FRPs) as a construction material. Sustainability of construction materials (second edition). Woodhead Publishing Series in Civil and Structural Engineering, pp 521–538
24. Rizkalla S, Tadros G (2003) FRP for prestressing of concrete bridges in Canada. Fibre reinforced polymer (FRP) composite materials for infrastructure. ACI Symposium Publication, pp 75–90
25. Marco Rossini and Antonio Nanni (2019) Composite strands for prestressed concrete: state-of-the-practice and experimental investigation into mild prestressing with GFRP. Constr Build Mater 205:486–498
26. Kopparthy SDS, Netravali AN (2021) Green composites for structural applications. Compos Part C Open Access 6:100169
27. NASCAR safety essentials to keep drivers protected. Accessed 2020. https://blog.nationwide.com/nascar-safety/
28. Sun Z et al (2023) Research progress and application of natural fiber composites. J Nat Fibers 20(2):2206591
29. Narayanan Nampoothiri E et al (2022) Experimental investigation on mechanical and biodegradation properties of Indian Almond–Kenaf Fiber-reinforced hybrid composites for construction applications. J Nat Fibers 19(1)
30. Khan T et al (2021) The effects of stacking sequence on the tensile and flexural properties of Kenaf/Jute fibre hybrid composites. J Nat Fibers 18(3)
31. Banana/coir biofibers and carbon/innegra fabrics and BN/MWCNT nanoparticles reinforced UV resistant polyester hybrid composites. Constr Build Mater 392:132014
32. Zhang X et al (2023) Bending and shear performance of a cross-laminated composite consisting of flattened bamboo board and Chinese fir lumber. Constr Build Mater 392:13193
33. Zhu Z et al (2023) Research on the road performance of self-adhesive basalt fiber geotextiles based on the background of long-life pavements. Constr Build Mater 392:131776
34. Ghosh SK (2015) Development of an innovative bituminized jute paving fabric (BJPF) along with its commercial field trials for potential application in the field of geotechnical construction with an eye towards global concern. J Nat Fibers 12(5)

35. Barbosa MT et al (2023) Lime-based mortars with added silica fume and bioproducts for restoration and preservation of heritage buildings. J Build Pathol Rehabil 8(37)
36. Sldozian RJA et al (2023) Mechanical properties of lightweight green concrete including nano calcium carbonate. J Build Pathol Rehabil 8(3)
37. Chandrasekhar Reddy K, Hemadri G (2023) Characterization of high-strength concrete by Preferential supplant of scrap tyre rubber powder-bambara nut shell ash constituent: utilization of novel waste materials in concrete. J Build Pathol Rehabil 8(14)
38. Safri SNA, Sultan MTH, Jawaid M, Jayakrishna K (2018) Impact behaviour of hybrid composites for structural applications: a review. Compos B Eng 133:112–121
39. Rostamiyan Y, Fereidoon A, Mashhadzadeh AH, Ashtiyani MR, Salmankhani A (2015) Using response surface methodology for modeling and optimizing tensile and impact strength properties of fiber orientated quaternary hybrid nano composite. Compos B Eng 69:304–316
40. Mansor MR, Sapuan SM, Zainudin ES, Nuraini AA, Hambali A (2013) Hybrid natural and glass fibers reinforced polymer composites material selection using Analytical Hierarchy Process for automotive brake lever design. Mater Des 51:484–492
41. Nunes F, Correia JR, Silvestre N (2016) Structural behavior of hybrid FRP pultruded beams: Experimental, numerical and analytical studies. Thin-Walled Structures. 106:201–217
42. Kim D-H, Kim H-G, Kim H-S (2015) Design optimization and manufacture of hybrid glass/carbon fiber reinforced composite bumper beam for automobile vehicle. Compos Struct 131:742–752
43. Surajit Biswas and Saroj Mandal (2022). Mechanical and micro-structural study of cement mortar with graphene oxide. *Journal of Building Pathology and Rehabilitation*, 7 (87).
44. Naslain RR (2005) SiC-matrix composites: nonbrittle ceramics for thermo-structural application. Int J Appl Ceram Technol 2(2):75–84
45. Effect of silicon carbide powder on temperature field distribution characteristics and microwave deicing efficiency of cement concrete containing magnetite (Fe3O4) powder. Constr Build Mater 392:132005
46. Fabricated CeO_2/ZrO_2 nanocomposite to improve thermal resistance, mechanical characteristics, microstructure and gamma radiation shielding of OPC composite cement pastes. Constr Build Mater 392:131971
47. Beardmore P, Johnson CF (1986) The potential for composites in structural automotive applications. Compos Sci Technol 26(4):251–281
48. Koronis G, Silva A, Fontul M (2013) Green composites: a review of adequate materials for automotive applications. Compos B Eng 44(1):120–127
49. Senokos E, Ou Y, Torres JJ, Sket F, González C, Marcilla R, Vilatela JJ (2018) Energy storage in structural composites by introducing CNT fiber/polymer electrolyte interleaves. Sci Rep 8(1):3407
50. Shirshova N, Qian H, Shaffer MS, Steinke JH, Greenhalgh ES, Curtis PT et al (2013) Structural composite supercapacitors. Compos A Appl Sci Manuf 46:96–107
51. Chan KY, Jia B, Lin H, Hameed N, Lee JH, Lau KT (2018) A critical review on multifunctional composites as structural capacitors for energy storage. Compos Struct 188:126–142
52. O'Brien DJ, Baechle DM, Wetzel ED (2011) Design and performance of multifunctional structural composite capacitors. J Compos Mater 45(26):2797–2809

Case Study and Applications of Composite Materials in Various Engineering Fields

Mechanical Properties of Sandwiched Layers of Natural Fibers of Sisal and Jute for Automotive Application

S. M. Sanjay Kumar and S. Tilak kumar

Abstract Due to their minimum weight relative to materials like steel, aluminium, and synthetic fibre based polymer, natural fibres are encouraged to be used in the automotive industry due to environmental concerns. A green composite is created for a variety of uses, including sports equipment, automobiles, and ships, among others. Due to their wide availability, eco-friendliness, and affordability, other natural fibres including jute, banana, flax, hemp, and sisal also have exceptional and satisfying qualities. Industrial items like sisal and jute fibres can use both structural and non-structural construction depending on the matrix. This work will help us to find out the mechanical properties of sandwiched composites as per ASTM standards and results obtained are discussed.

Keywords Natural composites · Mechanical properties · Sandwiched layer · Jute and sisal fibers

1 Introduction

A composite material is built from combination of two or more distinct materials with both physical and chemical properties that when joined to develop composite. The natural fiber has been considered due to its light weight and effective cost. The first phase of composites is called as matrix having a continuous character. Such matrix materials have a good binding property and fibers have the desired ability to arrange itself in position by transferring the loads to reinforcement.

However, nowadays, composites are well-established at high level and reached out from ordinary things to cutting edge applications. Most of composite material has legalized appearance at good quality as weight sparing items, and present estimation is to make them as money related spare.

S. M. Sanjay Kumar (✉) · S. Tilak kumar
Department of Mechanical Engineering, SJB Institute of Technology, Kengeri, Bengaluru 560060, India
e-mail: sanjay20376@gmail.com; sanjaykumarsm@sjbit.edu.in

S. B. Boppana et al. (eds.), *Structural Composite Materials*, Composites Science and Technology, https://doi.org/10.1007/978-981-99-5982-2_3

Composites made of natural fibre are more affordable and lighter than those made of synthetic fiber. In an experiment using varied conditions, Sivakandhan et al. [1] looked at the mechanical characteristics of sisal and jute fibres. When compared to jute and sisal hybrid epoxy composites, their research demonstrates that the examination of various hybrid composites offers outstanding mechanical characteristics. Animal cellulose fibres, which include bast, leaf, seed, and fruit fibres among others, are categorised as natural fibres. The sisal plant, officially known as Agave sisalana, is one of the sources of the sisal fibre that has recently made strides in the mechanical characteristics of sisal fibre reinforced polymer composites. Additionally, these fibres have been frequently utilised as reinforcing components in cementitious composites in recent years. This review article's main objective is to explore the mechanical characteristics of composites made of sisal and jute fibres sandwiched together.

2 Literature Review

Researchers [1–6] have made eco-friendly natural fabric sandwiched laminated composite using jute and linen fabric with reinforcement of epoxy. They examined the hybrid composite with different layers and found that linen fabric laminate which has five layers possessed more strength than other which has good thermal stability and better water absorption behavior and also has high flexural strength when compared to other materials, this can also been used as biodegradable materials for other house hold application. Researchers [1–6] have tested characterization of natural fiber reinforced composites between man-made fibers like glass fiber and natural fibers like sisal, pine apple, date palm. They had chopped the natural fibers into required length for making hybrid composites, and use electronic tensometer to examine the tensile strength and also the impact testing machine to measure the phenomenon governing the life span of the structure. Tests were according to ASTM standards and concluded that GFRC has high strength than FRC. It was also concluded that natural fiber reinforced composites with date palm is having high impact strength, which is suitable for different applications. Some researchers made experimental instigation on mechanical properties of aluminium sandwiched with sisal, kenaf, aloevera, jute, flax, natural fiber reinforced composites with different polymer resins like LY556 and GY250 as determined through various mechanical tests to ensure better bonding quality. AA6061 sheets were used for making reinforced natural fibers composites and it was examined that sisal with AA6061 with LY556 showed superior impact strength.

Sivakandhan et al. [1] in his study deals with the fabrication of hybrid composites and mechanical properties like tensile, flexural, compression strength and impact of sisal fiber and jute fiber combination. Epoxy resin (LY556 grade) and hardener (HY951) is taken as the matrix and binder. He made five different materials with different percentage ratio of randomly assigned chopped fibers of sisal and jute fiber with 35% of fiber and matrix ratio. Then he examined that sisal with jute fiber reinforced sandwich fiber with epoxy is increased by 32% than that of sisal fiber epoxy in axial and co-axial tensile strength and also has higher superior strength at flexural and as well as in impact strength than sisal fiber epoxy.

The researcher in his work proposed a detailed study by selecting treated and untreated jute fiber; one is woven and another is non-woven jute fabric, reinforced with the help of compression molding process, he made eco-friendly hybrid composites which with the application of green composites as of dias-deck assembly. It was also reported that some of natural fiber are too sensitive to attract moisture contents on fibers and concluded that water resistance coating is required and applied over the surface of composite materials and might also improve material life.

Balachandar et al. [5], in his work choosed Bamboo fiber, Sisal fiber, Hemp fiber and Sugar tree fiber as a natural raw fiber then he woven it with woven machine after woven mats are placed one on upon another to prepare hybrid composites. It was made by using reinforcing material like Epoxy resin with a suitable resin percentage as the matrix. The composite material is fabricated by hand layup technique. The fabricated composite material is tested for its mechanical properties such as tensile strength, flexural strength and impact strength.

Based on the literature [1–6], synthetic fibers, carbon fibers/glass fibers are widely used for building automotive parts, industrials applications, mechanical properties and are comparatively strong. Therefore natural fibers have received great interest in application of automotive industries for interior and exterior for examples particulates vegetal fibers reinforced composites, not only vegetal, jute, and sisal offer such benefits as reductions in weight and cost, and also they are recyclable and biodegradable easily.

Nowadays, in the automotive industries there is increase in pressure to fulfill performance demands and environmental aspects.

In the last decade, green technology in the field of materials science is given considerable importance.

In this work proposal, natural fiber like sisal and jute fiber and combined fiber mat are selected to perform different orientation of sandwiched fibers and fabricated by hand-lay-up techniques to obtain mechanical properties like flexural, tensile stress by the standard ASTM with the help of universal test machine and bonding quality of the material is analysed with the help of SEM (Scanning electron microscopy).

3 Materials and Methodology

Selection of raw material it is the basic step the researcher must be carefully while selecting raw material. All the properties of materials like mechanical, chemical, properties are to be analysed.

In this investigation, two natural fiber materials are selected, one is sisal fiber and the other being jute material and polymeric matrix material used is epoxy resin.

Sisal is one of the groups of natural fiber which is extracted from the plant leaves called agave sisalana, it is the family plants belonging to agave family. The sisal fiber are prepared from sisal leaves. The fibers one by one from leaves are extracted by some machine that was invented to extract the fibers threads from the leaves of the plant. And then the fibers threads may be treated with some chemical treatments to improve the chemical composition to improve percentage of cellulous, hemi-cellulous, lignin, wax, pectin. Here, we selected the sisal fiber woven mat to study the characteristic of mechanical properties of natural fiber with different orientation by some experimentation. The fibers mats were borrowed from go green product in Chennai.

The materials for sandwiched composites are sisal and jute fibre because of their acceptable mechanical qualities for the creation of hybrid composites. The matrix and binder for the composites are made of epoxy resin of grade LY556 and hardener HY951. Sandwich plates were made in the form of sheets for this procedure, and samples with various oriented layers were produced using the fibre and matrix ratio. This process was carried out utilising the hand layup method. Sisal Fiber is exceptionally durable with a low maintenance with minimal wear and tear.

Properties of Sisal Fibre:

- It is Recyclable.
- High specific strength weight ratio.
- The fibers are yield fibers that have more tensile strength so they used for ropes and other industrial applications.

Jute fiber is one of the groups of natural fiber which is extracted from the inner bark vascular tissue of plants of jute. It is the family plants belonging to mallow family malvaceae. The botanical name of jute fiber is *Corchorus olitorius* as shown in Fig. 4.4a, b. It falls into the **bast fiber** category (fiber collected from bast, the **phloem** of the plant, and then that fibers threads may treat with some chemical treatments to improve the chemical composition such that improve percentage of cellulous, hemi-cellulous, lignin, wax, pectin, which has follows as bio-fibers. Here we selected jute fiber mat type materials to study the sandwiched properties.

Properties of Jute fibre:

- 100% bio-degradable recyclable and thus environment friendly.
- The second most important and widely cultivated vegetable fiber after cotton.
- High tensile strength with low extensibility this helps to make best quality industrial yarn and fabric for application of packaging.
- Very versatile natural fibers that has been used in raw materials for packaging, textiles, non-textile, and agricultural and also industrial.
- Jute fiber are relatively high strength due to it has yield fibers.
- Jute fibers are cost effective and quite cheap (Table 1).

The following are the properties of materials (Tables 2 and 3).

Resin and Hardener

Epoxy resin is one of the stable fluids with relatively long shelf lives. It is mixed with an epoxy hardener, epoxy resins are low molecular weighted polymers which contain at least two epoxide group. The resin contains many grades of epoxy; LY556 epoxy

Table 1 Comparison between natural and synthetic fibers

Aspects	Property	Synthetic fibers	Natural fibers
Technical	Mechanical property	High	Moderate
	Thermal sensitivity	Low	High
	Moisture sensitivity	Low	High
Environmental availability	Resource available	Limited	Infinite
	Production	High	Low
	Recyclability	Moderate	Good

Table 2 Properties of sisal fiber

Properties	Sisal fiber
Elongation of break	2.75%
Orientation	Woven Fabric
Tensile strength	385–758 Mpa
Young's modulus	9–22 Gpa
Density	1.58 gm/cc
Diameter of each fibers	0.60 mm

Table 3 Properties of jute fiber

Properties	Jute fiber
GSM	400 GSM
Orientation	Plain woven fabric
Density	1.45 gm/cc
Elastic Modulus	10–55 Gpa
Tensile strength	399–800 Mpa
Diameter of each fiber	0.35 mm

and hardener HY951 to were used to prepare the composite material as (Figs. 1, 2; Table 4).

4 Mechanical Testing

The dimension details for testing are mentioned in Table 5.

Fig. 1 Fabricated composite plate-Plate 1 (Sisal-Jute-Sisal)

Fig. 2 Fabricated composite plate-Plate 2 (Jute-Sisal-Jute)

Table 4 Composite plate stacking order

Sl.No.	Sandwich plates	Stacking order
1	PLATE 1	Sisal-Jute-Sisal
2	PLATE 2	Jute-Sisal-Jute

Table 5 Dimension details of the samples [2]

Sl.No.	Type of test	ASTM standard followed	Measurement of the sample
1	Tensile test	ASTM D3039	$250 \times 25 \times 3$ (mm)
2	Flexural test	ASTM D790	$127 \times 13 \times 3$ (mm)

4.1 Flextural Test

Testing for flexural strength is used to assess a material's ability to deform under load. Using a Universal Testing Machine (UTM) in accordance with ASTM D790

[4], this three-point bend test aims to promote interlaminar shear-related failure. Flexural strength is determined using the equation.

$$\sigma_f = (3P_{max}L)/2bh^2 \tag{1}$$

where L is the specimen length between the two support points, b is the specimen breadth, h is the specimen thickness, and P_{max} is the maximum load at failure [3].

4.2 *Tensile Test*

Tensile testing is performed to determine the amount of strain or elongation needed as well as the force needed to break the test specimen using ASTM D3039 specimen.

5 Results and Discussion

For mechanical testing, including tensile and flexural tests, the samples were cut into slices in line with ASTM standards. The samples are illustrated in Fig. 3a, b (Figs. 4 and 5).

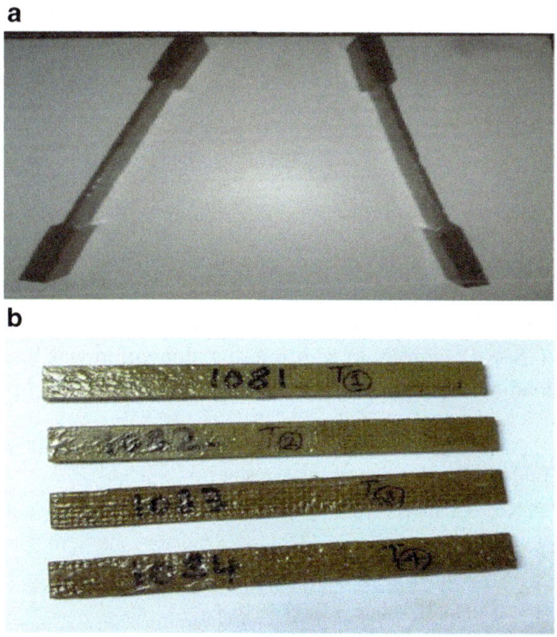

Fig. 3 Specimens prepared as per ASTM (D3039 and D790) standards

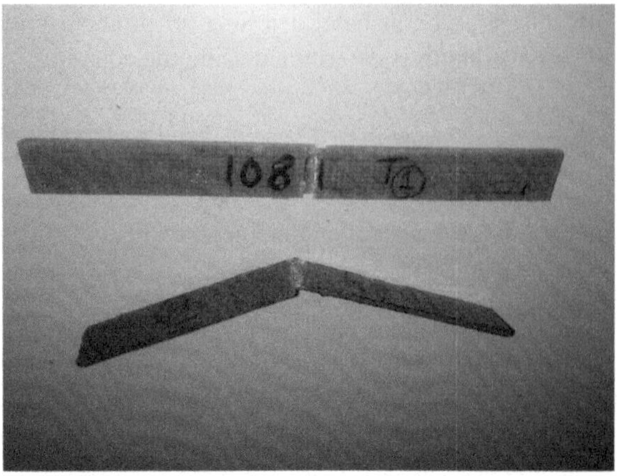

Fig. 4 Specimen after bending test

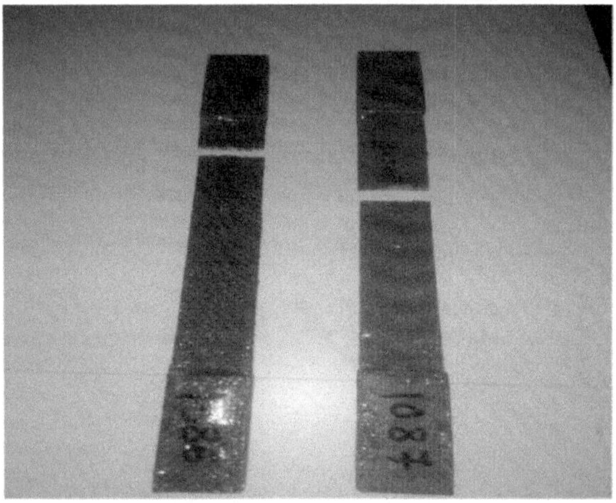

Fig. 5 Specimen after tensile test

Table 6 displays the results of the various tests that were performed. The tensile and bending test graphs are shown in Figs. 6 and 7, respectively. Elongation was limited to 5 mm in Fig. 6, with a maximum loading capacity of 1400 N for the (J + S + J) sample. In addition, the elongation was limited to 4 mm with a maximum loading capacity of 1700 N for the (S + J + S) sample in Fig. 7.

Table 6 Results of mechanical testing

Sl.No.	Samples type	Max. tensile strength (N/mm^2)	Flextural strength (N/mm^2)
1	Jute + Sisal + Jute (J + S + J)	16.2	71.79
2	Sisal + Jute + Sisal (S + J + S)	16.8	112.82

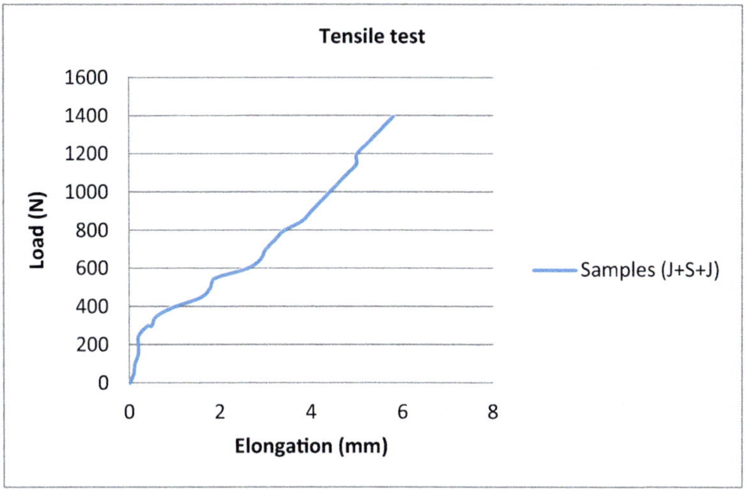

Fig. 6 Load versus elongation (tensile test)

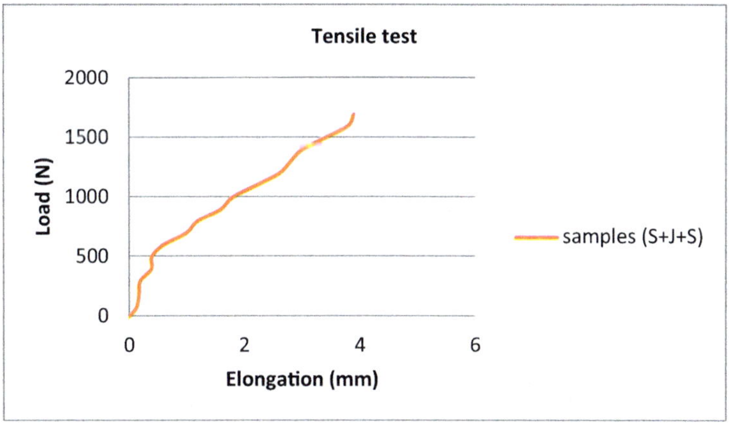

Fig. 7 Load versus elongation (tensile test)

According to the findings, samples of type (S + J + S) have greater tensile and flexural strength than samples of type (J + S + J) (Fig. 8).

6 Conclusions

- The mechanical properties of sisal and jute fibre composites are investigated in this proposed work.

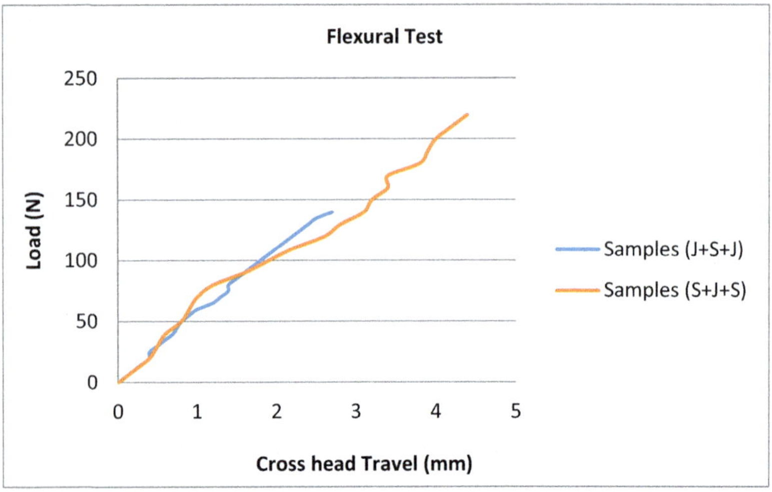

Fig. 8 Load versus cross head traveling (flexural testing)

- The tensile strength increased by 0.6 N/mm^2 and flexural strength increased by 41.03 N/mm^2 for sandwiched layers due to bidirectional orientation of sisal and jute fibers.
- Sisal fibres have more tensile strength when compared Jute fibres.

References

1. Sivakandhan C, Murali G, Tamiloli N, Ravikumar L (2019) Studies on mechanical properties of sisal and jute fiber hybrid sandwich composite. In: Article in press materials today: proceedings. https://doi.org/10.1016/j.matpr.2019.06.374
2. Das PP, Chaudhary V, Mishra S (2021) Chapter 1 emerging trends in green polymer based composite materials: properties, fabrication and applications. Springer Science and Business Media
3. Zhang P, Zhang C, Liu L (2022) Toughening 3D-printed Zr-based bulk metallic glass via synergistic defects engineering. Mater Res Lett 10(6):377–384
4. Tribological Applications of Composite Materials (2021) Springer Science and Business Media, pp 8–15
5. Balachandar M, Vijaya Ramnath B, Barath R, Bharath Sankar S (2019) Mechanical characterization of natural fiber polymer composites. Mater Today Proc 16:1006–1012
6. Green Composites (2021) Springer Science and Business Media, pp 25–32

Applications of Composites—A Case Study

K. S. Lokesh, C. G. Ramachandra, and D. Shrinivasa Mayya

Abstract The use of composite materials now a days designs new paradigm for all kinds of applications specially honours in 3 categories of transport systems such as roads, navy and flight vehicles which accommodates its integral usage in majority of the components that are built -in. Recent investigations also highlight the role of these light weight structures in energy harvesting and storage systems. This book chapter discuss about the various wide spread applications of all the core sectors of engineering which indulge the rigid structures which majorly corelates the structural and functional requirements. The present section also high light the case study on several applications including Aerospace and automotive and marine structures which depicts the resonance of utilizing different constituents which plays key role in performing the specific functions. The property derived application areas clearly gives the research spotlight on developing novel materials which intern found to be a better alternative for existing applications which demands for the availability of resources and process parameters. Hence this section also highlights the utilization of locally available resources as fillers to modify the matrix system and more likely to use for particular applications for which the specific properties are improved.

Keywords Case study on composites · Applications of composites

K. S. Lokesh (✉)
Department of Aeronautical Engineering, Srinivas Institute of Technology, Mangalore 574143, India
e-mail: lokeshvijay@sitmng.ac.in

C. G. Ramachandra
Department of Mechanical Engineering, Presidency University, Bangalore 560064, India

D. Shrinivasa Mayya
Department of Mechanical Engineering, Srinivas Institute of Technology, Mangalore 574143, India

S. B. Boppana et al. (eds.), *Structural Composite Materials*, Composites Science and Technology, https://doi.org/10.1007/978-981-99-5982-2_4

1 Introduction

The most considered composite category is reinforced plastics from fibres [1]. They are especially helpful related to this because of the extraordinary compositional properties of carbon and glass fibre conjugates [2]. FRC materials' low weight, great strength, durability, and form flexibility makes them a desirable alternative to steel and other materials. In the first half of the 1950s, new FRC materials were produced using carbon and glass fibre conjugates along with thermoset matrix system as a component. They are frequently used in the construction of autos, storage tanks, and boats [3]. In 2020, the global market for composite materials is predicted to reach \$96 billion, an increase of 40% from 2014. The most prevalent materials are glass and carbon fibres. Of the \$96 billion, carbon and glass fibre conjugates make up \$57 and \$30 billion market together. In attempt to reduce the unavoidable negative consequences, numerous studies have examined the supply due for different light weight member as amount of trash that is linked with it during the past few years. With an annual growth rate of 6.6%, the demand for these structures is likely to hit more than \$12 billion in coming days [4]. Figure 1 represents the global market volume of composites for the year 2020. Table 1 highlights the Case study on usage percentage of composites referring to various types of applications and Table 2 depicts the details of matrix/reinforcement/filler systems on applications based on property enhanced methods.

Applications comprising automotive, aeronautical, building and construction, energy harvesting and storing navy structures majorly utilize all categories of

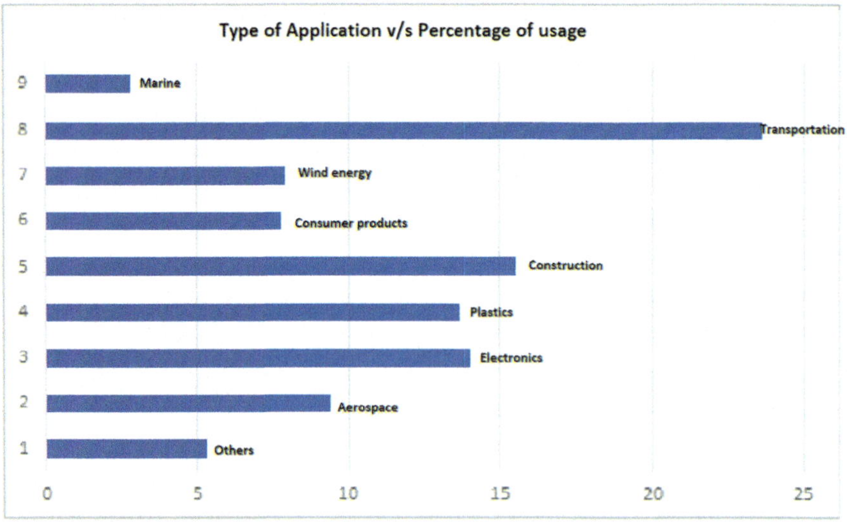

Fig. 1 Report representing the of global demand for composites for the marking year 2020 [5]

Table 1 Case study on usage percentage of composites referring to various types of applications

Application type	Type of composite used	Products made	Usage percentage/other benefits
Aerospace	Sandwich made from carbon, laminates of CFRP and fiberglass	Airbus commercial aircraft B787	50% of the total weight [6] Compared to aluminium, tensile and modulus strength is superior [7]
Automotive	CFRC	Carbon fiber components	Drastic reduction in the weight density of conventional car by 30% and CFRC also contributes to the regulating hazardous gas release through achieving fuel efficiency [8]
Marine	Carbon and glass fibre conjugates	Shipbuilding	Lighter weight, greater strength, and resistance to wear [9]
Construction	Hollow carbon composite tubes	Functional bridge for highway system	Flexibility and ease of application

Table 2 Details of matrix/reinforcement/filler systems on applications based on property enhanced methods [10–19]

Type of reinforcement and matrix system used	Type of production method	Fillers used	Recommended applications based on property derived
Fibre reinforced—Woven E-glass and chopped fibres with thermosets	Hand-lay-up method	Casio3, waste rubber, E-waste glass, mineral filler	Light weight structural requirements with good wear resistance
Particulate reinforced—Graphene particulate with Al-alloys	Stir casting method	Red mud particles, graphite particles, sic	Components made can be used for tribological applications with good resistance to hardness
Fibre reinforced—Fibre orientations fibres, fibre mat with thermosets	Hand lay-up method	Mineral fillers-waste and scrap glass	Thermal stability of the composites was achieved and recommended for application where strength to be retained once temperature is varied
Fibre reinforced—Jute/silk/CSM fibre with thermosets	Hand lay-up method	Fish scale, cocopeat powder, waste rubber	Light weight structures well suitable where strength to weight ratio is highly desirabl

Table 3 Composite components made for different passenger aircrafts [20]

Aircrafts name	Parts made of composites
F-14	Doors, skins of stabilizer, fairings
F-15	Fins, rudders, vertical tails, speed brakes, horizontal tails
F-16	Vertical and horizontal tails, fin leading edge
B-1	Doors, flaps, slats, inlets
AV-8B	Rudders, vertical and horizontal tails, ailerons, fin box
Boeing-737	Spoilers, wings
Boeing-757	Doors, rudders, elevators, ailerons, flaps
Boeing-767	Doors, rudders, elevators, ailerons, fairings

composites. The applications' specifics are covered in detail, along with case studies of a few categories.

2 Aerospace Applications

There is a prediction from international organization IATA that 11,000 aircraft will be discarded during the next ten years. Additionally, employing composites in aircraft has rapidly risen. Major aircraft producers like Airbus and Boeing have demonstrated the need for extensive composite uses in aero-industry, which is evident in the commercial aircraft they make. NASA is constantly searching for cutting-edge space solutions from composites producers for rockets and other spacecraft. New manufacturing processes and improved production yields were required due to the emergence of new competitors and the volume of parts to be produced. New problems require fresh approaches. In commercial, civilian, and military aerospace applications, heat resistive matrix systems are in the verge of making specific components for wings, fuselage parts and bulkheads whereas fibre glass, carbon laminates, and carbon sandwich composites are primarily used in building major components of aircraft. Antenna, radar, satellite structures, solar reflectors, and other items are manufactured for use in aerospace. The aircraft parts made of composites include surfaces of aerofoils, blades of compressors, engine bay doors, fan-blades, rotor-shafts in helicopters, turbine blades, turbine shafts and wing box structures (Table 3).

3 Automotive Applications

Due to global CO_2 pollution restrictions, composites will play a significant role in the Automotive and Road Transportation sector. Making an automobile lighter is one of the simplest ways to reduce emissions. Composites have been used in sports

Table 4 List of properties required with purpose of the requirement specifically to automotive applications [21]

Properties	Purpose
Low thermal expansion co-efficient	To know heat flow capability of material
Stability in dimension	Retention capability in dimension
Corrosive barrier	Resisting property variations against changing thermal environment
High value of toughness	Offer more resistance to sudden failure
Lower density	Less weight of the body

vehicles because their low weight enables the highest performance attainable. Recent improvements in mobility have made it possible for technologies like electric propulsion and self-guiding vehicles. To obtain longer ranges between recharging (EV'S), these cars need lightweight materials, which has increased demand for high-volume composite production. The automotive sector has a long history with composites and is the main market for them. Composites provide innovative vehicle designs in addition to making cars lighter and more fuel-efficient. Performance will be affected by design or production errors, which could result in a reduction in sales for the manufacturer. The advantages of composite materials over steel allow them to fulfil and even exceed the criteria of the automobile industry. The sandwich composites and laminates made of carbon and glass fibres are the most popular among the different types of fibres. Automotive needs dependable, coordinated systems with parts that can tolerate temperature changes, corrosion, and friction. In addition, these composites' unique qualities that make them suitable for automotive applications are quantifiable. Table 4 provides information on the nature and function of features that contribute to the distinct functions in the automotive industry.

4 Interior and Exterior Applications

A variety of automotive elements, including headlight positioning for placing front lights, under-the better barrier and temperature restricted assemblies, external body sections, and inside sized intricate assemblies frequently use composite materials. More specifically, components like air deflectors and spoilers and other integrated assembly things are made of composite materials. bumper beams and bumpers, cylinder head coverings, such as valve, rocker, and cam Window and roof frames, front grill opening panels, Heat shields (such as those for the engine and transmission), Pillars and covers, Headlamp housings for forward-facing headlamps.

5 Marine Applications

For many years, fibres impregnated with plastic categories are normally used in navy applications, including random and mass structures, superyachts, workboats, and recreational vessels. In less well determined areas, such as topside hatch covers, propellor shafts and bearings, FRPs have recently been used. One of the earliest notable applications for glass-fiber composites (GRP) was in the marine industry. It has completely changed how huge composite structures may be designed and produced in a number of industries. The UK employs a number of different fabrication techniques to create vessels, such as manual lay-up GRP, infusion method and better functional carbon conjugates for racing boats. In the offshore business, these structures are evident to be a driving kit for future market. Composite materials made of glass fibre (GF) and carbon fibre (CF) are currently very important in the development of maritime energy and marine engineering repair. The majority of installations substitute structural sections for conventional construction materials, which lowers the high freight weight, seawater corrosion, and issues associated with rebar materials [22]. One of the first industries to fully appreciate the advantages of composite materials, particularly their simplicity in use and long-term corrosion resistance due to the millions of cycles that a boat goes through once it begins to sail, is the marine industry. Hulls for the navigation and bulk heads for industry and recreational, and military ships and boats are among the marine composites' applications. For use in corrosive conditions, composites are perfect. Ceilings, struts pumping components are some of the major applicable areas. Owing to its qualities of lightness and resistance to corrosion.

The following are the sorted benefits of reinforced plastic categories:

- Adaptability to the environment, including resistance to rotting, corrosion, etc.
- The capacity to form seamless, intricately formed structures.
- The capacity to suit applied load.
- Superior structural aspect ratios.
- Identical steel structures typically weigh half as much as GRP marine constructions.
- Excellent longevity.
- Ease of repair and low maintenance.

Composites are used more frequently in racing yachts than in any other type of maritime structure. Due to unique needs, the materials used are not those typically used in marine construction. In order to sail with the most rate and barrier to the effects of oscillatory distractions in maritime ambience, they must be designed with the least amount of weight and the most amount of stiffness. Boat hulls as sandwiched foam like structures, frames, keels, masts, poles, and booms, as well as carbon winch drums and shafting, are frequently constructed from carbon based reinforced structures. In the various international sailing conditions, employing light weight structures improve functioning and reduce risk of navigational issues and other implications.

6 Construction

Fibre composite materials are now more frequently used in structural load-bearing applications by the construction industry, and they have proven to be a competitive and viable alternative to steel in reinforced concrete, for rehabilitation and retrofit of existing civil structures, and to a lesser extent, for new civil structures. Consider FRP composites for applications in civil engineering for a variety of reasons. Durability, corrosion resistance, cost, weight, material qualities, and ease of construction are the primary considerations for engineers when selecting a material to complete a task (Fig. 2).

FRP bars showcasing excellence while considering axial strength and offer better response to repetitive loading task i.e. After 107 cycles, long-fiber composites typically still have a significant amount of their short-term strength. Addition to lower mass as compare to steel reported they are desirable alternatives to structural materials like steel, which have a density of about 7850 kg/m^3, ranging from 1200 to 2600 kg/m^3.

The reason that why FRP rebars to be considered is that concrete component that could corrode from chemicals or chloride ions and for any concrete component that needs non-ferrous reinforcing because of electromagnetic factors. As an alternative to stainless steel, galvanised, or epoxy rebars are the suitable alternatives. And also, applications that demand thermal non-conductivity.

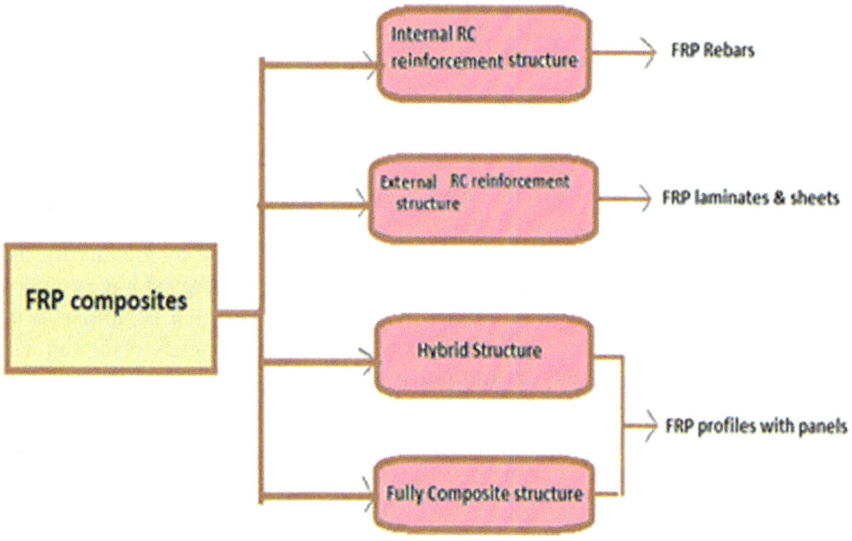

Fig. 2 Usage of FRP's in construction of RC structures

7 Electronics/Other House Hold Applications

Depending on the fibres and resins chosen, composites can be finely tuned to suit almost all of the requirements in electrical and electronic applications, including dielectric characteristics, insulation, even heat or electric conduction. Because of their excellent dielectric strength and arc resistance, modern Composite Materials are perfect for electrical and electronic applications. With the flexibility to create customised formulas, these materials have demonstrated success in supplying the necessary components to function in other hostile conditions. Frequent temperature loads, exposure to the elements, harsh weather, anti-static, and chemical assault are a few examples. The most difficult instances are mining settings and offshore exploration rigs. Lamp housings, terminal boxes, electrical enclosures, plugs, sockets, and energy distribution components are some of the most typical applications.

8 Energy Production

Through the creation of buildings that make it possible to harness renewable energy sources, light weight structures make a quintessential role in renewable energy industry. Composites are currently positioned to enable cost-effective solutions for large-scale projects because to factors including the less dens differentiable to crystalline compositions, lower transportation and construction expense and crucially, low price handling expense throughout its life cycle. Due to the lightweight and intricate air foil shape of wind turbine blades, composites are a market leader in this area thanks to moulds that are designed to efficiently produce blades with little effort.

For both land-based and offshore systems, increased turbine and rotor blade size is currently the focus of research and development. Some major components such as rotating sharp structure and devices for blow energy producers are among the manufactured parts. Multifunctional load preserved composites for sophisticated energy storage applications. Slices in the battery are anchored by locking mechanisms improving mechanical performance. Multifunctional energy storage composites tested experimentally exhibit electrochemical behaviour that is similar to the norm. Comparing MESC to pouch cells at 60% packing efficiency results in a 15 increase in mechanical stiffness. Additional uses for this stream include hydrogen fuel cells and aeronautical hydrogen tanks. natural composite battery cell.

9 Containers/Vessels

Chemical production facilities have been using composite ducting for more than 25 years. Composite materials saw a significant increase in utilisation in industrial applications during the 1970s. Composite pipe, duct, and tank installations are becoming more popular nowadays in above- and below-ground construction for commercial, governmental, and residential uses. As new chemical facilities are built and existing ones are expanded for chemical processing, cheaper price of gas is helping to later increase the demand. There are many such applications that container holds from utilization and goods with pollution prevention, strut operations with lining and treating polluted water and many more.

10 Sports and Recreation

As the economy grows and people's living levels rise, more and more contemporary people are relaxing at various sporting events. Additionally, the improvement and development of sports equipment is given significant weight in the development of contemporary athletic sports, which also places a strong emphasis on scientific training.

Due to its lightweight, high strength, great design flexibility, simple processing, and easy forming properties, fibre reinforced composite materials have found extensive use in sporting goods. Before the development of fibre reinforced composites, sports equipment was primarily made of materials like wood, steel, stainless steel, aluminium alloy, etc. Fibre reinforced composite material offers a clear advantage over these materials in the few areas like density and design aspects. Table 5 details about the based on their geometry of structures its application is considered.

Several reports suggested the specific type of reinforcement and matrix system where the yield of the research directs the corresponding method and material stream better suitable for wide spread applications. Table 6 details about one such consolidated data pertains to the works carried out with respect to various applications.

Table 5 FRP application in sports based on the geometry of the structure [23]

Geometry of the structure	Application type
Plate-like structure	Skis, surfboards, windsurfing, table tennis boards, slats and gliding wing spar etc.
Tubular structures	Tennis, badminton, fishing rods, golf clubs, baseball bats, hockey sticks, pole shaft, etc.
Sheet structure	All kinds of helmets, golf club heads, the hull structure of the various boat classes
Other structures	Match with a variety of vehicles, Sword, climbing ropes, various lines etc.

Table 6 Consolidated data pertains to the works carried out with respect to various applications [8, 9, 21, 23–30]

Method of manufacturing	Material system	Applications
Hand lay-up	*Matrix system*—Thermoset epoxy, polyester, polyaniline *Reinforcement*—Woven, knitted, stitched or bonded fabrics, graphene	Wind-turbine blades, boats, architectural shapes, secondary structure in aerospace Composites, automobile parts, dash board and deck
Spray layup	*Matrix system*—Thennoset, polyester *Reinforcement*—All type chopped fibers, flanks, particles	Enclosures, lightly loaded structural panels, e.g., caravan bodies, track fairings, bathtubs, shower trays, small dinghies, vent hoods, boats, shower unit
Prepreg layup	*Matrix system*—Thermoset epoxy polyester *Reinforcement*—Glass, carbon. Boron (roving and chopped strand mats, woven raving fabric)	Aerospace, spoils and leisure, automotive and motorsports, ballistic and defence marine
Filament winding	*Matrix system*—Epoxy, polyester, vinyl ester, phenolic *Reinforcement*—All types of fibres	Pipe lines, tubing, pressure vessels, tanks, gas cylinders, fire-fighters breathing tanks, tube light poles, aircraft fuselages, wing sections, radomes, helicopter rotor shafts
Resin transfer molding (RTM)	*Matrix system*—Thermoset, epoxy, polyester, vinyl ester and phenolic *Reinforcement*—Woven fabric, raving, chopped fiber mat	Bath and shower enclosures, cabinets, aircraft parts, automotive components, radomes and submarine sonar domes
Vacuum assisted resin transfer molding (VART M)	*Matrix system*—Thermoset resin-epoxy, polyester and vinyl ester *Reinforcement*—Woven fabric, roving, chopped fiber mat	Bridge sections and rail carriages, land-based and offshore utilityscale wind turbine blades
Pultrusion process	*Matrix system*—Thermoset and thermoplastic epoxy, polyester, vinyl ester *Reinforcement*—Continuous roving, fiber chopped strand mats, continuous filament mat	Bar and rod, pipe, tubing, ladder rails and rungs, and supports of many kinds, beams and girders used in roof structures, bridges, frameworks

(continued)

Table 6 (continued)

Method of manufacturing	Material system	Applications
Compression molding	*Matrix system*—Thennoset (epoxy, polyester, polyvinyl ester, phenolic), or thermo-plastic (polypro pylene, polyethylene nylon) *Reinforcement*—Chopped strand mats, woven fabric (UD/BD), short fibers, chopped fibers	Track parts, aircraft window frames, automobile panels, toys, electrical and aero-plane parts
Centrifugal casting	*Matrix system*—Thermoset resin—epoxy, polyester *Reinforcement*—Chopped fibers	Large diameter cylindrical pipes, telegraph poles, cylindrical components

11 Conclusion

Considering the different class of composites and their influence on property driven applications leading the industry requirements, the need and necessity terms in this regard are discussed briefly in this chapter. Various structural aspects out of available categories of composites are adhere with functional requirement. The data corresponds to various applications with respect to percentage utilization of composites depicts the overall market potential of composites. Prior to the profound use of advanced materials which are recently discovered, the usage percentage of existing composite systems and the benefits of increasing the percentage by weight of these material system in aircraft vehicles are also highlighted by considering the case study on reputed aircraft manufacturers. Advancements in materials usage in energy harvesting and energy storage generates keen concern on novel materials which are present and future materials for EV's and aerospace applications especially in multifunctional energy storage composites. The role of fillers on existing matrix systems aims to improve the property derived applications of composites thereby increasing the demands for these structures not only enhancing the structural stability but also improving the thermal and physical properties of composites.

References

1. The International Air Transport Association (IATA) (2020) Helping aircraft decommissioning. https://www.iata.org/en/programs/environment/aircraft-decommissioning/. Accessed on 11 Aug 2021
2. Pimenta S, Pinho ST (2011) Recycling carbon fiber reinforced polymers for structural applications: technology review and market outlook. Waste Manag 31:378–392
3. Molded Fiberglass Companies (2016) Moulding process. Retrieved Feb 2017, from http://www.moldedfiberglass.com

4. Naqvi SR, Prabhakara HM, Bramer EA, Dierkes W, Akkerman R, Brem G (2018) A critical review on recycling of end-of-life carbon fibre/glass fibre reinforced composites waste using pyrolysis towards a circular economy. Resour Conserv Recycl 136:118–129
5. Rubino F, Nisticò A, Tucci F, Carlone P (2020) Marine application of fiber reinforced composites: a review. J Mar Sci Eng 8:26
6. Murali Manohar D, Polym Eng. https://crescent.education/wp-content/uploads/2019/02/
7. Kumar N, Kannantha V, Pai R, Hebbale AM (2021) Development and experimental investigation of mechanical properties of graphene-based aluminum 6061 alloys. Mater Today Proc 46: 2421–2424 (Part 7). ISSN: 2214-7853. https://doi.org/10.1016/j.matpr.2021.01.303
8. Singh I (2013) Module—5 polymer matrix composites: processing index. Retrieved Feb 2017, from http://textofvideo.nptel.iitm.ac.in/112107086/
9. Velmurugan R (2017) Composite materials, Module II—manufacturing techniques, Department of Aerospace Engineering, Indian Institute of Technology, Madras, Retrieved Feb 2017, from http://www.nptel.ac.in/courses/101106038/mod02lec01.pdf
10. Raghavendra Pai K, Shrinivasa Mayya D, Naveen Kumar JR, Hebbale AM (2021) Experimental study on preparation and mechanical characteristics of jute/silk/coco-peat reinforced with epoxy polymers. Mater Today Proc 46:2764–2769 (Part 7). ISSN: 2214-7853. https://doi.org/10.1016/j.matpr.2021.02.511
11. Pinto T, Ravi SM (2017) Evaluation of mechanical properties and wear characterization of polymer composites under varying temperature conditions: a review. Int J Eng Inf Syst (IJEAIS) 1(4):64–68. https://doi.org/10.5281/zenodo.821168
12. Ummaji B, Gururaj P, Rayappa K, Yashavantha J (2018) Effect of red mud particles on scratch resistance of aluminum based metal matrix composites. Am J Aerosp Eng 5(1):24–29. https://doi.org/10.11648/j.ajae.20180501.14
13. Karthik BY, Avinash CG, Kamath G, Lobo S (2018) Experimental study on tribological characteristics of e-waste filled fibre reinforced plastics. Int J Sci Res Mech Mater Eng (IJSRMME) 2(1):11–16. ISSN: 2457-0435. http://ijsrmme.com/IJSRMME182203; vol 2|Issue 2|March–April 2019. www.ijsrmme.com
14. Gowspeer M, Pinto T, Effect of e-waste rubber on wear behaviour of glass fibre reinforced with epoxy composites 20–25. https://doi.org/10.7324/IJASRE.2018.32624
15. Pinto T (2018) Review on tribological parameters of fibre reinforced plastics modified with mineral fillers. Int J Sci Res Mech Mater Eng (IJSRMME) 2(5):49–56. ISSN: 2457-0435. http://ijsrmme.com/IJSRMME18257
16. Kannantha V (2017) Int J Eng Inf Syst (IJEAIS) 1(2):97–100. ISSN: 2000-000X
17. Pratibha PS, Pinto T, Kumar A, Chetan G (2017) Comparative study on tensile behaviour of e-glass (woven & chopped) fibre reinforced with epoxy composites. Int J Eng Sci Res Technol 06(03) 2017/3
18. Pinto T (2019) Examination of modulus of rupture of fibre mat bonded with potential adhesives. Int J Sci Res Mech Mater Eng (IJSRMME) 3(2):08–14. ISSN: 2457-0435
19. https://ijsrmme.com/IJSRMME19315
20. Summerscales J (1990) Manufacturing concepts for volume production of large composite components. Int Symp Compos Mater Text Reinforcement Use Build Constr Relat Appl 3:387–394
21. Maunder SK (2001) Composite manufacturing material, product and process engineering. Retrieved Feb 2017, from https://www.crcpress.com
22. Hihara LH, Bakkar A (2016) Corrosion of metal matrix composites, reference module in materials science and materials engineering. Elsevier
23. Hernández-Morenoa H, Douchin B, Collombet F, Choqueuse D, Davies P (2008) Influence of winding pattern on the mechanical behavior of filament. Compos Sci Technol 68(3–4):1015–1024
24. Manufacturing of Composite, Department of Aerospace Engineering, AE681 composite materials. Retrieved Feb 2017, from https://www.google.co.in
25. Huntsman International LLC. (2001–2017) Wet lay-up moulding processes. Retrieved Feb 2017, from http://www.huntsman.com/advanced_materials

26. Naveen Kumar JR, Prasad P, Hebbar N, Savitha MB, Navaneeth Gowda N (2021) A potential application of polyethyleneimine-reduced graphene oxide nanocomposite sensing film coated on interdigitated electrode prepared from copper-clad for Co_2-detection. Mater Res Innov 25(6):363–371. https://doi.org/10.1080/14328917.2020.1826674
27. Composite material and manufacturing by society of manufacturing engineering. Retrieved Jan 2017, from www.sme.org
28. Yang BJ, Ha SK, Pyo SH, Lee HK (2014) Mechanical characteristics and strengthening effectiveness of random-chopped FRP composites containing air voids. Composites 159–166 (Part 62)
29. Naveen Kumar JR, Kannantha V, Pinto T, Sampreeth U (2020) Experimental evaluation of substrate and annealing conditions on ZnO thin films prepared by sol-gel method. Mater Today Proc 24:201–208. ISSN: 2214-7853 (Part 2). https://doi.org/10.1016/j.matpr.2020.04.268
30. Lucintel (2016) Growth opportunities in the Indian composites market: 2016–2021 trends, forecast, and opportunity analysis. Retrieved Feb 2017, from http://www.lucintel.com/indian_composites_market_2016_2021.aspx

Various Methods of Processing MMCs, PMCs and Ceramic Matrix Composites

Processing of Composites with Metallic, Ceramic, and Polymeric Matrices

R. Samanta, B. Sengupta, G. Mandal, A. Wazeer, A. Das, and A. Sinha

Abstract One of the key engineering materials invented and explored by researchers in recent times is Composite materials because of their wide range of applications starting from transportation to aerospace, leisure industries and electronic industries. The development of various classes of composite materials requires derivations and evolution of several processing techniques to achieve successful incorporation and desired distribution of reinforcements throughout the matrix. In this chapter, an overview of the processing of various classes of composites viz. "Metal Matrix Composites" (MMCs), "Ceramic Matrix Composite" (CMCs), and "Polymer Matrix Composites" (PMCs) are presented along with recent trends in processing.

Keywords Composite · MMC · CMC · PMC · Processing

1 Introduction

In recent years, the evolution of composites has gained a massive momentum as these materials are increasingly acclaimed as the important and futuristic man-made materials. The advancement of composites has opened new avenues and new field of applications in various sectors. In general terms, composite materials are tailor-made materials suitable for specific applications and are defined as the combination of multiple distinct materials at a macroscopic level [1]. Various classes of composites such as MMCs, CMCs and PMCs are known to possess high strength-to-weight ratio,

R. Samanta · B. Sengupta · G. Mandal · A. Sinha (✉)
Department of Metallurgical Engineering, Kazi Nazrul University, Asansol, West Bengal 713340, India
e-mail: sinharijit@gmail.com

A. Wazeer
School of Laser Science and Engineering, Jadavpur University, Kolkata, West Bengal 700032, India

A. Das
Dept. of Aerospace Engg & Applied Mechanics, IIEST Shibpur, B.Garden, Howrah, West Bengal 711103, India

S. B. Boppana et al. (eds.), *Structural Composite Materials*, Composites Science and Technology, https://doi.org/10.1007/978-981-99-5982-2_5

hardness, tensile strength and other properties suitable for use in several applications [2]. The matrix materials support the reinforcements and effectively transfer load along the composites [3, 4]. These synthetic composites act as an adhesive binder that effectively supports fibers present in the matrix under compressive loads and it also grows shear stress-bearing capabilities by distributing load all over the matrix in the fiber lay-ups [5–7]. On the other hand, the development of several classes of composites significantly demands the knowledge of various processing techniques in order to grow in-depth understanding of the science behind the efficiency and effectiveness of the materials [8–10]. The present chapter highlights various conventional processing methods of MMCs, CMCs, and PMCs along with the recent trends in processing.

2 Processing of MMCs

MMCs are generally made by dispersion of a reinforcement material into the metal matrix. The surface of the reinforcement is effectively coated to achieve restricted oxidation in the matrix, e.g., in Al-matrix, carbon fibers are used in order to synthesize composite materials that attributes to higher strength and low density. Several techniques are used to form MMCs as shown in Fig. 1.

2.1 Solid State Processing

Solid state processing of MMCs methods can be further categorized as "Powder Blending and Consolidation", "Diffusion Bonding" and "Physical Vapour Deposition" as shown in Fig. 2. The processes are described as follows:

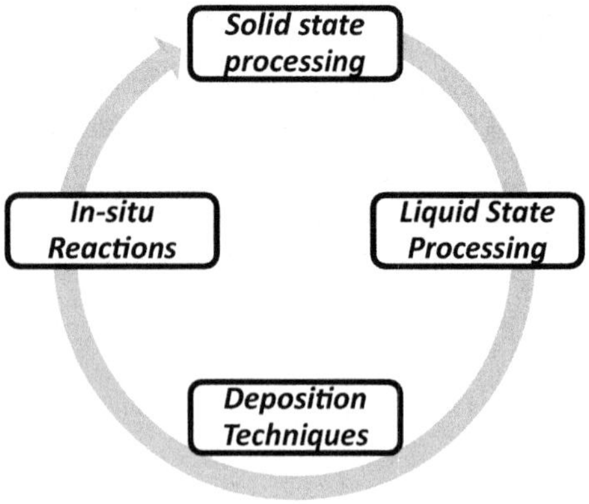

Fig. 1 Several processing techniques of metal matrix composites

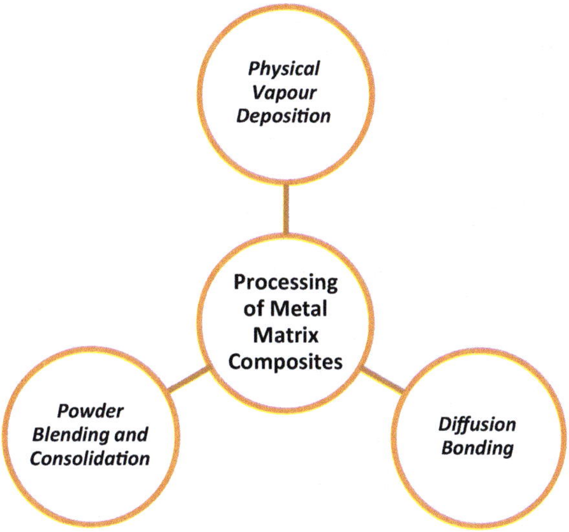

Fig. 2 Solid state processing of metal matrix composites

- **Powder Blending and Consolidation**: In this process, powdered metal alloys are blended with short fiber and ceramic particles in liquid suspension. This can also be carried out without using the liquid. After that the mix is processed using cold compaction, degassing and high-temperature consolidation [11]. In the powder blending and consolidation process, particles of certain volume fraction present in the matrix depending upon the type of powder and processing parameters help to strengthen the composites by the mechanism of dispersion-strengthening as shown in Fig. 3. Magnesium (Mg) and aluminium (Al) metal matrix composites are generally prepared using this method. Mechanical alloying (MA) is another solid state processing method in which the powders of the matrix and reinforcements materials are milled together inside a planetary ball mill followed by sintering.
- **Diffusion Bonding**: When two metallic surfaces are brought into contact and pressure is applied, inter-diffusion of atoms creates bonds between second phase particles and metal matrices as shown in Fig. 4 [11]. This fabrication method is used for developing composites with continuous or discontinuous fibers.
- **Physical Vapour Deposition**: In this process, fibers are continuously passed through a region having a high partial pressure of the material to be deposited. The matrix is deposited throughout a fibre preform. In Physical Vapour Deposition, the deposition rate is approximately 5–10 μm/min [12]. Sometimes composites are produced as a play by using hot pressing or hot isostatic pressing.

2.2 *Liquid State Processing*

This includes several methods like "Stir Casting", "Squeeze Casting" and "Infiltration Process" as shown in Fig. 5. The processes are described as follows:

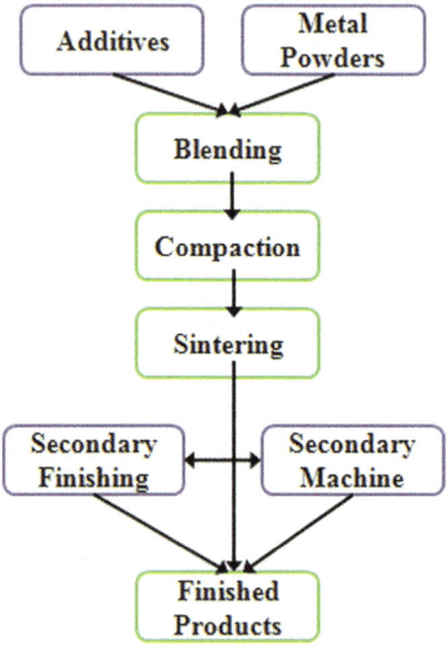

Fig. 3 Powder blending and consolidation process

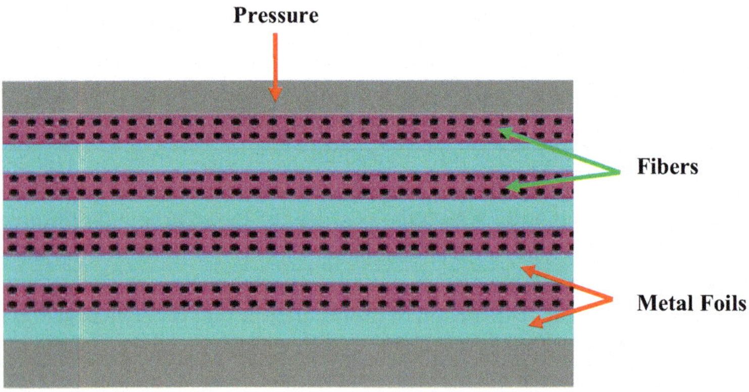

Fig. 4 Diffusion bonding

- **Stir Casting**: In case of stir casting, solid reinforcements are added to a molten metal matrix followed by casting in a mould. A common problem encountered in this process is distribution of the reinforcements throughout the matrix which, in many cases, may not be uniform [11]. Hence, the melt is continuously stirred with the help of a stirrer before casting. Mixing can also be improved by allowing the melt to cool in a semi-solid state.
- **Squeeze Casting**: In squeeze casting technique, an open die is used in which molten metal is directly inserted. The die is then closed. After that, the metal

Fig. 5 Several techniques of liquid state processing

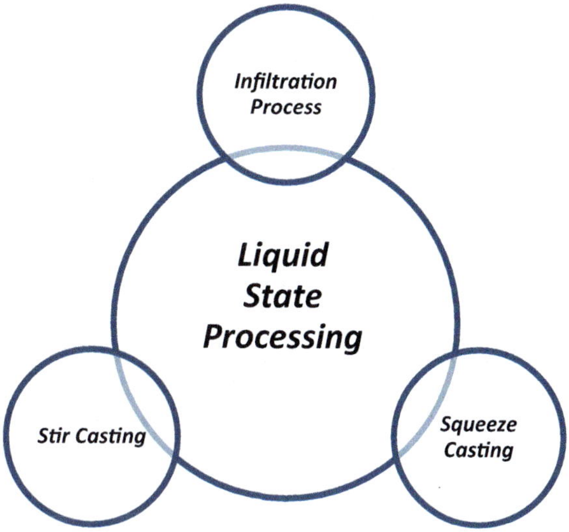

in molten condition comes into contact with the surface of the dies, and at high pressure, it starts transferring heat quickly. It ensures formation of very fine grain composites with negligible pores [13].

- **Infiltration Process**: In this process, liquid metals are infiltrated into a permeable form of whiskers or fiber reinforcements. Depending upon the porosity level, volume fraction of reinforcements is decided and in general, it is fixed as 10–70% [14]. In order to retain integrity, most of the time metal-based mixtures or silica is used as binders [11].

2.3 Deposition Techniques

A deposition layer of approximately 5–10% of the surface of the metal is generally created in this process with the help of whisker or fiber reinforcement, followed by the consolidation of depositions in full density [15]. In case of MMCs with long or continuous fiber reinforcement, metal spray is preferred which is sprayed on the fibers.

2.4 In-Situ Reactions

Here, the reinforcements are generated by the chemical reaction between the constituent components. As a result, the distribution of the reinforcements is uniform throughout the matrix and reinforcements are thermodynamically stable [16]. In-situ process also helps in generating a clean matrix-reinforcement interface [17].

3 Processing of Ceramic Matrix Composites (CMCs)

CMCs can be described as a special type of composite material in which only ceramic materials are used in both matrix material and refractory fibers. It can be produced by several techniques as shown in Fig. 6.

3.1 Dry Processing Including Pressing

Dry pressing can be described as axial compaction of comparatively loosely granulated ceramic powders of less than 3% moisture within an arrangement of punch or die. Under pressure, the powder spreads on the surface of the dies or punch faces to form a specific shape [18]. Powder compaction generally occurs within a rigid-walled die and usually between a bottom and top punch [19].

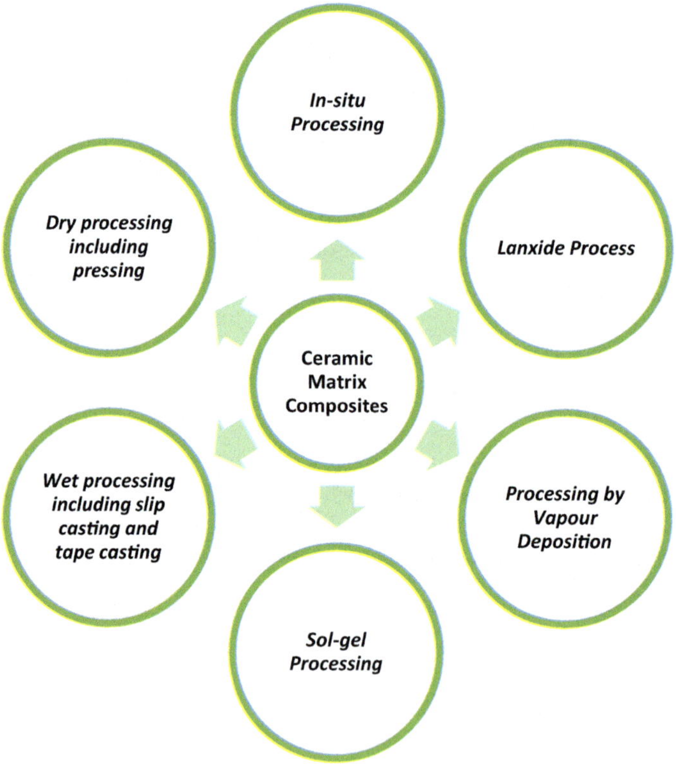

Fig. 6 Several processing techniques of ceramic matrix composites

3.2 Wet Processing Including Slip Casting and Tape Casting

Two widely used wet processing techniques of ceramic materials are slip casting and tape casting. Slip casting can be described as a ceramic forming process for pottery and other ceramics [20]. In this method materials which are of difficult shapes are prepared on a wheel. On the other hand, tape casting is used to manufacture thin tapes of ceramics and sheets using ceramic slurry [21]. Tape casting is also known as "knife coating", "doctor balding" and "shank shifting".

3.3 Sol–gel Processing

In sol–gel processing, two or more solutions (precursors) are mixed together in the stoichiometric ratio. At room temperature with continuous stirring, this solution becomes stabilized. After that a third solution is added to this stirred solution. In this process, a hydro-gel formation occurs that resulting the increase in viscosity of the solution [22]. Further, the gel is subjected to room-temperature aging for a couple of hours followed by washing using distilled water. Once the washing process is done the precipitate is dried for a couple of hours [23]. Sometimes, pressing and sintering is carried out after the sample is dried.

3.4 Processing by Vapour Deposition

In this process, through conversion or chemical reaction a solid layer is formed over a substrate [24, 25]. Deposition can be of two types: chemical vapour deposition (CVD) and physical vapour deposition (PVD). CVD is more common in case of CMCs. In CVD process, a solid generated in the gaseous state as a result of a chemical reaction is deposited over a heated substrate.

3.5 Lanxide Process

Here, a ceramic matrix is generated by the reaction between a molten metal and a gas. Growth of the ceramic matrix is generally occurred outwards from the surface of the metal. Lanxide process also can be described as a process where a composite material is generally synthesized through a controlled reaction impregnation [26]. In this process particle preforms or infiltrate ceramic fiber is generally used to produce lanxide materials [27]. One of the major advantages of this process is that the desired microstructure can be achieved by adjusting the process parameters.

3.6 In-Situ Processing

In-situ composites are known as a multiphase material manufacturing process in which the reinforcing phase is effectively synthesized within the ceramic matrix during the process of composite fabrication. In ex-situ composites reinforcing phase is separately synthesized and after that, it is inserted into the ceramic matrix using a secondary processing technique like powder processing or infiltration [28]. In in-situ processes, different kinds of reinforcement morphologies can be obtained and the reinforcements are mainly ceramic particles [29].

4 Processing of Polymer Matrix Composites (PMCs)

PMCs are composed of different kinds of continuous or short fibers which are generally bound together by a composite material matrix of both inorganic and organic polymers. It can be produced by several techniques as shown in Fig. 7.

4.1 Mechanical Methods Including Hand Lay-Up

In this process, resins are impregnated manually into the fibers using brushes or rollers. The fibers are in general in woven form or in the form of chopped strand mat. Liquid thermosetting resin is used. The process is cheap and suitable for a wide range of resins and fibers [30, 31]. Boats, blades of wind turbines, and architectural moulding components are processed by this method.

Fig. 7 Processing of polymer matrix composites

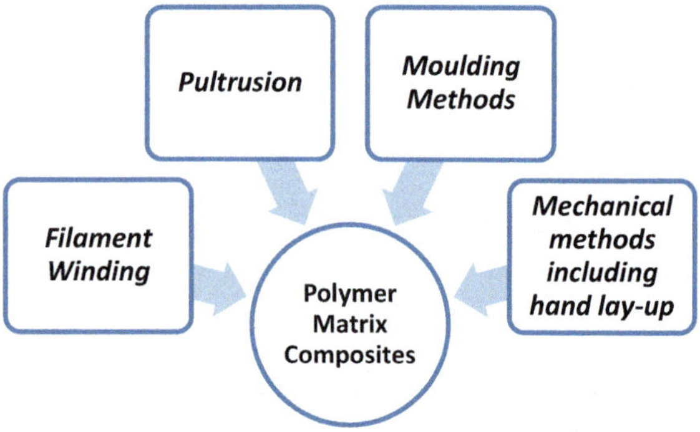

4.2 Moulding Methods

In this method, the material is subjected to hot pressing by placing it in the dies. It then flows into the mould and cured. In order to form a closed cavity mould fiber cloths are stacked up and then resins are injected into the mould. Train seats and internal complex components of aircraft are being made using this method [32].

4.3 Pultrusion

Pultrusion is basically a combination of pulling and subsequent extrusion. It is a manufacturing process of fiber reinforced polymer (FRP) composites. In this process, the fibers impregnated with resin are pulled through a heated die [33]. Subsequent cooling is also carried out to control the polymerization i.e. curing. The profile is then extruded through a die and cut into desired length. Good dimensional tolerance and high production rate can be achieved by this process.

4.4 Filament Winding

Filament winding process is widely used for manufacturing hollow, circular or oval-shaped materials. In this method, fiber tows are either passed through the resin bath prior to the process of winding [34] or impregnated fibers are wound on a mandrel of definite pattern. This process is economical and extensively used in the manufacturing of boat masts, pressure vessels, chemical storage tanks, and pipes [35].

5 Recent Trends

Several new processing techniques for MMCs have come up in recent years. These include "cryomilling" and "spark plasma sintering" (SPS) [36], "laser additive manufacturing" [37], "Accumulative Roll Bonding" (ARB) etc. In case of cryomilling, nanocrystalline powders are fabricated by mechanical attrition technique in a cryogenic medium like liquid nitrogen. SPS is an advanced consolidation technique of powder materials in which a pulsed direct current is used along with a uniaxial pressure. Processing of CMCs is also carried out using SPS. Compared to the conventional sintering, SPS can consolidate samples at a relatively lower temperature in a comparatively shorter duration. Emerging techniques for the processing of PMCs include additive manufacturing and processing using microwave energy [38]. In case of "laser additive manufacturing" or "laser melt deposition" (LMD) technique, a metallic substrate is taken on which a laser beam forms a pool of melt into which the reinforcement powders are fed. A nozzle is used to transport inert gas and powder into the small area around the focus of laser beam. Using this, layer by layer the composite structure is built up. ARB is a severe plastic deformation method in which two or more sheets are stacked together with the reinforcement powder placed in between the sheets. The sheets are then given a large amount of deformation by passing through the rolls which induces solid state bonding between the sheets. Recently PMCs are processed (curing, joining, moulding) using microwave energy in which uniform heating is expected to occur. As a result, the process becomes more efficient. Microwave hybrid heating is also carried out in which microwave energy is used to cure several layers of the PMC to build up a complete part.

6 Future Prospects of Composite Materials

An extensive discussion on the processing of MMCs, CMCs and PMCs has been carried out in this study. In foreseeable future, using these methods composites materials can be manufactured by altering parameters to get desired properties. Figure 8 shows some of the key advantages and areas that can change the course of manufacturing industry.

- **Ultra-Lightweight Composites**: Aluminum MMCs are mainly used in the development of electric and hybrid automobiles. It is also used in process of power transmission cables [39]. Ultra lightweight composites i.e. carbon nano-reinforced composites have been used in wind turbines and that is 30% lighter than existed

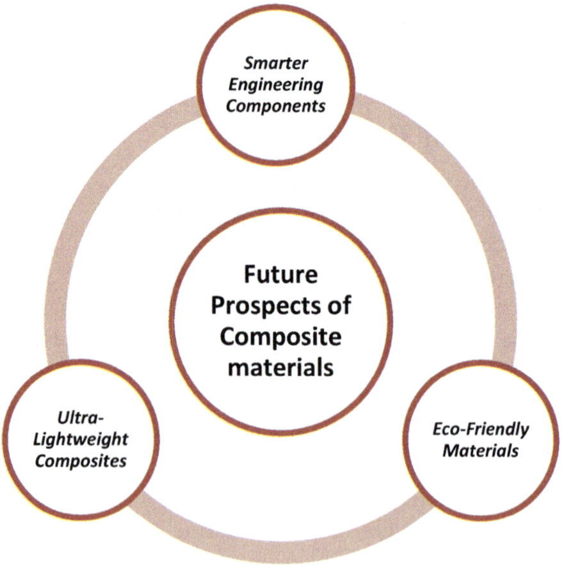

Fig. 8 Future prospects of composite materials

wind turbines [40]. As it is ultra-light the capacity for generating electricity is much higher.

- **Eco-Friendly Materials**: Advancements in green technology force manufacturing companies to utilize composite polymers more often in order to produce efficient and lighter materials that help to increase fuel efficiencies in aerospace and automobile industries. In order to sustain in the foreseeable future development of nano resins or fiber has become a necessity as it can be recycled and reused further [41].
- **Smarter Engineering Components**: In recent few years the evolution of composite materials changes the whole course of material applications and their usage in various sectors. Composite enables the intuitive design of structural and internal components previously design of such components was beyond the imagination of the designers [42]. Shaping conventional materials in such a way has become a revolution and that makes composite materials more inevitable.

7 Summary

Composite materials are increasingly becoming new state of the art engineering materials as they fulfil desired criteria for critical applications. In this chapter, a comprehensive study of the processing of various classes of composite materials has been summarized. Extensive discussions have been carried out regarding processing of MMCs, CMCs and PMCs. In case of MMCs, "solid state processing", "liquid state processing", "deposition techniques" and "in-situ reactions" have been described. For CMCs, "dry processing including pressing", "wet processing including slip casting and tape casting", "sol–gel processing", processing by vapour deposition, lanxide process and in-situ processing have been discussed. Outlines of PMC processing techniques like "mechanical methods including hand lay-up," "moulding methods", "pultrusion" and "filament winding" have been elaborated. Recent trends in various composite processing techniques are also highlighted in this chapter.

References

1. Reinhart TJ (1998) Overview of composite materials. Handbook of composites. Springer, Boston, MA, pp 21–33
2. Prashanth S, Subbaya KM, Nithin K, Sachhidananda S (2017) Fiber reinforced composites—a review. J Mater Sci Eng 6(03):2–6
3. Fiedler B, Gojny FH, Wichmann MH, Nolte MC, Schulte K (2006) Fundamental aspects of nano-reinforced composites. Compos Sci Technol 66(16):3115–3125
4. Chandra R, Singh SP, Gupta K (1999) Damping studies in fiber-reinforced composites—a review. Compos Struct 46(1):41–51
5. Soutis C (2005) Fibre reinforced composites in aircraft construction. Prog Aerosp Sci 41(2):143–151

6. Joshi SV, Drzal LT, Mohanty AK, Arora S (2004) Are natural fiber composites environmentally superior to glass fiber reinforced composites? Compos A Appl Sci Manuf 35(3):371–376
7. Ashbee KH (1993) Fundamental principles of fiber reinforced composites. CRC Press
8. Kim JK, Mai YW (eds) (1998) Engineered interfaces in fiber reinforced composites. Elsevier
9. Vassilopoulos AP, Keller T (2011) Fatigue of fiber-reinforced composites. Springer Science & Business Media
10. Zhao X, Wang X, Wu Z, Keller T, Vassilopoulos AP (2018) Effect of stress ratios on tension–tension fatigue behavior and micro-damage evolution of basalt fiber-reinforced epoxy polymer composites. J Mater Sci 53(13):9545–9556
11. Kumar A, Vichare O, Debnath K, Paswan M (2021) Fabrication methods of metal matrix composites (MMCs). Materi Today: Proc 46:6840–6846
12. Dey A, Pandey KM (2015) Magnesium metal matrix composite: a review. Rev Adv Mater Sci 42:58–67
13. Ye HZ, Liu XY (2004) Review of recent studies in magnesium matrix composites. J Mater Sci 39(20):6153–6171
14. Poletti C, Balog M, Schubert T, Liedtke V, Edtmaier C (2008) Production of titanium matrix composites reinforced with SiC particles. Compos Sci Technol 68(9):2171–2177
15. Trinh SN, Sastry S (2016) Processing and properties of metal matrix composites. Mech Eng Mater Sci Independent Study 10. https://openscholarship.wustl.edu/mems500/10
16. Kandpal BC, Kumar J, Singh H (2014) Production technologies of metal matrix composite: a review. Int J Res Mech Eng Technol 4(2):27–32
17. Cyriac AJ (2011) Metal matrix composites: history, status, factors and future. M.S. thesis, Oklahoma State University
18. Ewsuk KG, Arguello JG, Bencoe DN, Ellerby DT, Glass SJ, Zeuch DH, Anderson J (2003) Characterizing powders for dry pressing, sintering. Am Ceram Soc Bull 82(5):43
19. Abe H, Abe I, Sato K, Naito M (2005) Dry powder processing of fibrous fumed silica compacts for thermal insulation. J Am Ceram Soc 88(5):1359–1361
20. Pröbster L, Diehl J (1992) Slip-casting alumina ceramics for crown and bridge restorations. Quintessence Int 23(1)
21. Nishihora RK, Rachadel PL, Quadri MGN, Hotza D (2018) Manufacturing porous ceramic materials by tape casting—a review. J Eur Ceram Soc 38(4):988–1001
22. Singh LP, Bhattacharyya SK, Kumar R, Mishra G, Sharma U, Singh G, Ahalawat S (2014) Sol-gel processing of silica nanoparticles and their applications. Adv Coll Interface Sci 214:17–37
23. Zheng K, Boccaccini AR (2017) Sol-gel processing of bioactive glass nanoparticles: a review. Adv Coll Interface Sci 249:363–373
24. Nagel SR, MacChesney JB, Walker KL (1982) An overview of the modified chemical vapor deposition (MCVD) process and performance. IEEE Trans Microw Theory Tech 30(4):305–322
25. Dinata AA, Rosyadi AM, Hamid S, Zainul R (2018) A review chemical vapor deposition: process and application
26. BelyakovIvanov AV, Fomina GA (1997) A lanxide ceramic composite material. Glass Ceram 54(7):212–214
27. Xi-ya Z, Yue-hua T (2004) Fabrication of ceramic composites by directed metal oxidation. J Wuhan Univ Technol-Mater Sci Ed 19(1):48–50
28. Shehata F, Fathy A, Abdelhameed M, Moustafa SF (2009) Preparation and properties of Al_2O_3 nanoparticle reinforced copper matrix composites by in situ processing. Mater Des 30(7):2756–2762
29. Koczak MJ, Premkumar MK (1993) Emerging technologies for the in-situ production of MMCs. J Minerals Metals Mater Soc (TMS) 45(1):44–48
30. Davallo M, Pasdar H (2009) Comparison of mechanical properties of glass-polyester composites formed by resin transfer molding and hand lay-up technique. Int J ChemTech Res 1(3):470–475
31. Rydarowski H, Koziol M (2015) Repeatability of glass fiber reinforced polymer laminate panels manufactured by hand lay-up and vacuum-assisted resin infusion. J Compos Mater 49(5):573–586

32. Sabiston T, Inal K, Lee-Sullivan P (2020) Application of artificial neural networks to predict fibre orientation in long fibre compression moulded composite materials. Compos Sci Technol 190:108034
33. Gorthala R, Roux JA, Vaughan JG (1994) Resin flow, cure and heat transfer analysis for pultrusion process. J Compos Mater 28(6):486–506
34. Minsch N, Herrmann FH, Gereke T, Nocke A, Cherif C (2017) Analysis of filament winding processes and potential equipment technologies. Procedia CIRP 66:125–130
35. Mertiny P, Ellyin F (2002) Influence of the filament winding tension on physical and mechanical properties of reinforced composites. Compos A Appl Sci Manuf 33(12):1615–1622
36. Li M, Ma K, Jiang L, Yang H, Lavernia EJ, Zhang L, Schoenung JM (2016) Synthesis and mechanical behavior of nanostructured Al 5083/n-TiB_2 metal matrix composites. Mater Sci Eng, A 656:241–248
37. Jiang B, Zhenglong L, Xi C, Peng L, Nannan L, Yanbin C (2019) Microstructure and mechanical properties of TiB_2-reinforced 7075 aluminum matrix composites fabricated by laser melting deposition. Ceram Int 45(5):5680–5692
38. Naik TP, Singh I, Sharma AK (2022) Processing of polymer matrix composites using microwave energy: a review. Compos Part A: Appl Sci Manufact 106870
39. Wang JY, Chia KS, Liew JYR, Zhang MH (2013) Flexural performance of fiber-reinforced ultra lightweight cement composites with low fiber content. Cement Concr Compos 43:39–47
40. Wu Y, Wang JY, Monteiro PJ, Zhang MH (2015) Development of ultra-lightweight cement composites with low thermal conductivity and high specific strength for energy efficient buildings. Constr Build Mater 87:100–112
41. Korgel BA (2013) Composite for smarter windows. Nature 500(7462):278–279
42. Sengsri P, Ngamkhanong C, Melo A, Papaelias M, Kaewunruen S (2020) Damage detection in fiber-reinforced foamed urethane composite railway bearers using acoustic emissions. Infrastructures 5(6):50

Experimental Study on Synthesized Graphite Nano Particles Based PVA Nanocomposites

K. S. Lokesh, Thandra Paavan Kumar, C. G. Ramachandra, and D. Shrinivasa Mayya

Abstract The present work aims to develop Graphite Nano Particles (GNP's) based Poly Vinyl Alcohol (PVA) nano composites. Cold water soluble PVA is used in the present study which is blended fractionally with synthesized graphite nanoparticles in the ratio of 0.25, 05, 0.75 and 1% by weight by open mold method followed by autoclave curing to ensure proper curing of prepared samples. The synthesized graphite nanoparticles were subjected to UV–vis spectroscopy analysis, where in the highest peak was observed at 255 nm. Major contribution of this present investigation deals with the computational studies of composition elements to study various factors such as nuclear structure using the Density Functional theory (DFT). Stability analysis of each molecule is performed separately using the UB3LYP calculation method with the help of 6-31G basis set function. Addition to this Nuclear magnetic resonance (NMR) study and vibration analysis were performed using GAUSSIAN 9, Gauge-Independent Atomic Orbital (GIAO) method was used for NMR determination.

Keywords GNP's · PVA based composites · Molecular study

K. S. Lokesh (✉)
Department of Aeronautical Engineering, Srinivas Institute of Technology, Mangalore 574143, India
e-mail: lokeshvijay@sitmng.ac.in

T. P. Kumar
Department of Nanotechnology, Srinivas Institute of Technology, Mangalore 574143, India

C. G. Ramachandra
Department of Mechanical Engineering, Presidency University, Bangalore 560064, India

D. Shrinivasa Mayya
Department of Mechanical Engineering, Srinivas Institute of Technology, Manglore 574143, India

S. B. Boppana et al. (eds.), *Structural Composite Materials*, Composites Science and Technology, https://doi.org/10.1007/978-981-99-5982-2_6

1 Introduction

In the present day, polymer/layered-inorganic, polymer–clay nanocomposites are engineered every day for the future they promise in the advancement of present-day technologies, the notable ones include electro-chromic displays, sensors, catalysts, etc. The fact being that, nanocomposites provide significant proof in change at their behaviour and properties at ultra-low loads of such nanofillers as graphite nanoplatelets, and carbon nanotubes (CNTs). Graphite is a single layered material which is made up of single sheet of carbon which is one-atom thick. The structural foundation of graphite consists of silicate layers and their integration to their counterpart, the carbon nanotubes (CNTs) propose that graphite makes a better nanofiller in a wide array of scientific and technological aspects. Aspect ratio of a particular graphene sheet can be increased just by separating their graphite layers by a degree, this in turn can satisfy the high aspect ratio requirement for nanocomposites [1]. Graphite crystal when converted into nanoparticle size, we can observe edges that surround the particles. Graphite nanoparticles are made up of nanosized graphene sheets whose peripheries are formed with the help of s-dangling bonds, the stacking arrangement remains constant. These dangling bonds are responsible for the reaction between foreign chemical species such as hydrogen or oxygen with the actual bonds of the structure. In this sense, a nano graphite is made up by finite flat graphene sheets having open edges. There are many methods which have been engineered to design graphite nanoparticles and sheets, they include micromechanical cleavage, chemical vapor deposition (CVD), solvent thermal reaction, thermal desorption of Si from SiC substrates, and chemical routes via graphite intercalation compounds (GIC) or graphite oxide (GO) [2–5]. The best method which gave the maximum output is the chemical treatment of graphite. But, the only downside of this method is that it involves the use of undesirable solvents under extreme conditions. The process of chemical exfoliation involves chemical treatment which produces some impurities, on the other hand, mechanical milling is another suitable technique for the preparation of graphite nano powder and nano structured sheets [6]. In the solution method, poly vinyl alcohol is used as polymer matrix material for the synthesis of graphite nanocomposites. The resulting nanocomposites have been investigated for its barrier properties. Polyvinyl alcohol is a water-soluble synthetic polymer and it is preferred for their excellent hydrophilic interactions, biocompatibility, and non-toxic nature. The model properties of PVA include better adhesion, emulsification, thin film forming, and that became the reason to use it as a wrap sizing, paper-coating agent, adhesives, drug delivery agent, and as an important aspect in biomedical and material packing [7–9]. Physical and electronic aspects of nanoparticles synthesized from graphite from any other conventional methods are characterized and investigated with the help of Scanning electron microscopy, X-ray diffraction, Electron spin resonance (ESR), Raman scattering and magnetic susceptibility. They proved quite resilient in both the mechanical applications from the industry of packings to the armour designs of the defence and to the fields of the semiconductors such as sensors and actuators, The experimental findings suggest that the synthesized nano graphite arranges itself in a polyhedral fashion, whose faces consist of a stack like structure arranged in stacks of 3–6 planar graphene sheets with an in planar size of 7–8 nm [10]. Density of states for this structure is considerably enhanced due to the consistence of edge-inherited non-bonding *p* orbitals in their fermi level, which is consistent with the theoretical suggestions [11]. The current output of the composite is achieved by optimized dispersions, interfacial chemistry and the nano size morphology, the advantage here being enormous surface area per unit volume, which is a common property of a nanofiller. Irrespective of the degree of polymerization of PVA, which is enhanced by annealing, the tensile strength and dynamic modulus of PVA hydrogels, the water content in the hydrogels has decreased respectively. PVA has a high mechanical strength. Additionally, the hardness of PVA hydrogels can be improved by decreasing the water concentration in their composition. The hydrogel when prepared from PVA with a degree of polymerization of 11,000, annealed at 1400 C for 24 h, has resulted in its highest tensile strength of 20 MPa and the best dynamic modulus of 180 MPa can when it is [12].

A review on rapid innovation in nanotechnology in the recent years has enabled multiple developments in advanced metal matrix nanocomposites for structural engineering and functional device design and their fabrication. Carbon enriched materials, such as carbon nanotubes (CNT's), graphite, and graphene contain unique magnetic, mechanical, thermal, electrical properties [13]. Due to their lubricious nature, these carbon rich materials have increased attention and researchers started to synthesize light-weight, self-lubricating metal matrix nanocomposites with superior tribological properties and mechanical properties for multiple applications in aerospace and automotive industries [14]. From literature it was clear that the less work carried out on synthesis of the graphite nano particles blended with polyvinyl alcohol with rare computational studies, hence the present work aims to synthesize graphite nano particles and develop graphite based nano composite blended with PVA along with parallel computational studies carried out to investigate the nuclear structure using the Density Functional theory (DFT). Stability analysis of each molecule is performed separately using the UB3LYP calculation method with the help of 6-31G basis set function and also NMR analysis of the computed structure.

2 Materials

2.1 PVA (Poly-Vinyl Alcohol)

PVA is an excellent bio-based polymer which is non-toxic and odourless in nature, the most important properties of PVA include its chemical resistance and mechanical properties. PVA exists in various forms depending upon the function of the user, most of the time it is used in the form of fibre, powder, film without any separation of its monomer units. PVA based upon the length of its monomer unit 'vinyl acetate' varies its core properties such as molecular weight which generally ranges from 20,000 to

400,000, the degree of hydrolysis which varies in accordance with the change in its molecular weight. Generally, this range lies around 80–99%. Increase in molecular weight has proved a significant increase in the tensile strength, solvent and block resistance. On the other hand, the properties such as solubility, water sensitivity, flexibility. The best application of PVA is the 'packaging', the reasons being many but, the main one is the reaction of multiple hydroxyl groups on its surface, this makes PVA one of the most hydrophilic polymers with high moisture sensitivity, the resulting blends and composite materials, for this reason have become more prominent in the packaging applications. According to Grande et al., PVA may be used as a compatibilizer for other biopolymer blend systems. For example, Chitosan/PLA blends in the presence of glycerol as a plasticizer has been improved greatly in terms of miscibility and formability, PVA on the other hand acts a dispersion medium for 9 polymers. PVA and its blends/composites due to their high material performance has increased its dominance over the industries for packaging applications over the recent years.

The conventional laboratory synthesis included a series of treatments since PVA production on a large scale is very difficult when its monomer units are directly polymerized due to the disturbing by-products and their interferences over the main applicants. The saponification reaction was employed according to the following scheme shown above in Fig. 1, the reaction is interpreted in the following chemical reactions wherein reaction A is depicted as the alcohol conjunction of poly vinyl acetate, whereas the reaction B indicates the saponification of poly vinyl acetate; the reaction C results in the retrieval of methyl acetate. Industrially, acetylene or ethylene are used as base materials in the preparation of vinyl acetate within the presence of acetic acid and/or oxygen, the polymerization is initiated with the application of heat, this heat is used to activate the vinyl acetate in the presence of methanol solution, Poly vinyl alcohol is produced via the application of saponification, during the polymerization process more than 70% of the monomer units are converted to poly vinyl acetate. PVA is prepared by partial or complete substitution of ester groups from the poly vinyl acetate in the presence of hydroxyl groups in the hydrolysis process, in the presence of sodium hydroxide or anhydrous sodium methylate.

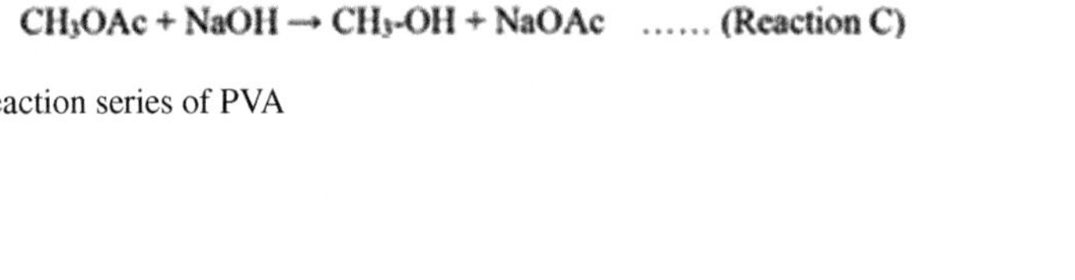

Fig. 1 Reaction series of PVA

2.2 Graphite Nanoparticles

The basic definition of a nanoparticle is that it is a particle of matter which is stable in the size range of 1–100 nm. The basic difference between micro particles (1–1000 μm), coarse particles (2500–10,000 nm), fine particles (100–2500 nm) and nano particles is their behaviour at their size ranges. At each of their size ranges, they have very different physical and chemical properties which include electronic, magnetic, optical and colloidal properties [15]. Graphite powder with a standard size of 28 μm and 99.75% pure is used as the base for this experiment. A Planetary Ball Mill (Retsch PM100) was used to carry out the ball milling process. The diameter of the balls used as the milling medium was 8 mm via the diameter, there were 20 balls used in the process along with a stainless-steel grinding jar of 50 ml capacity [16]. The ball to powder ratio (BPR) considered was 10:1 and the rotation speed of the jar was adjusted to 200 or 250 rpm in all the runs carried out. Pure and fresh was considered for each run to avoid and prevent sample mixing.

2.3 Physico-Chemical Properties

Graphite on the other hand is in crystalline form and is an analogue of carbon, the atoms are arranged in a hexagonal fashion. Under standard condition, it is the most stable form of carbon and it is a naturally occurring carbon-based structure. A study that worked on the composites made from graphite and nano-MgO proved better with the increased concentration of nanoparticles along with the support from the graphite base, through the aforementioned conclusions, it was established that the hybrid composite of aluminium, 5 weight percent graphite, and 2.5 weight percent nano-MgO has comparatively ideal mechanical and wear properties [17]. It is converted to diamond when exposed to high temperatures and pressures. The most basic use of graphite is seen in pencils and lubricants. Better properties of graphite can be described by using its good conductance towards heat and electricity. The high electrical conductivity of graphite makes it useful in the production of electronic products such as electrodes, batteries, solar panels, etc. Graphite is sp2 hybridized and the atoms are arranged in plane with each atom bound to 3 nearest neighbours with a bond angle of 120°. The individual layers arranged are called graphene. The graphite nanoparticles on the other hand consist of the following parameters with their melting point staging at 3550 °C and the boiling point at 4027 °C. The density of the particles on the conventional scale is 2.26 g/cm^3. The average particle size < 50 nm, the size here varies, depending on their concentrations but never cross 50 nm [18–21]. The thermal expansion of the graphite nanoparticles being 6.0 W/m–K and the thermal expansion being 4.9 μm/m–K, the young's modulus was found to be 21 GPa. These nanoparticles in some ways are similar to Si_3N_4 nanoparticles formed a well-compacted tribolayer that was retained across extended sliding distances, considerably influencing the hybrid nanocomposites' wear resistance [22].

2.4 Methodology

The present research aims to developing the GNP's filled PVA nano composites with varied GNP percentage of 0.5, 0.75, 1, and 1.5% by weight. The prepared graphite powder is analysed from the obtained values of wavelength confirms the presence of nano particles through UV spectroscopy. Samples are prepared using polycarbonate plates as a die with dimensions of 200 $\times$ 4 mm to accommodate the PVA and GNP's mixture added proportionally to get the cured samples when cured under oven for the period of 72 h.

2.5 Preparation Methods

The preparation method opted here is the ultrasound irradiation where the graphite structure is treated with variable ranges of ultrasound and the changes made to the structure was analysed by Transmission Electron Microscopy, laser light scattering, Raman spectroscopy for the structural modifications made. The crystalline graphite structure is subjected to irradiation with the frequencies ranging from 20 to 500 kHz, this analysis was performed in 3 varying solvent solutions: water (which is the best medium for cavitations), the surfactant (OMImBr) which is aqueous in nature and the solution containing sulphuric and nitric acids in mixed proportions, this resulted in the intercalation which led to exfoliation of the sample. The original basal particle of size 1681 m was ultimately reduced to 41 m after the ultrasonication, which delivered an insane energy density built up to 1.2 MW/m^2. Modifications in time for sonication and the nature of solvent has resulted in crystal structure, which affected the out of the plane thickness and the basal width as well. The ultrasonication promoted the disordering of the graphite tridimensionality stacking for all solvents used, with its strongest effect in acidic medium. With the use of a surfactant as sonication medium, the exfoliation of graphene flakes has been observed along with the formation of Turbostatic structures [17]. The next treatment which involved intercalation in acid medium before the application of ultrasound produced similar effects. Ultrasonic Processor (Sonics and Materials, 500 W Ultrasonic Processor-VC505) with 20 kHz converter, 350 W output which was adjusted with an ultrasonic probe (with amplitude of 431 m, and 19 mm diameter, Sonics and Materials) has been opted and used for the experiment. The probe is immersed inside a double jacketed reactor (cylindrical structure) with an internal diameter of 80 mm, containing 1.2 g of pure and fresh graphite powder treated with 60 ml solvent. In order to achieve precise temperature control, 50 °C constant temperature was maintained inside a double-jacket for the circulation of cooling liquid (with the help of a cryostat) inside the experiment setup to neutralize the temperature inside the reactor to ambient (250 °C). A 'rosette' type of reactor was used to carry out the sonication of graphite structures under the presence of sulfonitric-medium and water. The sonication of water suspensions containing GN4 graphite at 20 kHz which were set at a period of 1–22 h range [22].

Residual intercalation compound that was obtained by a consecutive treatment of water immersion and sulfonitric medium, this was subjected to ultrasonic irradiation at 20 kHz for 5 h. Sonication to GN4 graphite was performed at 20 kHz in a 4/1 (v/v) sulfonitric medium consisting of nitric acid (30% mass) and sulfuric acid (30% mass) for a time period of 40 min. A few suspensions (which are synthesized in the presence of water or in presence of OMImBr solutions) were subjected to irradiation at 500 kHz inside a homemade cup-horn reactor in the absence of cooling. The process of irradiation of GN4 in the presence of OMImBr solutions was carried out for a time period of 120 min at 500 and 20 kHz. The resulting KS4 samples was subjected to irradiation for 1 h at 20 kHz and for 10 h at 500 kHz inside water. The synthesized graphite nano particles from graphite powder are as shown in Fig. 2.

Fig. 2 **a** Dispersion of synthesized GNP's, **b** mixture of PVA blended with GNP's, **c** prepared and cured sample

2.6 UV Spectroscopy

UV spectroscopy is a structural spectroscopic study which is performed to identify the physical attributes of an unknown sample, the main principle used behind the UV studies is the Beer-Lambert law, this law proves states that for every solution in stable and meta-stable state, the concentration of the solute is directly proportional to the absorbance of the solution. The follow up theory states that whenever a molecule absorbs radiant energy from any region of the light spectrum, the atoms excite and due to the excess energy in the system, they move towards higher energy levels and return after a short interval leaving most of the excess energy in the form of heat or radiation, this emitted energy is often recorded as spectra, and is measured by a UV Spectro-photometer. Almost all of the chemical systems (compounds) absorb radiation either from UV region or visible region. The initiation of the process begins with a beam induced with power I, and it is shot onto the sample solution. Absorbance takes place when the sample molecules absorb some part of the incident radiation and the resulting beam is transmitted back, this transmitted beam is measured to calculate the ratio of absorbance by % T or %Transmittance. That is

$$\%T = 100 \times (I/I_o) \tag{1}$$

Absorbance is defined with the help of Beer-Lambert Law in the following way

$$A = -Log(I/I_o) = ECD \tag{2}$$

Here, E = Molar Absorptivity, C = Concentration of the Solute, MD = path length in cm. So, from the above equation

$$A = -LogT = ECD \tag{3}$$

For the thin film samples Eq. (2) is modifies as

$$A = -Log(I/Io) = ECT(\text{where T is the thickness}) \tag{4}$$

Therefore, absorbance is more useful compared to transmittance as it has a direct linear relationship with the concentration of the solute. Varian Cary 5000 UV–vis-NIR spectrometer was employed to observe the changes in the optical absorption spectra at various dilutions. Here, two empty optical cells were used to adjust the baseline correction. 0.15% polyvinyl alcohol solution, which was used as the dispersing agent in the original solution was taken as the reference. Transmittance recording mode spectra was adjusted from 300 to 1300 nm for pure dispersion samples and two other diffusions which were synthesized by the dilution of pure dispersion samples that are taken at 20 times and 40 times and fused with a neat PVA solution. The transmittance for the pure dispersion samples remained almost zero all through the range of wavelengths opted and the most important observation is that there were no

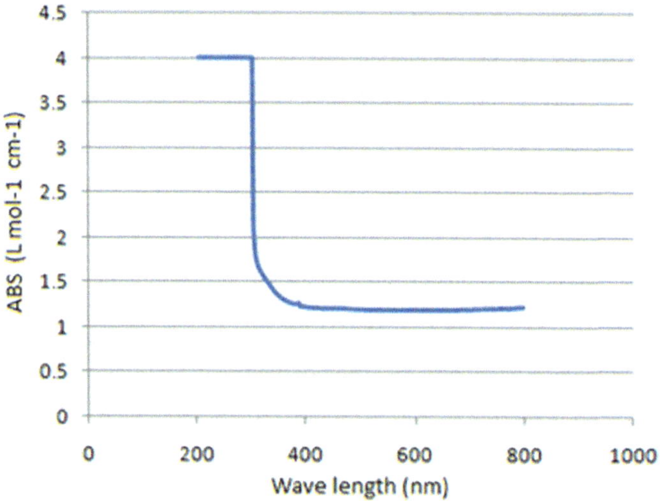

Fig. 3 UV-plot depicting the wavelength versus ABS

peaks in the spectra. The sensors were fabricated with the help of spray coating and dispersion [23]. UV–Vis spectroscopy with a range of 300 to 3000 nm was opted to read and record the transmittance data for every sample. The reference sample substrate opted was made with Polyester-imide. This reference sample was used to measure and obtain the transmittance values for the thin films. Experimental UV spectroscopy peak is observed and reported for the present synthesized graphite nano particles are shown in Fig. 3.

3 Results and Discussions

3.1 Computational Studies

The current computational modelling studies are performed using the Density Functional theory (DFT) for the investigation of the electronic structure (mainly the ground state) of the multi-atomic systems, molecules, and condensed phases. Using this theory, the properties of multi-atomic systems can be determined by functional, i.e., functions of another function, in the DFT, these are functions of the spatially dependent electron density. DFT studies is one of the most prominent, popular and versatile methods available in condensed matter physics, computational physics, and computational chemistry. It was first put on a firm theoretical footing by Walter Kohn and Pierre Hohenberg in the framework of the two Hohenberg–Kohn theorems (HK). The current study focuses on the structure of the PVA along with Graphite nanoparticle {which has been assumed as two graphite rings} and its prepared polymer which consists of PVA and Graphite nanoparticles which are treated in the ratio of 1:0.25, etc. To reduce the computational time cost we have used an input of monomer molecule of PVA and two rings of graphite molecules, the main factors of the computational studies which affects the performance is that the input components and the computational time, the specifications of the device which the work is performed too plays an equal role too. The input components however, can be optimized and taken by preferences; the computational time however, is dependent on the type of calculation and the no. of atoms taken from the molecule. The observed facts are that the more the no. of molecules the more the computational time. The ground state optimizations of PVA and graphite nanoparticle were carried out in Gaussian 9 software employing FOPT Simple calculation with the RB3LYP as the calculation method and with the 6-31G basis set. The following calculations has been performed separately for PVA, graphite nanoparticle and the polymer (PVA + Graphite np):

- Structure optimization—to refine the structure and remove any or all chemical restraints. (No change observed in structure as well as energy).
- Stability analysis—to check the stability of the molecule.
- Time dependent density functional theory calculation (TD-DFT)—To study the absorbance effects and plot a UV appropriately.
- NMR analysis—To plot the NMR graph for the polymer.
- Vibration analysis—To check the molecular oscillation of the computed structure and find any or all the imaginary molecular frequencies.

3.1.1 Stability Analysis

Stability analysis of each molecule is performed separately using the UB3LYP calculation method with the help of 6-31G basis set function [24]. This is the general basis set function used for every conventional computational calculation, the stability of each molecule used and the final polymer is presented below:

- PVA (poly vinyl alcohol)—a PVA monomer as a repeat unit was selected as the starting polymer material. In Fig. 4b, the stability is achieved by performing the above analysis. The resulting parameters we observed here is that there are no imaginary frequencies available. The spin type observed is singlet. The total potential energy of the molecule (E(RB3LYP)) being − 153.759419 a.u. (a negative value), this happens because when considering electrostatic potential we multiply charges, so for a − 1 and + 1 charge the results is always negative meaning attraction (but for two negative or two positive charges it's always positive, meaning repulsion). The dipole moment of the stabilized structure came to 2.1358 Debye.
- Graphite nanoparticles—Graphite nano particles are consistently formed as nanosheets when observed at nano scale, the original crystallographic structure of a graphite nanoparticle can be seen in Fig. 4a, and in Fig. 4c here we took 2 rings of one graphite nanoparticle sheet as representation, but when we increase the concentration to a few more rings, multiple properties may vary. The stability analysis is performed by using the above specifications and we observed that the

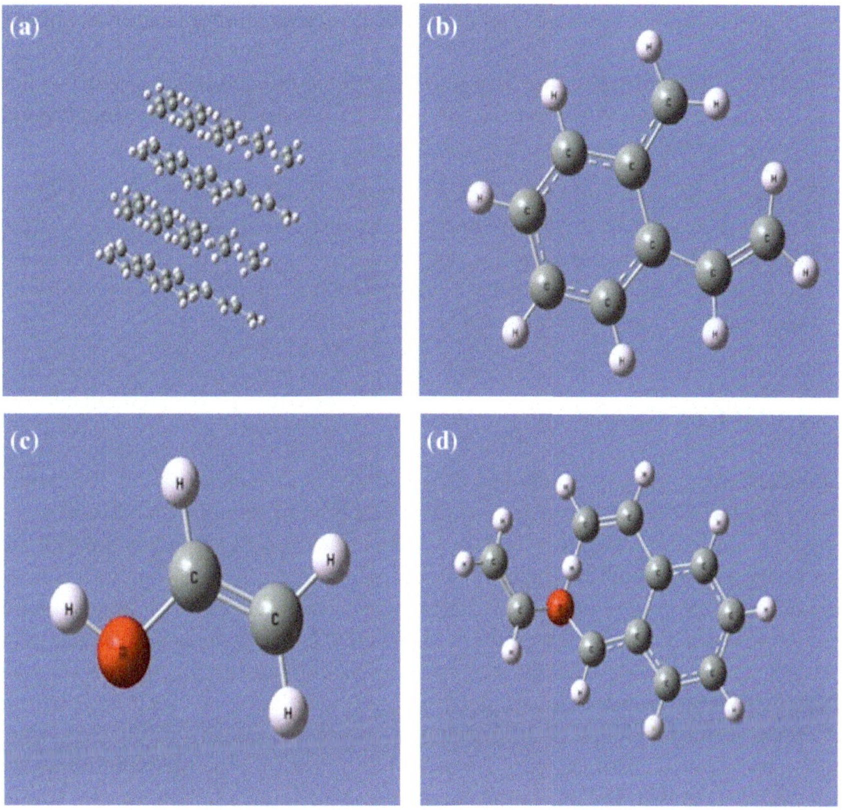

Fig. 4 **a** Graphite nano particle. **b** Graphite computed structure. **c** PVA structure. **d** Combined polymer (PVA + Graphite NP's)

spin type is doublet. The total potential energy of the molecule (E(UB3LYP) is –348.23038136a.u. The dipole moment of the stabilized structure is 0.3763 Debye.

- The polymer—The polymer taken here is the final product of the project and it is the mixture of PVA along with graphite nanoparticles. The stability analysis is performed by using the above specifications and we observed that the spin type is doublet. The total potential energy of the molecule (E(UB3LYP) is –497.38629372a.u. The dipole moment of the stabilized structure is 6.5416 Debye. The stability was a typical analysis done here; there was a slight confusion at the beginning, due to the bond interactions of the graphite nanoparticle to the PVA. So, we had to carry out 3 iterations

    (a) Attachment of graphite group to the oxygen atom—During this iteration the computational result gave a perfect crystal structure with the dipole moment being 6.5416 Debye units.
    (b) Attachment of graphite group to the hydrogen atom—during this iteration the structure gave a fatal error because hydrogen has only 1 valence electron

and by binding it to the graphite molecule became unstable and the molecule broke.

(c) Attachment of graphite group to the double bonded carbon atom—during this iteration the computational result gave a perfect crystal structure with the dipole moment of 2.5395 Debye units.

So, out of the 3 iterations performed the first iteration was considered for the further study because of its dipole moment, because the higher the dipole moment, higher will be the ionic character of that bond and lower will be the covalent character of that bond. If the ionic Character is higher, then the reactivity will also be higher, and ultimately the stability will be higher.

3.1.2 TD-DFT Calculation

Time-dependent density-functional theory (TD-DFT) is a quantum mechanical theory used to study the properties and dynamics of multi-atomic systems and many-body systems in the presence of time-dependant functions, these are potentials which can be represented with the help of electric or magnetic fields. The change in the properties of the system due to the interference of such potentials on molecules and micro solids can be analysed with the help of TDDFT. The features such as excitation energies and frequency response properties and photo absorption spectra are some examples. Gaussian software can determine the simulation of vibration electronic spectra described by three basic parameters (position, intensity and width), which can be used for the estimation of fundamental transition characteristics as well as quantitative analysis. The excitation energies and oscillator strength for each excited state were evaluated by QM calculations. Generally, absorption spectrum consists of a number of absorption bands plots of the molar extinction coefficient (ε) versus the excitation energy/wavelength (λ) in nm. The peak for the computed structure came around at 936.83 strength, where the peak is observed at 936.83 nm (Fig. 5).

3.1.3 NMR Analysis

Nuclear Magnetic Resonance is a most powerful tool for extracting the structure of many-body systems, this makes the tool a practical analysis method for the study of organic compounds, therefore the calculation of H and C chemical shifts, as well as coupling constants, it is helpful in the notation of measured signals on a spectrum to an actual functional group. NMR shielding tensors may be computed with the Continuous Set of Gauge Transformations (CSGT) method and the Gauge-Independent Atomic Orbital (GIAO) method. Magnetic susceptibilities may also be computed with both GIAOs and CGST. Gaussian also supports the IGAIM method (a slight variation on the CSGT method) and the Single Origin method, for both shielding tensor and magnetic susceptibilities. The NMR of the current sample is calculated using the SCF GIAO Magnetic shielding method where a graph is plotted

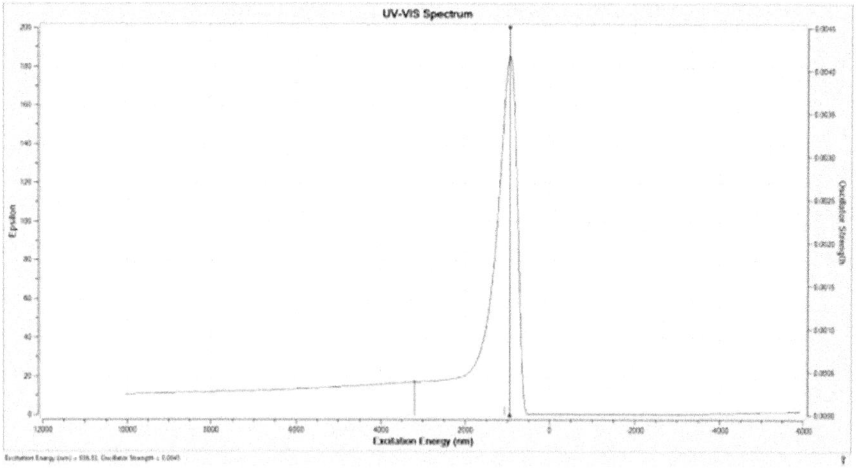

Fig. 5 Excitation energy against oscillations

on degeneracy versus shielding (ppm), with the degeneracy tolerance value set at 0.05. Figure 6a gives a glance at the NMR calculation of O shift, where the highest peak is observed 6th O atom at degeneracy = 1 and shielding = 337.204 ppm. The second Fig. 6b gives a glance at the NMR calculation of H shift, where the highest peak is observed on 20th H atom at degeneracy = 2 and shielding = 27.69. The third Fig. 6c gives us a glance at the NMR calculation of C shift, where the highest peak is observed at 10th C atom at degeneracy = 1 and the shielding value = 115.116 ppm.

3.1.4 Vibration Analysis

The vibration states of a particular molecule can be identified and probed in various ways. The best option available is the method of infrared spectroscopy, the vibration transitions generally demand certain amounts of energies which link to the infrared regions of the electromagnetic spectrum. Currently, infrared spectroscopy and Raman spectroscopy use visible light to the measure vibrational frequencies directly. These two techniques are complementary to each other and can provide useful information such as in the case of the rule of mutual exclusion for centrosymmetric molecules. In the present study, the sample has been analysed under 63 modes and the IR spectrum graph has been plotted with the resulting data, the highest peak was observed at 51th mode with the frequency at 1609.39 cm^{-1}. One outcome of this test is to check any imaginary frequencies of the computed sample, imaginary frequencies signify a local maximum on the potential energy surface, the presence of an imaginary frequency signifies that there is a negative Eigen value in the equation, while computing, which means that one of the square roots contains a negative number inside, which leads to an imaginary frequency state at one of its modes. Figure 7 shows the NMR analysis

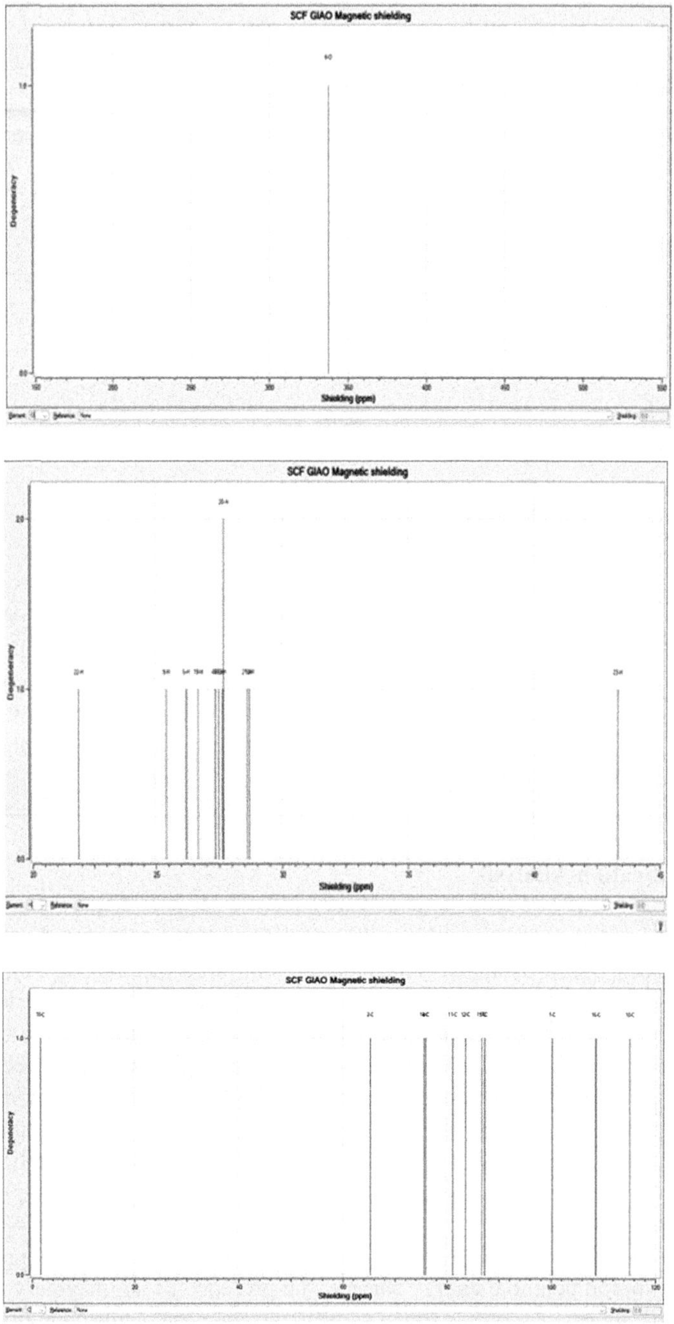

Fig. 6 **a** NMR analysis of the computed structure in Oxygen (O)-shift. **b** NMR analysis of the computed structure in Hydrogen (H)-shift. **c** NMR analysis of the computed structure in the Carbon(C)-shift

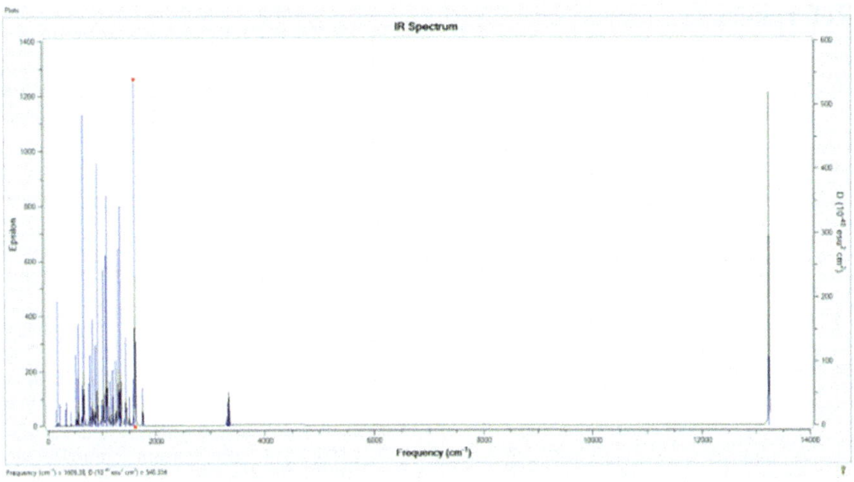

Fig. 7 NMR analysis of the computed structure in the Carbon(C)-shift

of the computed structure in the carbon(C)-shift. In the present study, however there were no imaginary frequencies observed in the computed structure shown in Fig. 7.

4 Conclusion

Developing the GNP's filled PVA nano composites with varied GNP percentage of 0.5, 0.75, 1, and 1.5% by weight. Synthesized graphite nano particles are analysed through UV spectroscopy where in the highest peak was observed at 255 nm. Computational studies were performed on the sample, by measuring important parameters such as total energies, absorbance, and resonance of each element and vibration energy of the sample. Stability analysis was done to various configurations of the same structure, out of which 3 structures have been selected; the final structure has been selected due to its optimum dipole moment. The spectrum analysis was done to check the absorbance nature of the computed structure by taking its position, intensity into consideration. The NMR study was performed to check the shifts at different degeneracies for each of the elements present in the structure and the highest peaks were observed at where the highest peak is observed on 20th H atom, 6th O atom and 10th C atom. Finally, the vibration studies are performed to check the presence of imaginary frequencies and none have been found so far. These studies prove that the produced structure is stable and strong in all of its aspects, when compared with the conventional ones.

References

1. Zhang J, Wang J, Lin T, Wang CH, Ghorbani K, Fang J, Wang X (2014) Magnetic and mechanical properties of polyvinyl alcohol (PVA) nanocomposites with hybrid nanofillers—graphene oxide tethered with magnetic Fe_3O_4 nanoparticles. Chem Eng J 237:462–468
2. Pandele AM, Ionita M, Crica L, Dinescu S, Costache M, Iovu H (2014) Synthesis, characterization, and in vitro studies of graphene oxide/chitosan–polyvinyl alcohol films. Carbohyd Polym 102:813–820
3. Samzadeh-Kermani A, Mirzaee M, Ghaffari-Moghaddam M (2016) Polyvinyl alcohol/polyaniline/ZnO nanocomposite: synthesis, characterization and bactericidal property. Adv Bio Chem 6(01):1. https://doi.org/10.1063/1.5092886
4. Hemalatha KS, Rukmani K (2016) Synthesis, characterization and optical properties of polyvinyl alcohol–cerium oxide nanocomposite films. RSC Adv 6(78):74354–74366
5. Deshmukh K, Ahamed MB, Deshmukh RR, Pasha SK, Chidambaram K, Sadasivuni KK et al (2016) Eco-friendly synthesis of graphene oxide reinforced hydroxypropyl methylcellulose/polyvinyl alcohol blend nanocomposites filled with zinc oxide nanoparticles for high-k capacitor applications. Polym-Plast Technol Eng 55(12):1240–1253. http://ijsrmme.com/IJSRMME18248
6. Chowdhury SR, Roy PS, Bhattacharya SK (2017) Green synthesis and characterization of polyvinyl alcohol stabilized palladium nanoparticles: effect of solvent on diameter and catalytic activity. Adv Nat Sci: Nanosci Nanotechnol 8(2):025002
7. Pooresmaeil M, Namazi H (2019) Preparation and characterization of polyvinyl alcohol/β-cyclodextrin/GO-Ag nanocomposite with improved antibacterial and strength properties. Polym Adv Technol 30(2):447–456
8. Aslam M, Kalyar MA, Raza ZA (2018) Polyvinyl alcohol: a review of research status and use of polyvinyl alcohol-based nanocomposites. Polym Eng Sci 58(12):2119–2132
9. Rani P, Ahamed MB, Deshmukh K (2021) Structural, dielectric and EMI shielding properties of polyvinyl alcohol/chitosan blend nanocomposites integrated with graphite oxide and nickel oxide nanofillers. J Mater Sci: Mater Electron 32(1):764–779
10. Khodiri AA, Al-Ashry MY, El-Shamy AG (2020) Novel hybrid nanocomposites based on polyvinyl alcohol/graphene/magnetite nanoparticles for high electromagnetic shielding performance. J Alloy Compd 847:156430
11. Łoś S, Duclaux L, Alvarez L, Hawełek Ł, Duber S, Kempiński W (2013) Cleavage and size reduction of graphite crystal using ultrasound radiation. Carbon 55:53–61
12. Katta JK, Marcolongo M, Lowman A, Mansmann KA (2017) (χ.χ.) friction and wear behaviour of PVA/PVR for articular cartilage replacement. 9–2017. https://ijsrmme.com/IJSRMME19315
13. Xia JY, Tang MT, Cui CHEN, Jin SM, Chen YM (2012) Preparation of α-Bi_2O_3 from bismuth powders through low-temperature oxidation. Trans Nonferrous Metals Soc China 22(9):2289–2294
14. VK et al (2019) Assessment of axial cramp resistance of graphene based aluminium nano composites by stir casting technique. Int J Sci Res Mech Mater Eng (IJSRMME) 3(2):01–07. ISSN: 2457-0435. https://ijsrmme.com/IJSRMME19314
15. Lokesh KS (2018) Determination of scratch resistance of graphene based nano composites. Int J Acad Eng Res (IJAER) 2(7):20–24. ISSN: 2000-001X
16. Moghadam AD, Omrani E, Menezes PL, Rohatgi PK (2015) Mechanical and tribological properties of self-lubricating metal matrix nanocomposites reinforced by carbon nanotubes (CNTs) and graphene—a review. Compos B Eng 77:402–420
17. Irhayyim SS, Hammood HS, Mahdi AD (2020) Mechanical and wear properties of hybrid aluminium matrix composite reinforced with graphite and nano MgO particles prepared by powder metallurgy technique. 7:103–115. https://ijsrmme.com/IJSRMME19314
18. Lokesh KS (2019) Experimental study on effect of silicon carbide and graphite particles on mechanical behaviour of alluminium-6061 alloys. Int J Sci Res Mech Mater Eng (IJSRMME) 3(1):10–17. ISSN: 2457-0435. http://ijsrmme.com/IJSRMME19311

19. Naveen Kumar JR, Prasad P, Hebbar N, Savitha MB, Navaneeth Gowda N (2021) A potential application of polyethyleneimine-reduced graphene oxide nanocomposite sensing film coated on interdigitated electrode prepared from copper-clad for carbon dioxide detection. Mater Res Innovations 25(6):363–371. https://doi.org/10.1080/14328917.2020.1826674
20. Lokesh KS, Chethan IC, Kumar N, Kannantha V (2018) Determination of compressive strength of graphene reinforced with aluminium-7075 metal matrix composites. Int J Mech Eng Technol 9(1):327–335
21. Naveen Kumar JR, Kannantha V, Pinto T, Sampreeth U (2020) Experimental evaluation of substrate and annealing conditions on ZnO thin films prepared by sol-gel method. Mater Today: Proc 24(Part 2):201–208. https://doi.org/10.1016/j.matpr.2020.04.268. ISSN 2214-7853
22. Hariharasakthisudhan P, Sathickbasha K, Surya Rajan B, Jeyanthram V (2020) Steady-state wear behaviour of multi-scale hybrid composite—AA6061/Al2O3 (μm)/Si3N4 (nm)/ Graphite—EN31 steel tribo system. J Tribol 142(1)
23. Wei Q, Wang Y, Rao Y, Jiang A, Zhang K, Lu T, Chen X (2019) Evaluating the effects of nano silica on mechanical and tribological properties of polyvinyl alcohol/polyacrylamide polymer composites for artificial cartilage from an atomic level. Polymers 11(1):76
24. Di Y, Li J, Ye T, Xiaoyun Y (2015) The preparation and mechanical properties of novel PVA/ SiO_2 film. J Exp Nanosci 10(15):1137–1142

Manufacturing Particulate and Fiber Reinforced Composites

Manufacturing Process of Fibre Reinforced and Particulate Reinforced Composites

K. S. Lokesh, C. G. Ramachandra, and J. R. Naveen Kumar

Abstract Composite structures meant for structural uniqueness and property enhancing members in the broad areas covering the areas of aerospace and automobile applications concerning the aspects relates to weight. Meantime manufacturing the light weight structures owing to their specific features measures the type of methods followed is another important factor. The present report discusses about the various manufacturing methods involved in producing the prominent category of composites which aims to establish the research aspect as well as commercial aspects. One such combination that composite manufacturers are ever look in to is fibre reinforced and particle reinforced composites. The present work also emphasizes the comparison of various manufacturing methods adopted with various benefits, limitations for the both manufacturing methods.

Keywords Manufacturing methods · Particulate composites · FRP composites

1 Introduction

Due to Considerable improvement in mechanical and physical properties, as well as their high performance, fibre and particle reinforced composite materials are seeing increased use in a variety of technological fields. Composite Materials created using either of these techniques are frequently utilised for a variety of structures like as aircraft, robots, tennis rackets, bicycles, manufacturing machinery, and so on. Fibres/particles serve as a load carrying medium in fibre and particle reinforced composite

K. S. Lokesh (✉)
Department of Aeronautical Engineering, Srinivas Institute of Technology, Mangalore 574143, India
e-mail: lokeshvijay@sitmng.ac.in

C. G. Ramachandra
Department of Mechanical Engineering, Presidency University, Bangalore 560064, India

J. R. Naveen Kumar
Department of Electronics and Communication, Srinivas University Institute of Engineering & Technology, Surathkal, Mangalore 574146, India

S. B. Boppana et al. (eds.), *Structural Composite Materials*, Composites Science and Technology, https://doi.org/10.1007/978-981-99-5982-2_7

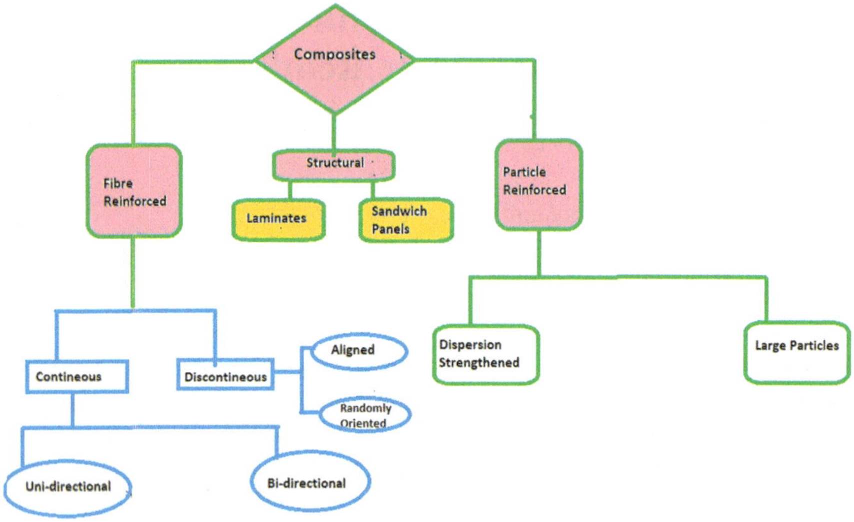

Fig. 1 Brief classification of composites

materials, while the matrix serves as a load transporting medium [1–3]. Different techniques, including filament winding, hand lay-up, stir processing etc. are used to create composite components including foam cored structures [4, 5]. They may need to be machined after fabrication to enable dimensional control for simple assembly and for functional aspects.

1.1 Classification of Composites

Composite structures are widely categorized in to fibre reinforced, structural and Particulate Reinforced Composites in which fibres are further divided in to continuous and discontinuous with uni and bi-directional group and randomly oriented and aligned manner respectively. Whereas the structural composites are categorized with laminates and sandwich panels and particulate reinforced in to dispersion strengthened and large particles. Figure 1 shows the brief classification of 3 categories of composite structures.

1.2 Fibre Reinforced Composites Production Techniques

Depending on the product's use and the type of reinforcement used, many procedures exist. Figure 1 depicts the classification of composite manufacturing processes. Fibre reinforced polymer composites are made using a variety of methods. Refer to

According to the types of reinforcement employed, the matrix system used, the application, and the cost of production, Table 1 compares various manufacturing procedures, the product's quality, and the production rate.

1.2.1 Hand Lay-Up Method

The hand lay-up open moulding technique can be used to create a variety of composite goods, from very small to extremely large. Despite the modest manufacturing volume per mould, several moulds can be used to create significant production quantities. The simplest composites moulding hand lay-up which is as shown in Fig. 2 which offers inexpensive equipment, straightforward processing, and a variety of part sizes. Modifications to the design are simple. Equipment requires a minimum investment. Good production rates and consistent quality are achievable with trained operators.

1.2.2 Spray-Layup Technique

When spraying a mixture of chopped fibres, resin, and catalyst onto a mold's surface, a spray gun is employed. Open mould and a semi-automatic method are used here. Air trapped inside fibres is squeezed out using spray rollers. Figure 3 depicts the process of developing FRP composites by spray lay up method.

1.2.3 Filament Winding

The process of filament winding is typically applied to the production of hollow, spherical, or prismatic objects, such as pipes and tanks. A customised winding machine is used to wound continuous fibre tows onto a rotating mandrel. The aerospace, energy, and consumer product industries frequently use filament wound parts. The filament winding machine receives continuous fibre tows through a fibre supply system, and then winds them onto a mandrel in a present, repeating geometric pattern. A fibre delivery head that is fastened to a moveable carriage on the filament winding machine directs the tow location. The winding angle, or angle at which the tow is in relation to the mandrel axis, can be adjusted to create the necessary strength and stiffness. Once there are sufficient amounts of two layers, the resulting laminate is cured on the mandrel. The mandrel shape and laminate thickness dictate the overall size and shape of the completed item. Method of Producing cylindrical materials as shown in Fig. 4.

1.2.4 Resin Transfer Moulding

Figure 5a designates the RTM method, for the compression of the curved like surface, RTM uses a resin transfer technique with vacuum assistance and a flexible solid

Table 1 Distinguishing features of various manufacturing methods for the production of FRP composites

Method of manufacturing	Details of the methods followed	Material system	Benefits	Limitations
Hand lay-up [6, 7]	Fiber layers are laid up to the appropriate sample thickness, then they are thoroughly impregnated with resin	*Matrix system*—thermoset epoxy, polyester, polyaniline *Reinforcement*—fabrics that are knitted, bonded, woven, or sewn	Process is simple, inexpensive, and quick	Fibre volume fraction will be less
Spray layup refer Fig. 3 [7–10]	When spraying a mixture of chopped fibres, resin, and catalyst onto a mold's surface, a spray gun is employed. Open mould and a semi-automatic method are used here. Air trapped inside fibres is squeezed out using spray rollers	*Matrix system*—thermoset, polyester *Reinforcement*—all type chopped fibers, flanks, particles	Shape complexity may be achieved using this process, which also produces medium to big components more quickly, at a lower cost, with greater versatility, and with ease	It can produce volatile chemical substances, dangerous air pollution, and being heavy and resinous
Prepreg layup [11, 12]	Cured pre-impregnated resin textiles are used in the procedure. The tool is coated with pre-impregnated resin, and the final step is curing the component, either at ambient temperature or at a higher temperature	*Matrix system*—thermoset epoxy polyester *Reinforcement*—carbon with glass. Boron (roving and chopped strand matting, woven roving fabric) (roving and chopped strand mats, woven roving fabric)	This method is quicker and offers a good fibre volume ratio	Compared to hand layup, cost is enormous. Prepreg should be handled carefully, stored correctly, and the resin's pot life checked before use
Filament winding refer Fig. 4 [13, 14]	To create cylindrical pieces, a rotating mandrel is wrapped under tension in the right pattern with roving or filaments from bobbins	*Matrix system*—epoxy, polyester, vinyl ester, phenolic *Reinforcement*—all types of fibres	As a high fibre volume fraction is possible, the process is automated	Because of the initial cost is high, mass production is preferred

(continued)

Table 1 (continued)

Method of manufacturing	Details of the methods followed	Material system	Benefits	Limitations
Resin transfer molding (RTM), refer Fig. 5 [15]	A closed injection moulding method is the RTM. In the RTM method, dried cloth is first placed in a closed mould and then resin is injected into it under high pressure	*Matrix system*—thermoset, epoxy, polyester, vinyl ester and phenolic *Reinforcement*—roving, woven fabric, and mat made of chopped fibre	The product that comes out of this close-mould process is well finished on both sides	Because each part needs a unique cure and custom tooling, production time and costs are slightly higher
Vacuum assisted resin transfer molding (VART M) refer Fig. 5b	A different RTM procedure—vacuum bag, release or distributor film, and sealing tape are used to wrap the fabric after it has been placed on a mold. before being vacuum-bagged	*Matrix system*—thermoset resin-vinyl ester, polyester, and epoxy Woven fabric, roving, and chopped fibre mat serve as reinforcement	The method of achieving smoother thickness was made simpler by using flexible, simple moulding components The resin flows into and out of the system	The area of interest for further research in VARTM is control over part thickness variation
Pultrusion process refer Fig. 6 [16]	The continuous roving is pulled from the bobbins and directed into the resin tank, where it is mixed and runs through the pre-heater to take on the desired shape before going through the heating die for the final curing	*Matrix system*—thermoset and thermoplastic continuous roving, fiber-chopped strand mats, polyester, vinyl ester, and epoxy reinforcement	To create consistent cross sections, the method is utilised in mass production	Not cost-effective for batch or unit production

(continued)

Table 1 (continued)

Method of manufacturing	Details of the methods followed	Material system	Benefits	Limitations
Compression molding refer Fig. 7 [16]	Temperature, pressure, and time are the three key variables in the compression moulding process, and they all depend on the size, shape, and thickness of the pieces as well as the required curing cycle	*Matrix system*—thermoset (epoxy, thermo-plastic materials (polypropylene, polyester, polyvinyl ester, phenol) polyethylene nylon, polyether ether ketonetc) *Reinforcement*—woven fabric (UD/BD), short fibers, chopped fibers, and chopped strand mats	It requires less trimming and finishing, has excellent part-to-part repeatability, and is less expensive	For the best product quality, there must be ideal adjustment of pressure and temperature between the top and bottom moulds mold
Centrifugal casting [17]	Resin and chopped fibres are spun into cylindrical shapes using centrifugal force	*Matrix system*—Thermoset resin-epoxy, polyester *Reinforcement*—chopped fibers	This process flow is uninterrupted and best fit for mass production	Centrifugal force must be produced by the component rotating

Fig. 2 Hand lay-up method

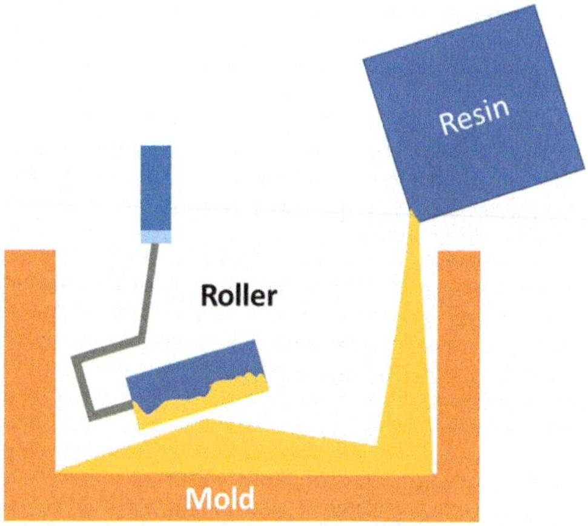

counter tool. This procedure results in improved strength-to-weight properties, higher laminate compression, and a high glass-to-resin ratio. Two finished surfaces are on RTM pieces. The mould is filled with reinforcement mat or woven roving before being sealed and clamped. Up until the mould is completely filled, air is displaced and vented at the edges as low-viscosity resin that has been catalysed is poured in under pressure. Typically, composite or nickel shell-faced composite construction is used to create the moulds for this low-pressure system. In order to produce uniform homogenous cross section of the final product vacuum bagging shown in Fig. 5b

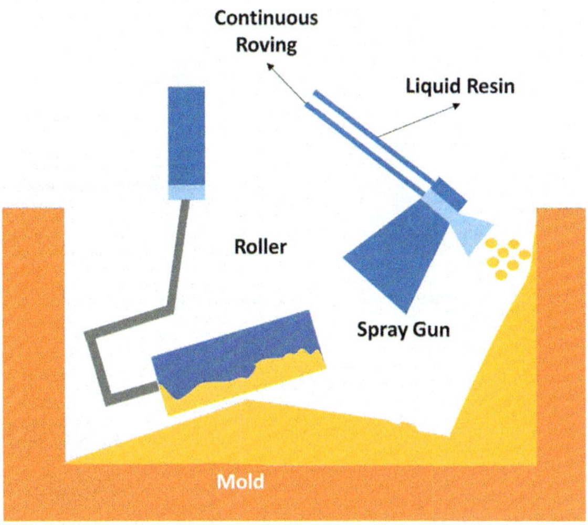

Fig. 3 Spray lay-up method

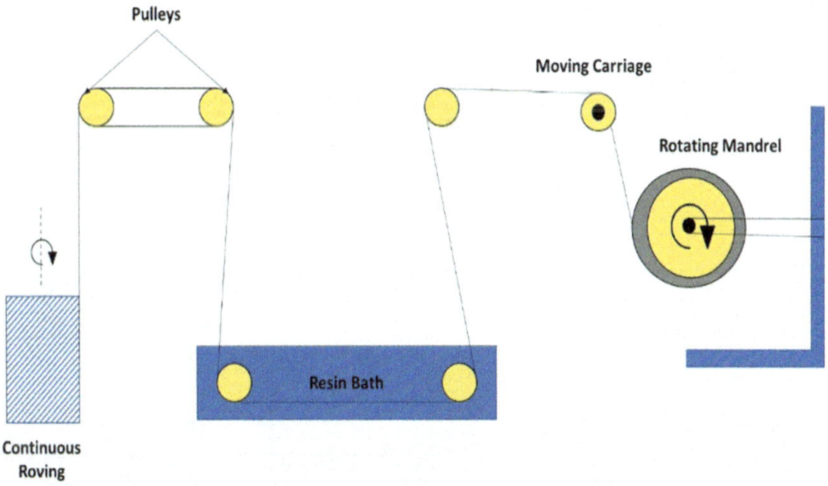

Fig. 4 Filament winding process

method is adopted where bag is compressed by vacuumizing the air inside the bag thereby creating the pressure inside to the layup to obtain the nice surface which forms the inner contour of the mold used.

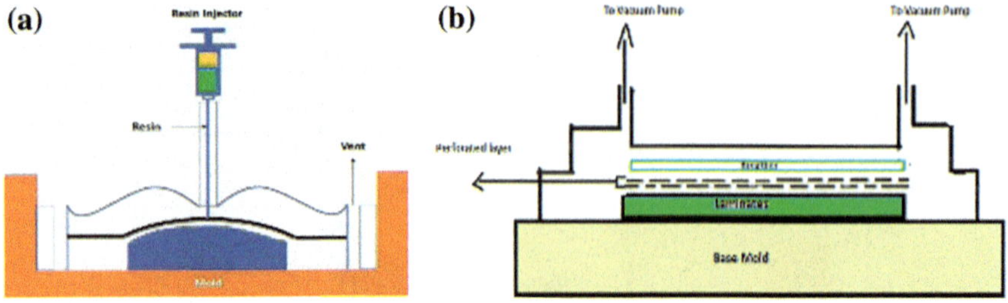

Fig. 5 **a** and **b** Represents resin transfer moulding and vacuum bagging method

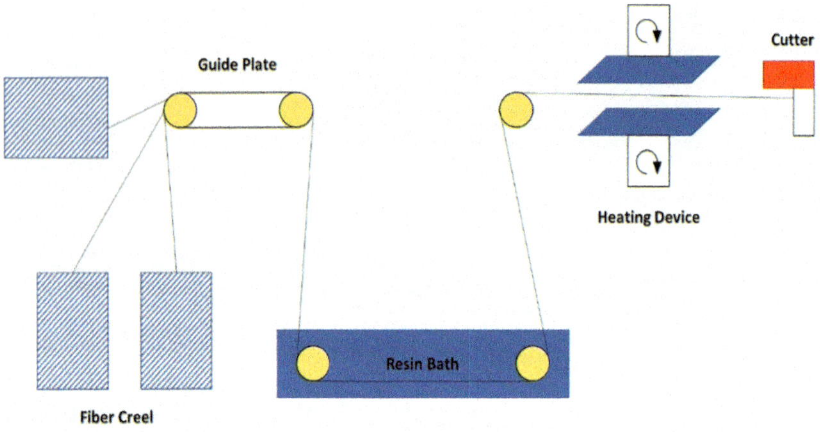

Fig. 6 Pultrusion process

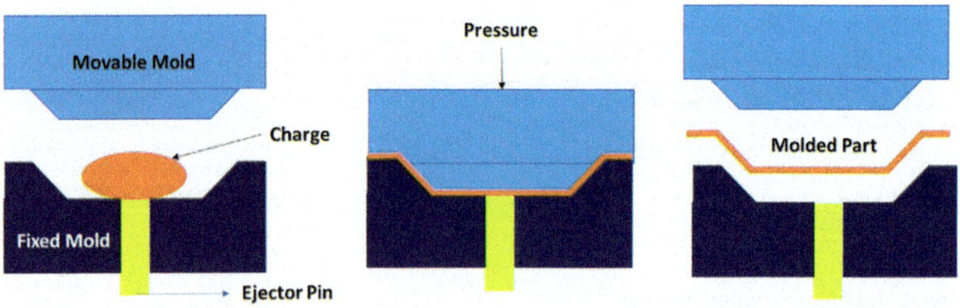

Fig. 7 Compression molding

1.2.5 Pultrusion Process

The continuous roving from the bobbins is directed to be gathered and combined in the resin tank, where it then flows through the pre-heater to take on the desired shape before passing through the heating die for the final curing. In contrast to extrusion, this method involves pulling a portion using a system-provided pulling mechanism. A cutting saw is used to shape the finished product into the required shape at the conclusion of the procedure. The method is employed in mass manufacturing to create continuous cross sections, but it is not cost-effective for individual or batch production. Typical manufacturing process flow is as shown in Fig. 6.

1.2.6 Compression Molding

A measured quantity of moulding material that is usually preheated (referred to as a charge) is compressed into the desired form using two heated moulds during the manufacturing process known as compression molding. Figure 7 represents the production of FRP parts by compression molding method. Following are the steps in the sequential compression moulding process:

1. There are many ways to make tooling, including machining, die casting, and 3D printing which demonstrates creating a part.
2. Depending on the particular machine or device you're using, this may entail cleaning the mould, turning on the heat, and other preparation procedures.
3. Prepare the charge by choosing the material type and calculating the necessary quantity.
4. Position the charge from the middle of the mold at the bottom.
5. Close the top mould, apply pressure, and wait for your component to form to compress the portion. Heat is another common manufacturing technique used during compression, which softens the raw materials and helps hasten production.
6. Remove the finished piece.
7. Resin flash around.
8. Before the final assembly, the part may need to be cleaned and the edges manually cut off or removed.

1.2.7 Centrifugal Casting

High material soundness components are produced by centrifugal casting. As a result, it is the technology of choice for many high-reliability products, including the compressor casings for jet engines, hydro wear rings, and numerous military items. It has been demonstrated that forgings and fabrications are inferior to a more affordable to provide shapes and lower manufacturing costs and less machining requirements. The first step in the centrifugal casting process is the pouring of molten metal into a heated, spinning die. Depending on how the intended part is set up, the die may be orientated either on a vertical or horizontal axis. Though the different production methods yield different outcomes with different compositional arrangements, there also exists the specific methods could yield different outcomes in conjunction with utilization of different reinforcements, fillers and matrix systems. In this view, few reports reveals that the effects of layup method to draft the considerable outcomes while processing it with different reinforcements and fillers. One such method is

Table 2 Role of production methods to yield significant outcomes through various fillers on type of reinforcement combined with polymer/metal matrix composites [18–31]

Type of reinforcement	Type of production method	Fillers used	Results observed
Fibre reinforced-woven E-glass and chopped fibres	Hand-lay-up method	Casio3,waste rubber, E-waste glass, mineral filler	Significant improvement in wear resistance
Particulate reinforced-micro and nano particulate with al-alloys	Stir casting method	Red mud particles, graphite particles, Sic	Hardness and, compressive strength is improved greatly
Fibre reinforced-fibre orientations fibres, fibre mat	Hand lay-up method	Mineral fillers-waste and scrap glass	Tensile and flexural properties are enhanced greatly for different loading of fillers
Fibre reinforced-jute/silk/ CSM fibre	Hand lay-up method	Fish scale, Cocopeat powder, waste rubber	Tensile and flexural properties are improved
Particulate reinforced-polyethyleneimine	Thin film deposition	Reduced graphene oxide	Accuracy in sensitivity of detecting CO_2 gas is achieved

hand-up method which is adopted to process different reinforcements with different fillers generated considerable enhancement in the mechanical results. Table 2 represents the Role of Production methods to yield significant outcomes through various fillers on type of reinforcement combined with polymer/metal matrix composites.

2 Particle Reinforced Composites

2.1 Methods of Manufacturing

One significant kind of material is particulate composites, which have a polymers and metal matrix system. These substances are typically utilised as building materials, high-performance engineering substances, or protective organic coatings. Cutting tools made of metal matrix composites reinforced with ceramic particles and high-temperature environments can both benefit from ceramic matrix composites. Normally particulate composites are prepared by various methods including In-Situ reaction precipitation, Powder metallurgy technique, Liquid In-filtration method etc. Some of the familiar manufacturing methods for producing particulate reinforced composites are discussed as follows. Table 3 depicts the various manufacturing methods for the production of Particle Reinforced Composites.

Table 3 Distinguishing features of various manufacturing process for the production of particulate reinforced composites

Method of manufacturing	Details of the methods followed	Benefits	Limitations
Agitation-preparation	Here, a specific method of mixing results in the direct addition of particle reinforcement into the molten metal melt, dispersing the particles throughout the melt of the aluminium alloy	Practically all casting techniques may be used to create it, and it has low equipment needs, a straightforward manufacturing process, and broad particle type and size adaptability	It is vital to maintain control over the mixing speed to ensure consistent mixing of the particles in the matrix and prevent clogging
In situ production method	Chemical processes inside a solid or liquid matrix can provide reinforcement elements for composites	Strong resistance to wear and corrosion reinforcements with minimal sensitivity to thermal shocks and temperature changes and thermodynamically stable reinforcements	The mechanical properties of in-situ composites created using fluid and strong state powder handling processes depend on a number of complex factors, including the weight rates of reactants, response time, sintering temperature, tension of hot–cold squeezing, and warming cooling rates
Powder metallurgy method	Using metallurgy in powder the powder metallurgy procedure entails three steps: mixing, compaction, and sintering	Using metallic powder, the powder metallurgy process is composed of three phases: mixing, compaction, and sintering	Because the pieces are made from powders and the procedure is not efficient, the difficult and labour-intensive, intricate shapes and interior pores is not entirely eliminated

(continued)

Table 3 (continued)

Method of manufacturing	Details of the methods followed	Benefits	Limitations
Spray deposition method	Cost savings are the key advantage of spray deposition. The process of spray deposition can be used to create materials that cannot be produced traditionally	Mainly cost savings are the advantage of spray deposition. Spray deposition is a method for producing materials that cannot be produced traditionally	Mainly cost savings are the advantage of spray deposition. Spray deposition is a method for producing materials that cannot be produced traditionally
Infiltration method	A molten matrix metal is used to fill the spaces between the prepared dispersion phase inclusions (ceramic particles, fibres, or weaves)	Components with complicated shapes and a potential for high fibre volume fractions	Porosity in the composite cast can result from solidification that occurs before complete infiltration if the fibre temperature is too low. On the other hand, a fibre temperature that is too high results in a lengthy solidification process

2.1.1 Agitation Preparation Method

Method for preparing agitation the primary idea behind the agitation method, sometimes called the whirlpool method, is to feed the using a specialised mixing technique, particle reinforcement is added directly to the molten metal melt, causing the particles to disperse throughout the alloy melt. Mechanical agitation, high-energy ultrasonic composite method, and electromagnetic stirring method can all be used to separate this, according to the increased particles and aluminium liquid mixing method. Practically all casting procedures may be used to create the mechanical agitation method, which has cheap equipment requirements, an easy procedure, and extensive particle type and size adaptability.

2.1.2 Preparation by In-Situ Reaction Method

Chemical reactions inside a solid or liquid matrix can produce the reinforcement elements that make up composites. This method of making composites (in situ) has several great benefits, including a clean interface between the reinforcement and matrix, excellent compatibility, steady thermodynamic reinforcements, and a strong bonding force between the reinforcement and matrix as well as inexpensive fabrication costs. Figure 8 represents the simplified production method for developing solidified particulate composite through in-situ production route.

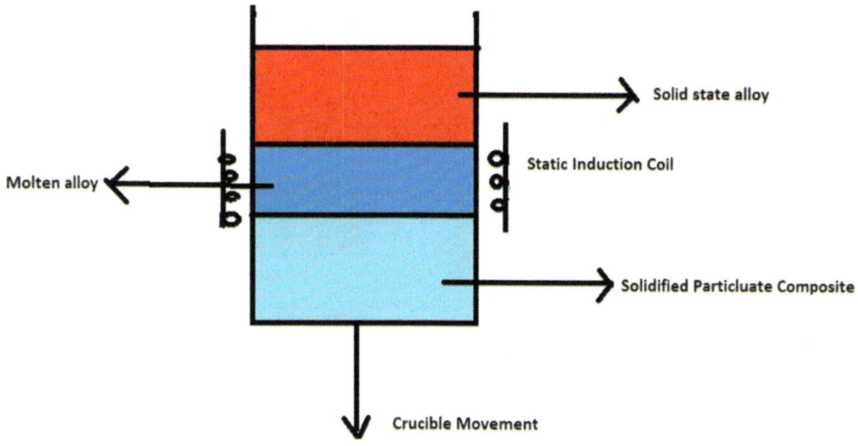

Fig. 8 In-situ reaction process

2.1.3 Extrusion Casting

Casting by extrusion Constrict Casting. The high-pressure solidification of the composite improves the wettability of the molten metal and the reinforcing particles and defects such looseness and porosity are also eliminated, considerably enhancing composites' strength and plasticity. Consequently, the developed material systems have greater quality.

2.1.4 Powder Metallurgy Method

With the use of extreme pressure, powdered metals and alloys are pressed into a rigid die during the production process of powder metallurgy, which produces extremely exact and accurate parts. As a result of the design and use of technological developments, powder metallurgy has developed into an essential process for the production of bushings, bearings, gears, and a variety of structural components. Utilizing powder metallurgy few steps make up the powder metallurgy procedure. Minimal interfacial reactivity, adjustable enhanced phase content, uniform enhanced phase distribution, steady performance, and ability to machine using traditional techniques are some of the benefits of the products. Figure 9 represents the Steps involved in powder metallurgy technique.

2.1.5 Spray Deposition Method

It includes atomizing molten metal, but the in order to form billets for later forging, spray is collected on a substrate. Rather than being left to solidify as powder. During the atomization process, the metal stream is filled with the reinforcing particles. Spray

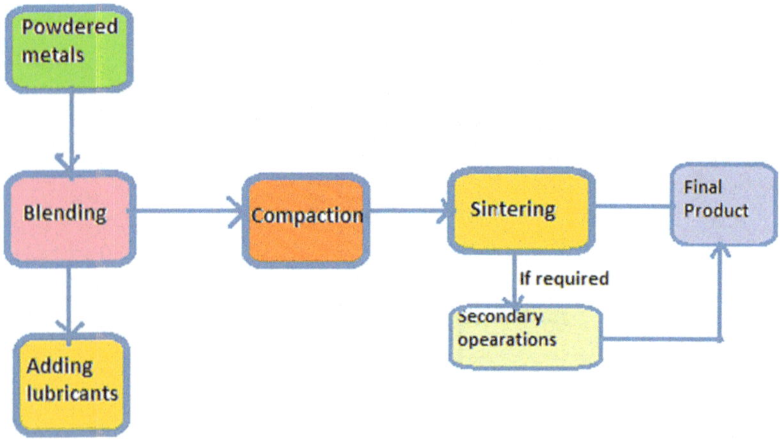

Fig. 9 Steps involved in powder metallurgy technique

deposition appears to have a bright future. Al alloys, Cu alloys, superalloys, stainless steels, and high Cr alloy steels are just a few of the materials that are processed in this method. The fundamental advantage of spray deposition is cost reduction. Among the materials that can only be produced using spray deposition and cannot be produced traditionally.

2.1.6 Infiltration Method

By soaking prepared dispersion phase (ceramic particles, fibers, or weaves) in a molten matrix metal, which fills the space between the dispersed phase inclusions, composite materials are created in the liquid state. Infiltration processes can be driven by external pressures (gaseous, mechanical, electromagnetic, centrifugal, or ultrasonic) applied to the liquid matrix phase or by the capillary force of the dispersed phase (forcing infiltration). The liquid infiltration method first creates a prefabricated portion by adhering the reinforcement phase to it with glue. The metal is then hydraulically pressed into the gap of the prefabricated part and solidifies to create the composite material.

Figure 10 depicts the method of manufacturing particulate composite by infiltration method where powdered composite is infiltrated through liquid metal there by initiating the formation of embryo of final composite with the help of hot crucible.

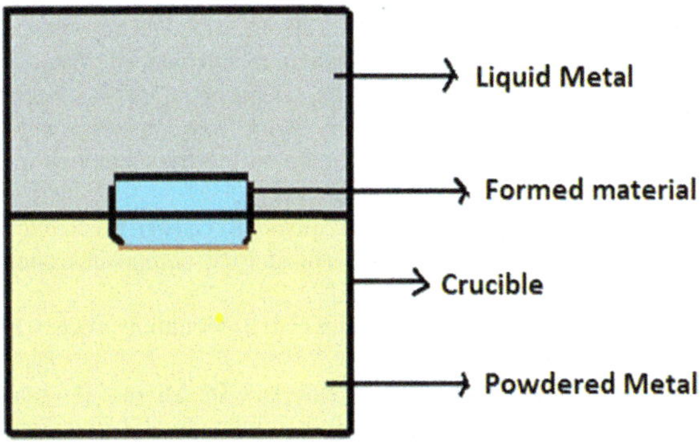

Fig. 10 Infiltration method of producing composite

3 Conclusion

Composite designs implied for primary uniqueness and property improving individuals in the expansive regions covering the areas of aviation and vehicle applications concerning the perspectives connects with weight. Interim assembling the light weight structures attributable to their particular elements estimates the kind of techniques followed is another significant variable. The current report examines about the different assembling techniques engaged with delivering the conspicuous classification of composites which expects to lay out the exploration perspective as well as business angles. One such blend that composite producers are at any point search in to is fiber supported and molecule built up composites. The current work additionally concluded with stresses that correlates with different production techniques states it from conventional to advanced level of production methods take off with different advantages, restrictions for the both research and commercial aspects.

References

1. Lokesh KS (2018) Evaluation of toughness on varied thickness of chopped strand mat/PU-foam sandwich structures. Int Res J Eng Technol (IRJET) 05(09). e-ISSN:2395-0056lp-ISSN: 2395-0072
2. Lokesh KS (2018) Synthesis and study on effect of thickness on 3-point bending strength of sandwich composites. Int Res J Eng Technol (IRJET) 05(08). e-ISSN:2395-0056, p-ISSN: 2395-007
3. Thomas Pinto, & Ravi S.M. (2017). Evaluation of mechanical properties and wear characterization of polymer composites under varying temperature conditions: a review. Int J Eng Inf Syst (IJEAIS) 1(4):64–68. https://doi.org/10.5281/zenodo.821168
4. Chetan IC, Thomas PU (2017) Influence of graphene on hardness number of aluminium-7075 based metal matrix composites. Int J Innov Sci Res Technol 2
5. Chethan IC, Kumar N, Kannantha V (2018) Determination of compressive strength of graphene reinforced with aluminium-7075 metal matrix composites. Int J Mech Eng Technol 9(1):327–335

6. Huntsman International LLC (2001–2017) Wet lay-up moulding processes. Retrieved February 2017, from http://www.huntsman.com/advanced_materials
7. Manufacturing of composite. Department of Aerospace Engineering, AE681 Composite Materials. Retrieved February 2017, from https://www.google.co.in
8. Composite material and manufacturing by society of manufacturing engineering. Retrieved January 2017, from www.sme.org
9. Yang BJ, Ha SK, Pyo SH, Lee HK (2014) Mechanical characteristics and strengthening effectiveness of random-chopped FRP composites containing air voids. Composites: Part 62:159–166
10. Molded Fiberglass Companies (2016) Moulding process. Retrieved February, 2017, from http://www.moldedfiberglass.com
11. Velmurugan R Composite materials. In: Module II—manufacturing techniques. Department of Aerospace Engineering, Indian Institute of Technology, Madras. Retrieved February 2017, from http://www.nptel.ac.in/courses/101106038/mod02lec01.pdf
12. Zhu H, Wang H, Ge L, Chen S, Wu S (2007) Formation of composites fabricated by exothermic dispersion in Al-TiO_2-B_2O_3 system. Trans Nonferrous Met Soc China 17:590–594
13. Hernández-Morenoa H, Douchin B, Collombet F, Choqueuse D, Davies P (2008) Influence of winding pattern on the mechanical behavior of filament. Compos Sci Technol 68(3–4):1015–1024
14. Lucintel (2016) Growth opportunities in the Indian composites market: 2016–2021 trends, forecast, and opportunity analysis. Retrieved February 2017, from http://www.lucintel.com/indian_composites_market_2016_2021.aspx
15. Maunder SK (2001, December 27) Composite manufacturing material, product and process engineering. Retrieved February 2017, from https://www.crcpress.com
16. Singh I (2013, January 7) Polymer matrix composites: processing index. Retrieved February 2017, from http://textofvideo.nptel.iitm.ac.in/112107086/
17. Summerscales J (1990) Manufacturing concepts for volume production of large composite components. In: International symposium on composite materials with textile reinforcement for use in building construction and related applications 18, vol 3. pp 387–394
18. Ummaji B, Gururaj P, Rayappa K, Yashavantha J (2018) Effect of red mud particles on scratch resistance of aluminum based metal matrix composites. Am J Aerosp Eng 5(1):24–29. https://doi.org/10.11648/j.ajae.20180501.14
19. Karthik BY, Avinash CG, Ganesh Kamath SL (2018) Experimental study on tribological characteristics of E-waste filled fibre reinforced plastics. Int J Sci Res Mech Mater Eng (IJSRMME) 2(1):11–16. ISSN : 2457-0435. http://ijsrmme.com/IJSRMME182203. March–April 2019. www.ijsrmme.com
20. Pinto T (2019) Examination of modulus of rupture of fibre mat bonded with potential adhesives. Int J Sci Res Mech Mater Eng (IJSRMME) 3(2):08–14. ISSN: 2457-0435. https://ijsrmme.com/IJSRMME19315
21. Hihara LH, Bakkar A (2016) Corrosion of metal matrix composites. In: Reference module in materials science and materials engineering. Elsevier
22. Bharath Kumar S, Ramachandra C (2016) G3, effect of E-waste filler on tensile and flexural behavior of E-glass/epoxy composites. Int J Eng Trends Technol 02(05)
23. Pinto T, Shrinivasa Mayya D et al (2022) Effect of wollastonite filler on the experimental and microstructural analysis of epoxy composite reinforced with E-glass fibre. J Inst Eng India Ser D. https://doi.org/10.1007/s40033-022-00347-9
24. Raghavendra Pai K, Shrinivasa Mayya D, Naveen Kumar JR, Hebbale AM (2021) Experimental study on preparation and mechanical characteristics of jute/silk/coco-peat reinforced with epoxy polymers. Mater Today: Proc 46(Part 7):2764–2769. ISSN 2214-7853. https://doi.org/10.1016/j.matpr.2021.02.511
25. Naveen Kumar JR, Prasad P, Hebbar N, Savitha MB, Navaneeth Gowda N (2021) A potential application of polyethyleneimine-reduced graphene oxide nanocomposite sensing film coated on interdigitated electrode prepared from copper-clad for carbon dioxide detection. Mater Res Innovations 25(6):363–371. https://doi.org/10.1080/14328917.2020.1826674

26. Pai R, Bangarappa L, Shrinivasa Mayya D, Naveen CR, Pinto T (2022) Hybridization effect on water absorption and flexural properties of E-glass/banana fibre/epoxy composites. Mater Today: Proc 52(Part 3):1841–1845. ISSN 2214-7853. https://doi.org/10.1016/j.matpr.2021.11.491
27. Hunasikatti S, Naveen Kumar J, Ramegowda NKJ, Kannantha V, Pai R (2019) Effect of E-waste rubber on mechanical behavior of glass a fiber reinforced with epoxy composites. AIP Conf Proc 2080:020003. https://doi.org/10.1063/1.5092886
28. Naveen Kumar JR, Kannantha V, Pinto T, Sampreeth U (2020) Experimental evaluation of substrate and annealing conditions on ZnO thin films prepared by sol-gel method. Mater Today: Proc 24(Part 2):201–208. ISSN 2214-7853. https://doi.org/10.1016/j.matpr.2020.04.268
29. Kumar N, Kannantha V, Pai R, Hebbale AM (2021) Development and experimental investigation of mechanical properties of graphene-based aluminum 6061 alloys. Mater Today: Proc 46(Part 7):2421–2424. ISSN 2214-7853. https://doi.org/10.1016/j.matpr.2021.01.303
30. Shanmugam Nagarathna B, Paavan Kumar T, Mohanraj T, Srivatsav SYS, Kumar ABK (2022) Experimental and statistical evaluation of the mechanical performance of (Jute and Cocopeat) plant and (Silk) animal-based hybrid fibers reinforced with epoxy polymers. J Nat Fibershttps://doi.org/10.1080/15440478.2022.2073501
31. Shanmugam BK, Panduranga BP, Naveen Kumar JR, Harish H (2021) Experimentation and prediction analysis on the mechanical performance of fish scale and coconut shell powder-based composites. J Nat Fibers.https://doi.org/10.1080/15440478.2021.1958410

Manufacturing of Particulate and Fiber Reinforced Composites: A Review

Samuel Dayanand and Satish Babu Boppana

Abstract Composite materials, which have greater strength, less weight, and better attributes, are progressively replacing traditional materials in use. The globe is investigating the use of particle and fibre reinforced composites in all applications, including toys, instrumentation, medicine, the building sector, air, land, and sea transportation. Particle and fibre reinforced composites can be manufactured in a heterogeneity of methods depending on the application and type of reinforcement utilized. As the trade improves in need for light-weight materials with higher strength for particular purposes, composites reinforced with different fibres of natural or synthetic materials are finding more importance. To identify the particle and fiber reinforced composite material for important applications, an outline of a large assortment of fibres, its classification, their attributes, functionality and different fibre composite manufacturing procedures is offered. Polymer fiber reinforced composites exhibit superior and superior qualities such as great damping property, durability, impact, stiffness, flexural strength, and resistance to corrosion and wear. The production of various particle and fibre reinforced composites components using diverse manufacturing procedures has been extensively covered in this chapter. The utilitarian attributes of different fibres that are promptly accessible all over the world, and the fabricating methods to manufacture the composite materials, should be concerted on to decide the improved attribute of the material for the expected application.

Keywords Particle · Fibres · Composites · Manufacturing

S. Dayanand (✉)
Department of Mechanical Engineering, Government Engineering College, Ganagavathi 583227, India
e-mail: samueldayanand@gmail.com

S. B. Boppana
Department of Mechanical Engineering, School of Engineering, Presidency University, Bengaluru 560064, India
e-mail: satishbabu3@yahoo.co.in

S. B. Boppana et al. (eds.), *Structural Composite Materials*, Composites Science and Technology, https://doi.org/10.1007/978-981-99-5982-2_8

1 Introduction

Rapid evolution in the manufacturing unit has generated an ultimatum for materials that are stronger, denser, stiffer, cheaper, and more viable [1, 2]. Composite materials suit one of the important materials with such enhanced and increased strength that they may be used in a wide range of ventures. Steel and aluminium have largely been replaced with composite, which has frequently demonstrated improved performance. By using composites, it was shown that the weight of steel and aluminium may be reduced by 60–80% while maintaining or improving their qualities [3, 4]. The word "composite" cites to a material fabricated of two basic components: reinforcement and matrix [3, 5]. Three categories, polymer matrix composite, metal matrix composite, and CMC (Ceramic Matrix Composite), can be used to classify composite materials. Fiber reinforced plastic (FRP) composites are made when fibres are mixed with resin or plastic [6–8]. A composite can have a metallic, ceramic, or polymeric matrix and reinforcement. When adding reinforcement to various forms of matrix material, a variety of mechanical property considerations are involved. For instance, continuous fibre reinforced polymer matrix composites give a considerable increase in strength and stiffness over unreinforced polymer matrix composites [9, 10]. Metal Matrix Composites (MMC's) have better temperature resistance and stiffness than the parent matrix, but less ductility. In order to boost toughness without sacrificing the parent ceramics, numerous desirable qualities, such as stiffness, wear resistance, and refractoriness, reinforcement of ceramic materials is used. Composites may be constructed to have certain physical characteristics in addition to their mechanical characteristics [11]. For instance, MMC's can be manufactured with nearly perfect dimensional firmness throughout a broad temperature range which is advantageous in space shuttle applications. Particulate composites come in a wide diversity of structure and sizes [12, 13].

Very fine particles of 0.25 μm and chopped fibres (like glass), hollow spheres, platelets or novel materials like carbon nanotubes can all be part of the particulates. In each instance, the matrix serves as the binding medium required for structural application purposes while the particles offer desirable material qualities. Particulate composites have a number of benefits. They strengthen the material by adding reinforcement to the matrix material. Specific material qualities may be provided by the reinforcement and matrix mix. For instance, adding conductive reinforcements to a plastic can result in slightly conductive plastics. Particulate composites can frequently be produced using more affordable, conventional techniques like injection moulding [5–7, 14, 15].

The various parts created with FRP materials have exceptional advantages over the traditional materials, together with high strength to minimum weight ratio, dimension stability, good impact resistance, good corrosion resistance, durability, and ease of moldability among others. Various production techniques are used for processing composites. Numerous processes evolved and expanded over time. Consideration is given to fibre and particle reinforcing of materials [16]. The variety of reinforcement types and their influence on mechanical characteristics and processing are among the universal issues covered. It addresses the numerous problems with reinforcing different matrices, such as the higher specific strength and stiffness of continuous fibre reinforced polymers and particle reinforced metals and the improvement in firmness of reinforced ceramics. The negative effects of reinforcing various matrices, which can be related to cost, fabricating challenges, as well as loss of specific features, are addressed. Axial particles inserted into a matrix material make up fiber-reinforced composites (FRCs). To fabricate a material with a higher specific modulus and higher specific strength is the goal of fiber reinforced composites. The fibers can transfer load that has been put on matrix and strength can be achieved. At the end, interfacial bonding is crucial for FRCs. Traditional FRC examples include wood and fibre glass [17, 18].

Due to their favorable performance and mechanical and physical characteristics, fibre and particle reinforced composite materials are finding more and more applications in a variety of technological fields. A wide range of structures, including aero sectors, robots, tennis rackets, bicycles, industrial machines have frequently utilized composite materials. Reinforcement fibers and particles serve as a load-carrying capacity medium and the matrix serves as a load-transporting media in fibre and particle reinforced composite materials. In order to maximize each phase's strengths and reduce its flaws, two or more phases are used [17, 19–21].

Since continuous fibre reinforcements affect the most types of matrix materials' attributes, they will be the subject of attention. The world focuses on structural applications since this is currently the main application sector and polymer composites account for the vast majority of the composite market. First, a general discussion of some of the most important features of using reinforcements will be given, followed by sections on different types of reinforcement, fundamental mechanical behaviour, and processing [22]. This study also provides the several basic manufacturing methods accessible to the FRP sector and contrasts them in terms of their benefits, drawbacks, applications, types of fibres and resin systems, needed part quality, cost, and production rate. Other important considerations for determining the correct procedure include component size, shape, and desired qualities [23].

1.1 Types of Reinforcements

The reinforcement phase of a composite material can take on several forms, such as continuous, short fibre, platelet, sphere, spheroid, and is categorized according to its shape, geometric arrangement, aspect ratios and concentrations. Continuous fibre reinforcement can be of uni-axial, bi- or multi-directional in plane, or 3-dimensional. When applied perpendicular to the direction of the fibres, the continuous fibre reinforcement types offer the best mechanical qualities among the several reinforcement kinds. Even if their nature is locally diverse, the behaviour of particulate systems is nonetheless primarily isotropic [10–14, 22, 24]. A significant amount of anisotropy is present in continuous fibre systems, meaning that the attributes parallel to the fibre direction and those along the fibre direction differ significantly. Depending on the anisotropy, composites can be very helpful in some applications, however laminates, which are made up of layers of fibres oriented at various angles, may be built if a more isotropic behaviour is desired. Other varieties of continuous fibre systems include built on woven fabric, which offers reinforcement consisting of two orthogonal directions or non-fabrics, which are made up of layers of fibres in different orientations held together by through-thickness stitches. The through-thickness reinforcing provided by the stitches in non-crimp fabrics enhances the characteristics (strength and toughness) in this direction. In comparison to their non-woven equivalents, composites with woven fabric reinforcement offer better processing flexibility (particularly with regard to drape) while preserving appropriate mechanical qualities [24–28].

Continuous fibre composites typically get their strength and its stiffness from fibre characteristics. The strength and firmness of fiber composites as a rule get from the fiber reinforced attributes. The amazing attributes of the strands are evident. The greater part of the strands are fused into polymer matrix. Bigger width SiC filaments, produced by compound fume testimony methods are the most widely recognized continuous fortifications for MMCs. For few matrix frameworks, overwhelmingly silicon carbide filaments, created by pyrolyzing a polymer forerunner are the most concentrated on fortifications however interest in oxides strands is expanding [5–7, 29–31].

Of the different reinforce types, those in view of constant filaments give the best mechanical attributes, when stacked and lined up with the fiber heading. Particulate frameworks, albeit nearby heterogeneous materials in nature. They are transcendently isotropous in their mechanical and physical way of behaving. Short fiber frameworks are frequently planar isotropic. Constant fiber frameworks display significant anisotropy. This anisotropy material can be exceptionally useful in specific applications yet in the event that a more isotropic way of behaving is wanted, layers of filaments situated at various points are included. Different sorts of constant fiber framework depend on woven material, which gives reinforcement in two symmetrical headings, or non-pleat textures, that contain layers of strands at different direction kept intact with increase in density. The fastens present in non-pleat textures give through-density reinforcement, which works on the attribute's principle (strength and durability) towards this path. Composites in view of woven texture reinforce offer more prominent adaptability in handling (particularly as to wrap), contrasted with their non-woven partners, while keeping up with satisfactory mechanical attributes [24, 30–32].

Fiber reinforcement can be gotten from providers in various structures, for example consistent or cleaved rovings, whirl mat, slashed strand mat, woven textures, non-pleat textures or even weaved pre-structures, customized to the state of the possible part. These filaments then must be joined with the lattice somehow or another. On the other hand, for persistent fiber materials, material might be accessible in a transitional structure, where it has proactively been joined with the framework. For PMC's, 'pre-pregs' are a halfway structure, wherein the fiber reinforcement has been joined. Layers of 'pre-preg' can be rested up with the reinforce at explicit directions and stiffed at temperature and strain to give an inflexible overlay. Handling of fibre reinforced endlessly, involves careful assessment. The interaction includes drawing heaps of filaments along a slurry. For some of the MMC's there is likewise a halfway structure gotten from putting varieties of strands between dainty metallic thwarts (this should be possible basically through a fiber twisting activity) preceding hot squeezing. Ceaseless rovings of fibers can be cleaved to give spasmodic fiber reinforcement [3, 30, 31, 33].

A predetermined number of various materials are accessible in bristle structure. The high angle proportion (for example a couple of micrometers in breadth and several micrometers in length) of silicon carbide might be integrated into matrix's, normally ceramic. This kind of reinforce will in general create better durability attributes than lower angle proportion fortifications yet issues connecting with the particular wellbeing dangers related with whiskers and handling troubles might have forestalled more broad utilization of them. Possibly, any material in the form of particulate structure can be utilized as the second stage in spasmodic composites yet similarity all through the handling steps prompts numerous blends being unsatisfactory. Further, a few blends give no benefits above the base matrix material so practically speaking, the quantity of mixes that are accessible financially are restricted. The mass of well-known of these will be talked about in the pertinent materials segments [30, 34–36].

2 Role of Particulate Reinforcements

Particle reinforced composites (PRC's) are less viable in fortifying than fiber reinforce. PRC's chiefly accomplish increase in stiffness, however they additionally can accomplish expansions in toughness and strength. Regardless, the upgrades are less than the accomplished FRC's. The fundamental advantage of PRC's is their minimal expense and simplicity of creation and framing contrasted with FRC's. Specifically, PRC's finds various applications where elevated degrees of wear-resistances are needed. The reinforce material is implanted into a matrix. The reinforce doesn't necessarily serve a simply underlying errand (building up the compound), but at the mean time is used to change actual attributes like wear resistances, thermal conductivity and friction coefficient. The reinforce can be either continuous, or broken. Broken MMC's can be isotropic in nature, and can be operated with quality metal working methods, like expulsion, forging or rolling. Likewise, they might be machined using traditional procedures, however generally would require the use of poly crystalline diamond tooling [1, 5–7, 25, 30, 31, 37, 38].

Consistent reinforcements uses single filament strands, like carbon fiber or SiC particle. However, the wires/strands are lodged into the base matrix in a specific course, the outcome, is an anisotropic construction wherein the ordering of the material impact its strength. It is noted that, the primary MMCs involved boron (B) fiber as reinforcement. Broken reinforcement utilizes whiskers or particulates. The most well-known reinforced materials in this classification are alumina and SiC particulate. Ceramic materials provide various advantages in different ventures. They give higher wear, intensity and erosion resistances along with high rigidity, dielectric strength, modulus of versatility and volume resistivity. These materials likewise offer low thermal extension than metals or plastics, and a more drawn out specimen life at unique plan aspects and resistances [13, 14, 29–31, 34, 35, 37, 39, 40].

2.1 Size and State of the Particulates in Composites

For a particle reinforced composite, the calculation and size of particles are significant in deciding composite attributes like mechanical attributes. Particle scattering and dissemination in a composite assume a basic part in deciding composite attributes. For superior execution of particulate bio ceramic-polymer composites, particle agglomerates should be separated during composite handling into essential particles (i.e., the littlest particulate bits of the minor part existing in as-manufactured or as-gotten ceramic powder), which show adequate scattering and sensible appropriation in the polymer matrix. Scattering particles from the dense state isn't sufficient as the particle reaching the upper layer in the composite will give break commencement destinations, hence causing untimely failure of the composite when the composite is under mechanical load. In a perfect world, particles present in the composite ought to be in a scattered state. Subsequently, uncommonly planned handling gear or potentially great creation methods are frequently required, which produces shear powers sufficiently huge to conquer different particle grip powers during composite handling with the goal that particle agglomerates can be diminished to tiny particles and can be equitably dispersed in the composite. By and large, there is a basic measure of particulate reinforce for the composite, just above which, wanted attributes can be gotten for the composite. In such a circumstance of profoundly filled polymers, handling of the composites turns into a test, in some cases outlandish. Surface treatment of bio-ceramic particles might ease handling troubles however may not be guaranteed to prompt improved particle scattering [10, 30, 37, 41–44].

Aluminum Oxide or alumina: It has great electrical and mechanical attributes, wear resistances and corrosion resistances. It has somewhat unfortunate thermal shock resistances. It is utilized as an electrical cover for various electronic and electrical ventures, along with flash attachment separators and electronic items substrates. It is additionally utilized in synthetic, clinical and wear resistance applications [1, 13, 29, 30, 45, 46].

Zirconium Oxide: Particulate zirconium oxide (ZrO_2) has the most noteworthy crack strength of any high level specialized clay. It's known for strength, corrosion resistances and mechanical attributes. ZrO_2 is having thermal extension coefficient that is extremely near steel, which makes it perfect for use in applications of steel. A few ecological circumstances can make the material unsteady, making it no longer have its mechanical attributes. Its moderately lower hardness and higher weight likewise limit, its wide use in wear ventures [1, 31, 47].

Silica (SiO_2): Silica is an amazing barrier pertaining to thermal property and has basically zero thermal extension. It has great synthetic protection from liquid metals yet is restricted by its exceptionally low strength. It is utilized for various hard-headed and glass type ventures [1, 30, 48].

Titanium Diboride: Particulate titanium diboride (TiB_2) is an electrically leading artistic and can be machined utilizing unconventional electrical discharge machining methods. It is an extremely harder material; nonetheless, it has poor mechanical attributes. Its significant uses are in metallurgical attributes applications, including liquid aluminium. It is likewise utilized for some restricted wear attributes applications, for example, ballistic protection to stop enormous measurement shots [1, 30, 49].

Aluminum Diboride (AlB_2): Aluminum diboride (AlB_2) is a synthetic compound produced using the metal aluminum and the metalloid boron. AlB_2 particle is found to be one amongst mixtures of aluminum and boron, the other is AlB_{12}, which are both generally alluded to as aluminum boride. High immaculateness, submicron and nano powder structures are associated with the compound. Borides are hard, high-softening materials with metal-like conductivity. They are utilized in semiconductors, superconductors, diamagnetic, paramagnetic, ferromagnetic, turbine cutting edges and rocket spouts. Borides have as of late been found to be superconductive and super incompressible [50, 51].

Zirconium Diboride (ZrB_2): Zirconium diboride (ZrB_2) is an exceptionally covalent unmanageable ceramic material with a hexagonal gem structure. ZrB_2 is an ultra high temperature ceramic (UHTC) with a dissolving point of 3246 °C. ZrB_2 is a covalent bond fortified with a (HCP) hexagonal crystal structure. Because of it's hardness held up to higher temperatures and flexible modulus, great abrasion, and creep resistances, it is thought of as the base material for a wide scope of mechanical applications. For example, heater components, higher temperature cathodes, rocket motors, thermal insurance structures for space shuttles vehicles with temperature capacities at high temperature and uncompromising wear applications. Great thermal conductivity, idleness to liquid metals, good thermal shock resistances, and higher electric conductivity, permit forming of ZrB_2-based ceramic through EDM. This will also help in the forming of ZrB_2-based materials into complex designs [1, 52].

Boron Carbide: Boron carbide (B_4C) is one of the strongest materials following diamond, and has extraordinary wear resistances. It is utilized widely for indignant covering and impact spouts. B_4C is likewise a neutron safeguard, going with it an essential decision for control bars and other atomic applications [1, 53].

Graphite (Gr): Graphite comprises of sheets of three-sided planar carbon material. The singular layers are named graphene material. In each and every layer, the carbon particles are organized in a honey-comb matrix with a good bond length and the distance between each plane is 0.336 nm [2–4, 30–34, 54]. The same can be used in powder form.

Silicon Carbide: Particle Silicon carbide (SiC) has exceptional wear attributes and thermal shock resistances. It has great mechanical attributes, particularly at higher temperatures. SiC is a good material for semiconductor having excellent electrical resistivities. It is utilized broadly for mechanical adhesive due to its compound and wear resistances [13, 14, 55].

Tungsten Carbide: Tungsten Carbide (WC) related cermets or clay metals have wider use in cutting devices and other metal-framing instruments. Unadulterated WC particle can be made utilizing a higher temperature using hot isostatic squeezing method. WC material has extremely higher hardness, strength and wear resistances and is utilized for grating water fly spouts; nonetheless, its weight restricts its utilization in numerous advantages [2–4, 56].

Graphene: Graphene has some of the stand-out attributes: higher surface zone, high electron transportability, high Young's modulus with an incredible conductivity. In view of its thermal conductivity, it is a fair competitor for the stronghold of an aluminium matrix to update the thermal conductivity [50, 57].

Aluminum Nitride: Aluminum nitride (AlN) has an exceptionally higher thermal conductivity being used as an electrical insulator. It is used as an optimal material for different applications in thermal and electrical administration circumstances [58].

Boron nitride: Boron nitride (BN) is of hexagonal shape and a pasty white color material which is frequently named as white graphite particle. It has extraordinarily higher temperature resistances in dormant climates yet can't be utilized above a temperature of 500 °C temperature in an atmosphere air. It is utilized as a higher temperature protector and in mix with TiB_2 in numerous ferrous and aluminium metallurgical usage approach [1–4, 59].

Silicon Nitride: Silicon nitride (Si_3N_4) has the good blend of thermal, electrical and mechanical attributes of any high-level specialized clay material. Its higher strength and durability go with the material for car and used in bearing automotive demands [1, 15, 60].

3 Fabrication of Particulate Reinforced Composites

The composite is prepared by joining powdered materials in a matrix. A number of methods are available for preparing these [1–5, 30–34, 61].

3.1 Powder Metallurgy (PM)

Powder handling techniques related to deformity handling are utilized to create particle or polymer fiber reinforced composites. This normally includes cold squeezing and hot squeezing to create fundamentally particulate reinforced MMCs. The base matrix and the reinforce powders are mixed to deliver a clear homogeneous circulation; it is normally utilized for higher liquefying point matrix and dodges isolation impacts and fragile response item development inclined to happen in fluid state processes. This technique grants to acquire intermittently particulate reinforced aluminium MMCs with the most noteworthy mechanical attributes. Aluminium MMCs are utilized for armor applications yet stay restricted for enormous scope creation. PM process is fundamentally comprising of three stages. The principal stage gives the readiness of the powder and it accompanies the progressive stages, after the first stage the powder items are combined as one with the reinforce. The next interaction is the course of union, amid which, the powder form of worked blend are joined jointly by sintering [31, 33, 61, 62].

3.2 Diffusion Bonding

It is a typical strong state handling strategy for joining comparable or different metals. The dissemination of particles between clean metallic surfaces at a raised temperature, prompts holding. The chief benefits of this strategy are the capacity to deal with a wide assortment of metal matrices and influence of fiber direction and volume portion. Amidst, the disadvantages are longer handling times, higher handling temperatures and an impediment on the intricacy of shapes that could be created. There are numerous variations of the essential dispersion holding process, albeit every one of them include synchronous use of strain and higher temperature [1–4, 31, 33, 61, 63]

3.3 Stir Casting

The least difficult and most financially utilized method is known as 'vortex strategy'. It is the result of effortlessness, minimal expense of handling, adaptability, generally efficient for huge estimated parts to be ready as well as creation of close to net molded parts. The vortex procedure includes the presentation of pre-treated particulates into the vortex formed of liquid composite made by the turning impeller [1–6, 31, 33, 55–58, 61, 64].

An intriguing late improvement with regards to mix projecting is a two-step blending technique. In this cycle, the base matrix material is heated to over its fluids temperature; so, the metal is completely dissolved. The soften is then chilled off to a required temperature in between the fluids and solid focuses and retain in a semi-strong state. Then the preheated particulates are included and blended. The slurry is once more heated to a completely fluid state and blended completely. This two-step blending method has been utilized in the creation of aluminium. Amongst all the deep rooted metal composite manufacture strategies, the said one is the most affordable [55–58, 61, 64].

3.4 Squeeze Casting

This is also referred to a role as fluid metal forging. The liquid metal filled the base portion of the pre-thermalled die. As the metal begins to harden, the upper top half applies strain. How much strain subsequently applied is fundamentally not exactly utilized in forging. Curing can be utilized with this cycle to shape openings and breaks. The porosity is very low and the basic mechanical attributes are moved along. Both the materials from ferrous and nonferrous can be delivered utilizing this strategy [1, 31–34, 61, 65].

3.5 Compo Casting (CS)

It is a fluid state method in which, the reinforce particulates are included to a melt that is to be solidified. It is noted that the strong particles previously formed in the semi strong slurry can precisely entangle the reinforcing particulates, forestall their gravity isolation and decrease their formation of cluster. These will bring about better dispersion of the reinforce particulates. The low porosity observed in the prepared castings has been credited to the good wettability in between the lattice and the reinforce particulates along with the low volume shrinkages in the composites [31, 61, 66].

3.6 In-Situ Synthesis

The composites arranged by ex-situ strategy experiences thermodynamic unsteadiness among lattice and fortifications, accordingly restricting their surrounding and high temperature mechanical attributes [4]. Ex-situ process has downsides like agglomeration, unfortunate wetting and heterogeneity in microstructure. As of late, another course is created for manufacture of composites through in-situ blend. In-situ combination is a cycle where in the fortifications are formed in the framework by the managed metallurgical responses (exothermic response), which shows a spotless matrix/reinforce interface, that prompts better enhancement in mechanical attributes of the composites. In situ composites are multiphase materials where the building up stage is blended inside the lattice during the composite manufacture [31, 46, 51, 61, 66, 67].

3.7 Advanced Shear Innovation

The above traditional strategies produce agglomerated structures displaying lower strength and malleability due to non-wettability of reinforce by matrix composites and thickness contrasts between the two materials and absence of effective blending innovation to accomplish a uniform dissemination of fine-size reinforce inside the lattice. Thus, the presentation and maintenance of building up particles in lattice is very troublesome [31, 61, 66].

To defeat this, a new rheo-handling strategy called the melt conditioning using advanced shear technology (MCAST) process has been created for assembling close net-shape MMCs with homogeneous dissemination reinforce in the matrix. The key thought is to apply adequate shear pressure (τ) on particle bunches implanted in the fluid metal to beat the typical rigidity of the group [26–32, 66–68].

3.8 Distributive Blending

Distributive blending utilizes ordinary mechanical mixing to pre-blend the Al compound with building up particles. The blending hardware is same as regular mix cast arrangement. A controlled argon climate will be kept with a heater all through the entire investigation to forestall liquefy oxidation. The reinforce particles are moved gradually and ceaselessly into the liquefy which would be precisely mixed at 600–800 rpm. After all the reinforce is brought effectively into the fluid metal, the composite blend would be permitted to set in the pot and accordingly thermalled to the preset dissolving temperature, and afterward mixed, however it produces agglomerated structures. Agglomerated structures will be formed in stale zones (e.g., close to pot walls) because of absence of adequate shear force in distributive blending. To separate the agglomerates, it is essential to apply a satisfactory shear pressure which defeats the typical strong power or rigidity of the bunches. The course of dispersive blending under escalated shearing creatively takes on a high-shear dispersive blending activity of the twin-screw component to the errand of conquering the firm power of agglomerates. The twin-screw system is utilized for the MCAST cycle comprising of a couple of co-turning, completely intermeshing, and self-cleaning screws. The screws have extraordinarily planned profiles which bring about high-shear rate and focused energy of disturbance. The fundamental capability of the twin screws is to separate the agglomerates and groups implanted in the fluid liquefy under a high-shear pressure and scatter the particles consistently under the extreme focus of disturbance [31, 61, 66–68].

3.9 Ultrasonic Assisted Projecting

As of late, metal matrix nano composites research is going on, in a more prominent degree. It beats numerous restrictions like flexibility, low crack strength and machinability.

Anticipating, as a liquid stage technique, is outstanding for its ability to convey as cast light weight portions of metal matrix nanocomposites with incredible support dissemination. Regardless, nanosized particles present irksome issues: it is exceptionally difficult to disperse them reliably in liquid metals by virtue of their lamentable wettability in metal framework and their gigantic surface-to volume extent, which successfully impels agglomeration and gathering. To achieve uniform dispersing and movement of nanoparticles in aluminum system nanocomposites, researchers encouraged a creative technique that joined solidifying processes with ultrasonic cavitation-based dissipating of nanoparticles in metal melts. Transient cavitation could convey an implosive impact adequately ready to isolate the clustered fine particles and disperse them more reliably in liquids [31, 61, 66–69].

3.9.1 Friction Stir

The above existing strategies are not reasonable for creating surface composites with particulate reinforce, due to arrangement of unfavorable stages and furthermore it structures interfacial response among reinforce and metal matrix. The above issues can be abstained from by setting up the composite underneath the melting place of substrate.

Friction stir processing, is one of the newer and promising thermo mechanical handling procedures that changes the microstructure attributes and good mechanical attributes of the material in single pass to accomplish most extreme execution with low creation cost significantly quicker utilizing a basic and modest device. Its rule is same as rubbing mix welding, in which a non-consumable pivoting device with an exceptionally planned pin and shoulder is dove into the connection point between the two plates to be welded and navigated along the line of the joint. The contact brought about by the pivoting instrument thermals up the materials around the pin to a temperature beneath the softening point and disfigures plastically. The turn of the device "blends" the material and reinforce together and brings about a combination of the two materials [12, 31, 55, 61, 66–68, 70].

4 Fiber Reinforced Composites (FRPs)

Depending on the fibers that are present in the composites, FRPs can be categorized. The one that has longer fibers relate to diligent fiber reinforced composites, on the other hand, the composites that have shorter fiber strongholds are arranged under non-uniform fibers, while some have dual kinds of fibers. Fibers can be put mono directionally in the network development of reliable FRPs, and can take stress and load applied on the composite framework, the fibers that are broken ought to need longer length to absorb proper load while restricting split advancement. The processing method and course of fibers define the various attributes and basic approach to acting of composite material. Enhancement in attributes, for instance, impact sturdiness and shortcoming strength ought to be noticeable while dealing with artificially treated typical fibers. Strands related to glass and basalt in the uniformly distributed stage were regularly used in composite materials. Important attributes of typical polymer fiber composites have likely applications in the state-of-the-art business, as experts as of now are obliged towards the headway of innocuous to the biological system materials in light of serious environmental guidelines [16, 23, 25, 71–74].

4.1 *Natural Fiber Reinforced Composites*

Huge attributes of natural polymer fiber reinforced with polymer fiber composite (NFPCs) have possible ventures in the cutting edge manufacturing, as analysts presently are constrained towards the advancement of harmless ecosystem materials because of severe ecological regulations. There are different strands available for composite materials and they are basically delegated typical or designed fibers. Further, late assessments have disclosed sensational material attributes, when two strands are joined, blending in with a grid material to approach a combination composite material. Natural fibres are particularly simple to get and are available in nature. They uncover a couple of wonderful material attributes, insignificant cost to unit volume, higher strength and express firmness. Composites are built of regular fibre and fortifications appear to convey a few different attributes over manufactured filaments, like decreased weight, cost, harmfulness, natural contamination, and recyclability. These monetary and ecological advantages of natural fibres composites made them prevalent over manufactured FRPs for current advantages. Contingent upon the sort, normal strands have comparative designs with various structures. The consideration of longer and shorter regular strands in thermoset lattices have showed elite execution application [8, 10, 19, 28, 30–32].

4.2 *Synthetic Fiber Reinforced Composites*

Artificial strands that are conveyed by compound mix are named synthetic fibers and additionally named normal or inorganic considering their substance. Glass fibers composites are for the most part comprehensively involved amid all of the produced strands as they provide eminent strength and firmness, warm consistent quality, insurance from impact, substance, disintegration, and wear attributes. Regardless, the machining operations of glass fiber-built up polymers is reasonably difficult. Glass fibers also convey the damage of evacuation around the completion life. In any case in specific advantages, more immovability is essential, carbon fibers are used as opposed to GFs. Graphene strands are one more sort of unrivaled execution of carbonaceous fibers that give back higher inflexibility with redesigned electrical conductivity when diverged from carbon strands. A couple of updated attributes of graphene fibers show their chance in various advantages, similar to light weight conductive connections and wires, knittable super capacitors, micromotors, day-light based cell materials, actuators, etc. [8, 10, 17, 19, 28, 30–32].

4.3 Hybrid Filaments Reinforced Composites

Thermoplastic reinforced composites with normal fiber, by and large, show unfortunate strength execution when contrasted with thermoset composites. Consequently, to gain advantages of plan adaptability and reusing potential outcomes, these normal fiber reinforced composites are blended with limited quantities of engineered fibers to make them more alluring for specialized applications.

Particle Reinforced Composites (PRP) contrasted with FRC, PRC isn't that compelling through material strength and crack resistances property. In any case, ceramic, metal or inorganic particulates confine the twisting and give great material firmness. As of late, PRCs are likewise getting a touch of consideration because of their isotropic material attributes and cost-viability. Besides, these reinforced composites are fabricated utilizing comparable procedures utilized for stiff material. PRCs are utilized for common applications like streets and substantial designs, where a serious level of wear resistance is normal, one such example is concrete that goes about as a folio material, total span of coarse stone or rock as an additional material gives good stiffness and hardness. [3–6, 8, 11].

Contrasted with FRC, particulates reinforced composite (PRC) isn't that effective through material strength and crack resistances property. Be that as it may, metal, ceramic or inorganic particulates limit the distortion and give great material stiffness. As of late, PRCs are likewise getting a touch of consideration because of their isotropic material attributes and cost efficacy. Also, these reinforced composites are fabricated by comparative procedures utilized for stiff material [5, 6, 8, 59, 60, 64].

4.4 Manufacture Strategies for Fiber Reinforced Composites

Treatment of FRP composite incorporates gathering of fibers and subsequently building up fibers, using network material through numerous strategies. The work incorporates twisting around, winding around, plaiting, and sewing of fibers.

In the present scenario, automated facilities, using an advanced computer might also help the way in which the fiber angle can be varied. The volume of fibers to be used while preparing composites can also be known easily [31, 32, 75, 76].

4.4.1 Hand Layup Method

It is the one of the most notable and extensively used open shape composite preparation technique. From the get go, fiber executes are kept through a definite shape; a coat would be enforced in basic removal process. Resin is poured while using a brush. A roller is finally used to send the resin while packing on the surfaces for securing a redesigned relationship in between the ever-evolving thin layers of the build-up and the grid materials [8, 10, 17, 19, 28, 30–32, 75].

4.4.2 Spray up Process

This method is nearly same as the hand rest up. In any case, it utilizes a handgun that splashes tar and cleaved filaments on a shape. All the while, a roller bar is used to mix these strands into the matrix material. It is an open form sort of technique, where slashed strands give superior comparability and very quicker than hand layup [8, 10, 17, 19, 28, 30–32, 76].

4.4.3 Resin Transfer Molding (RTM)

Vacuum implantation is one more new turn of events, where fibers are put on a shape and a chamber is organized between a vacuum pack and a mould. Vacuum forces make the material to be siphoned into the chambers around the fibers to join the cover structural property. For the enhancement in the strength of material composites, fibers need to undergo surface treatment [8, 10, 17, 19, 28, 30–32, 77].

4.4.4 Vacuum Bag Molding

Generally, the vacuum sack shaping method would be made by the hand recharge method. Overlay is first made by employing the hand recharge methodology, subsequently, the same is used under vacuum pack. Air present would then be taken care while pneumatic force packs the part. Different evened out composites were formed with multi scale fortresses of carbon fibers with the help of a vacuum terminating process, which killed potential outcomes of unmistakable porosity and less than ideal impregnation of twofold strongholds, with extensions in flexural, and inter laminar shears attributes, independently. The fiber support mat coordinated at the base piece of the shape and pre heated pitch is directed under strain through an injector. Various blends of fiber materials with its heading, along with three dimensional fortresses, can be done by RTM process [8, 10, 17, 19, 28, 30–32, 78–80].

4.4.5 Pultrusion

Pultrusion cycle involves strands of constant fibers getting through a pitch shower. The fibers are additionally combined using a die that is relatively warm. The persistent interaction is valuable one in the manufacture of composites involving steady cross segment that has extensive length. The method also empowers creation using robotization that often reduces the price of manufacturing [16–18, 30–32].

4.4.6 Injection Molding

The method can manufacture composite specimens, with higher accuracy and at extremely low process durations. In a common place infusion shaping cycle, fiber composites as pellets are taken care of through a container, and afterward they are instructed by a screw with a thermalled barrel. When the necessary measure of material is liquefied in a barrel container, the screw infuses the material into a spout into the form, where it is to be cooled and gets the ideal shape. Infusion shaping is viewed as very proper manner for thermoplastic embodiments of electronic items expected in clinical enterprises. Enhancement in fiber matrix similarity and consistency in the scattering of fibers in the lattice material is accomplished [19, 20, 30–32, 79, 80].

4.4.7 Electrospinning Process

The arising nano technology has incited scientists to search out new nano-scale fiber producing methods for composite assembling. An electro static, fiber manufacture method named electro spinning process, utilizes electrical powers to create continuous filaments of two nano-meters to a few micrometers. Natural polymer arrangement catapulted into spinneret structures a persistent fiber, which is gathered. It serves enhanced mechanical and physical attributes; adjusting over process confines. Some applications involve its use in wound recuperating, tissue designing frameworks, drug delivery, as a layer in bio-sensors, beauty care products, and so on [22, 30–32, 81].

4.4.8 Computerized Assembling Process

It is a steady collaboration that present self-computerization, which prompts reduced amount. Fiber winding is very useful to make axi-symmetric, along with some non-cylindrical symmetry composite parts, for instance, pipe turns. Guided by a couple of pulleys, steady 'prepreg' thin sheets, rovings together with single filament would be built such that it is subjected under pitch shower while its being received under a turning mandrel. Then, ensuing to solicit number of layers in mandrel, which has the best condition of the thing, is set for alleviating at the lab temperature. The prepared mechanical filament winding method is outfitted with a cutting-edge robot. The benefits are better method monitoring, accuracy, repeatability while displacing human executive [30, 82, 83].

4.4.9 Additive Manufacturing (AM)

AM offers an elevated degree of mathematical intricacy for the creation of completely modified objects as it exploits PC helped planning and furthermore wipes out the necessity of moulds, which reduces cost and assembling process. Additive Manufacturing relates to main advances in composite assembling since the technique gives extensive reach while determining fiber orientation, its direction and volume fraction. It can cross over plan thought in the final item rapidly without burning through material and process duration, which makes it good model for prototype and personal [30, 31, 84–88].

5 Execution of FRP

There are a few variables, compared to composite material elements and assembling methods, that influences the FRP's composite execution.

5.1 Interphase

The area over the fiber/powder relating to a structure of a matrix is generally pertained to interphase. While stress is caused at inter-phase, move from cross section to fiber occurs under stacking. Thusly, surveying contents of composite and various attributes of its materials, is of prime importance. Further, knowing the approach of acting of interphase is critical [8, 10, 30–32, 42, 71, 83].

5.2 Pre-treatments

Alkalization, preheating, utilization of silane pairing specialist to adjust fiber over the matrix brings about the enhancement of bond at the point of interaction and combination of the matrix to the fibers [8, 10, 30–32, 42, 71, 83].

5.3 Fiber Volume

It is the percentage of volume of fiber in the whole volume of composite material. It is dependent on weight and density of matrix and fibers. Usually, it is found that when the volume of fiber increases, the modulus of elasticity of the composite was also found to increase [8, 10, 30–32, 42, 71, 83].

5.4 Fiber Direction

The composite's strength and stiffness is based on fiber's direction and thus its completely relevant to the mechanical properties of the prepared composite. Directions associated with composites are random, multi-directional and sometimes uni-directional [8, 10, 30–32, 42, 71, 83].

5.5 FRP Supplanting Traditional Material

When a fiber reinforced composite, is considered that has the blend of unquestionable base matrix, offering a redesign in attributes of materials compared to polymers, pure metals or mixtures, FRP reinforced composites become most compatible for needed application. Strands as support in a grid of a composite development go probably as a stack conveying part. While the major material of matrix can keep fibers in their normal position and bearing, it similarly works with pressure movement. FRP reinforced materials have been seen as better compared to metals for various advantages. Other than the striking wear attributes, polymeric composites offers flexibility in multipurpose functioning and serve as newer tri-biological materials. For vehicle and aeronautics ventures, MMC composites with carbon fibers is overriding existing ascast metals since it gives fantastic mechanical, good electrical attributes alongside redesigned wear and erosion insurance at higher load applications. The most broadly perceived kinds of FRP composites used as support in the significant plans are CFRP composites. These FRP's show extraordinary security from shear and also flexural stresses. For the significant plans to persevere in an unforgiving environment, build up materials ought to be non-corrosive and non-magnetic. FRP's bars have these attributes and can be considered relevant for the various RC structural property over ordinary steel support. Expansion in versatility with increase in damping extent has been taken note. Besides, there was reduce in weight saving because of a decline in material thickness. Hybrid composite with jute material alongside fibers of carbon offer monetary, doable choices compared to the CFRP composites, uncovering uncommon damping attributes [8, 10, 30–32, 42, 71, 83].

6 Difficulties in Using FRC

For the composite, to become serious with various metals, it is critical to diminish cost alongside required ensured strength, practicality and dependability. A few significant difficulties with FRPs are absence of information and normalization, nonbiodegradability of manufactured filaments, high unrefined substance price, little endeavors being developed of items and applications. Once in a while, it has been seen that individuals don't share their insight for improvement to the general public and attempt to keep their syndication. This is the initial step for research that the exploration ought to be moral and ought to be advancement for the general public. Artificial engineered material accessible for composite isn't biodegradable and establishes issues for climate. In India, significance of exploration isn't that basic as different nations and consequently, advancement of composite innovation is turning out to be extremely tedious and expensive cycle [8, 10, 30–32, 42, 71, 83].

A critical test in making FRC reinforced material is the shortfall of fiber-framework depiction discernment. Increasing demand of FRC, because of the automotive business, prompts more critical bet of placing assets into normal substances [8, 10, 30–32, 42, 71, 83, 89, 90].

7 Conclusions

This review manages not just investigating different particle and fiber reinforced composites, yet additionally it features ongoing turn of events and future market prerequisites. Composite materials are uncovering different updates specifically material attributes since their improvement to some degree. An attempt is made all over the world to know about possible alternative for traditionally used materials particularly through composites. All through late numerous years, strongholds involving fibers or particulates that pertains for structural development showed exceptional results, considering notable selection relevant to most noteworthy advantages. Various categorization of composite materials, close by the attributes of their element parts, were analyzed for finding the use of composites in day to day activities. Fiber-supported composite material would often be considered as supreme uplifting reinforced composites due to its strength associated with the vast majority of purposes from most noteworthy fields. Composite materials are made with different different strategies, although every strategy is relevant for explicit material. Success of the material is based upon the blend of type alongside volume of matrix as each composite has different genuine attributes, namely stiffness, melting point, unbending nature, etc. Composites, offered increase in strength of material, but decrease in terms of weight was appreciable. Further assessment is required for knowing newest plans involving a mix of various varieties that embrace recent synthesizing strategies. It has been seen, there is immense effort in synthesizing fiber and particle supported composite which consolidates various applications. The present world has been keen enough to know the importance of economically and effectively synthesizing fiber and particle reinforced composite and implement the same in a majority of diverse applications.

References

1. Alavala CR (2016) Thermoelastic behavior of nanoparticulate BN/AA5050 alloy metal matrix composites. Int J Eng Adv Res Technol 2(1):6–8
2. Suresh NSV, Murthy, ShenBag N (2012) Aluminium-Titanium di boride metal matrix composites-challenges and opportunities. Proceedia Eng 38:89–97
3. Ajay Kumar S, Prabhu Kumar A, Balu Naik B, Ravi B (2018) Production and investigation on mechanical properties of TiC reinforced Al7075 MMC. Mater Today Proc 5(9):17924–17929
4. Boppana SB (2020) In situ synthesis of titanium carbide in pure aluminium. J Mater Sci Chem Eng 8:1–10
5. Alavala CR (2016) Influence of temperature on particulate fracture of AA2024 alloy/titanium oxide nanoparticulate metal matrix composites. Int J Sci Eng Appl Sci 2(4):1–6
6. Attar S, Nagaral M, Reddappa HN, Auradi V (2015) A review on particulate reinforced aluminum metal matrix composites. J Emerg Technol Innovative Res (JETIR) 2(2). (ISSN-2349-5162)
7. Reddy M, Narasimha BG, Anthony, Xavior M (2013) A review on processing of particulate metal matrix composites and its attributes. Int J Appl Eng Res 8(6)
8. Al-Mahaidi R, Kalfat R (2018) Fiber-reinforced polymers and their use in structural rehabilitation. In: Rehabilitation of concrete structures with fiber-reinforced polymer. Elsevier Science, Amsterdam, The Netherlands, pp 15–20
9. Biswas B, Hazra B, Sarkar A, Bandyopadhyay NR, Mitra BC, Sinha A (2018) Influence of ZrO_2 incorporation on sisal fiber reinforced unsaturated polyester composites. Polym Compos 40:2790–2801
10. Singh JIP, Singh S, Dhawan V (2017) Effect of curing temperature on mechanical attributes of fiber reinforced polymer composites. J Nat Fibers 15:687–696
11. Sozhamannan GG, Balasivanandha Prabu S, Venkatagalapathy VSK (2012) Effect of processing parameters on metal matrix composites: stir casting process. J Surf Eng Mater Adv Technol 2:11–15
12. Hascalik A, Orhan N (2007) Effect of particle size on the friction welding of Al_2O_3 reinforced 6061Al alloy composite and SAE 1020 steel. Mater Des 28:313–317
13. Bedir F (2007) Characteristic properties of Al-Cu-SiCp and Al-Cu-B4Cp composites produced by hot pressing method under nitrogen atmosphere. Mater Des 28:1238–1244
14. Unlu BS (2008) Investigation of tribological and mechanical properties Al_2O_3–SiC reinforced Al composites manufactured by casting or P/M method. Mater Des 29:2002–2008
15. Arik H (2008) Effect of mechanical alloying process on mechanical properties of a-Si_3N_4 reinforced aluminum-based composite materials. Mater Des 29:1856–1861
16. Correia JR (2013) Pultrusion of advanced fibre-reinforced polymer (FRP) composites. In: Advanced fibre-reinforced polymer (FRP) composites for structural applications. Woodhead Publishing, Sawston, UK, Cambridge, UK, pp 207–251
17. Verma D, Joshi G, Dabral R, Lakhera A (2019) Processing and evaluation of mechanical attributes of epoxy-filled E-glass fiber–fly ash hybrid composites. In: Mechanical and physical testing of biocomposites, fibre-reinforced composites and hybrid composites. Woodhead Publishing, Sawston, UK, Cambridge, UK, pp 293–306
18. Joshi SC (2012) The pultrusion process for polymer matrix composites. In: Manufacturing techniques for polymer matrix composites (PMCs). Woodhead Publishing, Sawston, UK, Cambridge, UK, pp 381–413
19. Leong YW, Thitithanasarn S, Yamada K, Hamada H (2014) Compression and injection molding techniques for natural fiber composites. In: Natural fibre composites. Woodhead Publishing, Sawston, UK, Cambridge, UK, pp 216–232
20. Werner VMK, Krumpholz R, Rehekampff C, Scherzer T, Eblenkamp M (2019) Thermoplastic encapsulations of a sensor platform by high-temperature injection molding up to 360 °C. Polym Eng Sci

21. Gonz a lez-Lopez ME, Perez-Fonseca AA, Manro quez-Gonz a lez R, Arellano M, Rodrigue D, Robledo-Ortiz JR (2018) Effect of surface treatment on the physical and mechanical attributes of injection molded poly(lactic acid)-coir fiber biocomposites. Polym Compos 40:2132–2141
22. Bhardwaj N, Kundu SC (2010) Electrospinning: a fascinating fiber fabrication technique. Biotechnol Adv 28:325–347
23. Habeeb MN, Ashour AF (2008) Flexural behavior of continuous GFRP reinforced concrete beams. J Compos Constr 12:115–124
24. Gunge A, Koppad PG, Nagamadhu M, Kivade S, Murthy KVS (2019) Study on mechanical attributes of alkali treated plain texture banana fabric reinforced biodegradable composites. Compos Commun 13:47–51
25. Abed F, El-Chabib H, AlHamaydeh M (2012) Shear characteristics of GFRP-reinforced concrete deep beams without web reinforcement. J Reinf Plast Compos 31:1063–1073
26. Davis DC, Mensah TO (2017) Fabrication and fatigue of fiber-reinforced polymer nanocomposites-a tool for quality control. In: Mensah TO, Wang B, Bothun G, Winter J, Davis V (eds) Nanotechnology commercialization: manufacturing processes and products. Wiley, Hoboken, NJ, USA
27. Plummer CJG, Bourban PE, Månson JA (2016) Polymer matrix composites: matrices and processing. Ref Modul Mater Sci Mater Eng
28. Ishikawa H, Takagi H, Nakagaito AN, Yasuzawa M, Genta H, Saito H (2014) Effect of surface treatments on the mechanical attributes of natural fiber textile composites made by VARTM method. Compos Interface 21:329–336
29. Bharath V, Auradi V, Nagaral M, Dayanand S, Boppana SB (2021) Impact of alumina particulates addition on hardness and wear behaviour of 2014 al metal matrix composites by vortex method. IOP Conf Ser Mater Sci Eng 1013:012018
30. Rajak DK, Pagar D, Menezes PL, Linul E (2019) Fiber-reinforced polymer composites: manufacturing, attributes and applications. Polymers 11(10):1667–1676
31. Boppana SB, Palani Kumar K, Ponshanmugakumar A, Dayanand S (2022) Different natural fiber reinforced composites and its potential industrial and domestic applications: a review. In: Bio-fiber reinforced composite materials. Springer, pp 51–73
32. Chung DDL (2017) Polymer-matrix composites: structure and processing. In: Carbon composites: composites with carbon fibers, nanofibers, and nanotubes. Elsevier Science, Amsterdam, The Netherlands, pp 161–217
33. Taha MA (2001) Practicalization of cast metal matrix composites (MMCCs). Mater Des 22(2001):431–441
34. Nagaral M, Deshapande RG, Auradi V, Boppana SB, Dayanand S, Anilkumar MR (2021) Mechanical and wear characterization of whiskers boron carbide-reinforced Al2024 alloy metal composites. J Bio Tribo-Corros 7:19
35. Sekine H, Chent R (1995) A combined microstructure strengthening analysis of SiCp/Al metal matrix composites. Composites 26:183–188
36. Abbasipour B, Niroumand B, Monir SM (2010) Compo-casting of A356-CNT composite. Trans Nonferrous Met Soc China 20:1561–1566
37. Pei XQ, Friedrich K (2016) Friction and wear of polymer composites. In: Reference module in materials science and materials engineering. Elsevier, Amsterdam, The Netherlands
38. Chattopadhyay SK, Khandal RK, Uppaluri R, Ghoshal AK (2010) Bamboo fiber reinforced polypropylene composites and their mechanical, thermal, and morphological attributes. J Appl Polym Sci 119:1619–1626
39. Rashid MA, Mansur MA, Paramasivam P (2005) Behavior of aramid fiber-reinforced polymer reinforced high strength concrete beams under bending. J Compos Constr 9:117–127
40. Balasubramanian K, Sultan MTH, Rajeswari N (2018) Manufacturing techniques of composites for aerospace applications. In: Sustainable composites for aerospace applications. Woodhead Publishing, Sawston, UK, Cambridge, UK, pp 55–67
41. Jamir MRM, Majid MSA, Khasri A (2018) Natural lightweight hybrid composites for aircraft structural applications. In: Sustainable composites for aerospace applications. Woodhead Publishing, Sawston, UK, Cambridge, UK, pp 155–170

42. Awan FS, Fakhar MA, Khan LA, Zaheer U, Khan AF, Subhani T (2018) Interfacial mechanical attributes of carbon nanotube-deposited carbon fiber epoxy matrix hierarchical composites. Compos Interface 25:681–699
43. Meola C, Boccardi S, Carlomagno G (2017) Composite materials in the aeronautical industry. In: Infrared thermography in the evaluation of aerospace composite materials: infrared thermography to composites. Elsevier Science, Amsterdam, The Netherlands, pp 1–24
44. Ahmad N, Bilal I, Khattak S (2018) Polyester usage in manufacturing of electrical and mechanical products and assemblies. Polyest Prod Charact Innov Appl
45. Boisse P (ed) (2015) Advances in composites manufacturing and process design. Woodhead Publishing: Sawston, UK, Cambridge, UK
46. Samuel D, Boppana SB, Palanikumar K, Ramesh S, Auradi V (2021) Role of heat treatment on hardness of Al 6061-AlB_2 metal matrix composites. Int J Surf Eng Interdiscip Mater Sci 9:26–39
47. Marques AT (2011) Fibrous materials reinforced composites production techniques. In: Fibrous and composite materials for civil engineering applications. Woodhead Publishing, Sawston, UK, Cambridge, UK, pp 191–215
48. De P, Pathak SK (2012) Influence of varying SiO_2 % on the mechanical properties of al based MMC. Trans Indian Inst Met 65(6):741–745
49. Suresh S, ShenBaga N, Moorthi V (2012) Aluminium-titanium diboride (Al–TiB_2) metal matrix composites: challenges and opportunities. Proceedia Eng 38:89–97
50. Boppana SB, Dayanand S, Vedashantha Murthy B, Nagaral M, Telugu A, Kumar V, Auradi V (2021) Development and characterization of Al6061-Al_2O_3-grpaphene hybrid metal matrix composites. J Compos Sci 5(6):155. https://doi.org/10.3390/jcs5060155
51. Boppana SB, Dayanand S, Ramesh S, Auradi V (2020) Effect of reaction holding time on synthesis and characterization of AlB_2 reinforced Al6061 metal matrix composites. J Bio Tribo-Corros 6:88
52. Mustafa N, Basir NNA, Ibrahim RE, Mahat MM (2019) Characterization of in situ zirconium diboride (Zrb2) reinforced by aluminium-copper (Al-Cu) metal matrix composites. Int J Eng Technol 7:406–410
53. Sharma DK, Sharma M, Upadhaya G (2019) Boron carbide (B4C) reinforced aluminum matrix composites (AMCs)
54. Poddar S, Kothavade NS (2013) Analysis of properties of aluminium-graphite metal matrix composites. Int J Eng Res Technol (IJERT) 2(11). ISSN: 2278-0181
55. Dario B, Gionluca B, Davide C, Livan F (2017) Al-SiC metal matrix composite production through friction stir extrusion of aluminum chips. Procedia Eng 207:419–424
56. Choudhary S, Singhal S, Srivastava UM, Jain P, Yadav V (2012) Tungsten-carbide composite: a review. Int J Sci Eng Res 3(9). ISSN 2229-5518
57. Natrayan L, Yogeshwaran S, Yuvaraj L, Kumar MS (2019) Effect of graphene reinforcement on mechanical and microstructure behavior of AA8030/graphene composites fabricated by stir casting technique. In: Proceedings of the international conference on inventive material science applications, vol 2166. ICIMA, Coimbatore, India, pp 020012
58. Wahab MN, Daud AR, Mariyam G (2009) Preparation and characterization of stir cast-aluminium nitride reinforced aluminium metal matrix composites. Int J Mech Mater Eng 4:115–117
59. Vasudeva R, Periyaswamy P, Bejaxhin Bovas Herbert A, Naveen E, Ramanan N, Aklilu T (2022) Wear behavioral study of hexagonal boron nitride and cubicboron nitride-reinforced aluminum MMC with sample analysis, Hindawi. J Nanomater 7816372:10
60. Aravind S, Raghupathi P (2019) Characterization of aluminium hybrid MMC reinforced with Si_3n_4 and BN. J Adv Res Dyn Control Syst 11:1106–1111
61. Narendra P, Amit C (2017) Fabrication methods of particulate reinforced aluminium metal matrix composite—a review. Mater Today: Proc 5(2):5933–5939
62. Meignanamoorthy M, Ravichandran M (2018) Synthesis of metal matrix composites via powder metallurgy route: a review. Mech Mech Eng 2:1

63. Aravinda T, Niranjan HB, Boppana SB, Dayanand RS (2021) Solid state diffusion bonding of Al2024 sheets. J Mines, Metals Fuels 455–458
64. Nagaral M, Bharath V, Auradi V (2013) Effect of Al_2O_3 particles on mechanical and wear properties of 6061Al alloy metal matrix composites. J Mater Sci Eng 2:2169–0022
65. Dhanshekara M, Senthil Kumar VS (2014) Squeeze casting of aluminium metal matrix composites-an overview. Procedia Eng 97
66. Arunachalam R, Krishnan PK, Rajaraman M (2019) A review on the production of metal matrix composites through stir casting—furnace design, properties, challenges, and research opportunities. J Manuf Process 42:213–245
67. Lakshmi S, Lu L, Gupta M (1998) In situ preparation of TiB_2 reinforced Al based composites. J Mater Process Technol 73(1–3):160–166
68. Rohatgi PK, Gupta N, Daoud A (2008) Synthesis and processing of cast metal-matrix composites and their applications. ASM Handbook 15:1149–1164
69. Li X, Yang Y, Cheng X (2004) Ultrasonic-assisted fabrication of metal matrix Nanocomposites. J Mater Sci 39(9):3211–3212
70. Baffaria D, Buffaa G, Campanellaa D, Fratini L (2017) Al-SiC metal matrix composite production through friction stir extrusion of aluminum chips. Procedia Eng 207:419–424
71. Rafi MM, Nadjai A, Ali F (2007) Experimental testing of concrete beams reinforced with carbon FRP. J Compos Mater 41:2657–2673
72. Altalmas A, El Refai A, Abed F (2015) Bond degradation of basalt fiber-reinforced polymer (BFRP) bars exposed to accelerated aging conditions. Constr Build Mater 81:162–171
73. Ozbakkaloglu T (2013) Compressive behavior of concrete-filled FRP tube columns: assessment of critical column parameters. Eng Struct 51:188–199
74. Ozbakkaloglu T, Vincent T (2014) Axial compressive behavior of circular high-strength concrete-filled frp tubes. J Compos Constr 18:04013037
75. Elkington M, Bloom D, Ward C, Chatzimichali A, Potter K (2015) Hand layup: understanding the manual process. Adv Manuf Polym Compos Sci 1:138–151
76. Perna AS, Viscusi A, Astarita A, Boccarusso L, Carrino L, Durante M, Sansone R (2019) Manufacturing of a metal matrix composite coating on a polymer matrix composite through cold gas dynamic spray technique. J Mater Eng Perform 28:3211–3219
77. Sussane L, Marcheti M (2012) Advanced composite materials by RTM for aerospace applications. In: Composites and their properties
78. Yalcinkaya MA, Guloglu GE, Pishvar M, Amirkhosravi M, Sozer M, Altan MC (2018) Pressurized infusion (PI): a new and improved liquid composite molding process. J Manuf Sci Eng 2018(141):011007
79. Mitschang P, Hildebrandt K (2012) Polymer and composite moulding technologies for automotive applications. In: Advanced materials in automotive engineering. Woodhead Publishing, Sawston, UK, Cambridge, UK, pp 210–229
80. Matveenko VP, Kosheleva NA, Shardakov IN, Voronkov AA (2018) Temperature and strain registration by fibre-optic strain sensor in the polymer composite materials manufacturing. Int J Smart Nano Mater 9:99–110
81. Wang G, Demei Y, Kelkar AD, Zhang L (2017) Electrospun nanofiber: emerging reinforcing filler in polymer matrix composite materials. Prog Polym Sci 75:73–107
82. Park CH, Lee WI (2012) Compression molding in polymer matrix composites. In Manufacturing techniques for polymer matrix composites (PMCs). Woodhead Publishing, Sawston, UK, Cambridge, UK, pp 47–94
83. Shirvanimoghaddam K, Hamim SU, Akbari MK, Fakhrhoseini SM, Khayyam H, Pakseresht AH, Ghasali E, Zabet M, Munir KS, Jia S (2017) Carbon fiber reinforced metal matrix composites: fabrication processes and attributes. Compos Part A—Appl Sci 92:70–96
84. Gonzalez-Henriquez CM, Sarabia-Vallejos MA, Rodriguez-Hernandez J (2019) Polymers for additive manufacturing and 4D-printing: materials, methodologies, and biomedical applications. Prog Polym Sci 94:57–116
85. Chua CK, Leong KF (2017) 3D printing and additive manufacturing: principles and applications. In: Fifth edition of rapid prototyping, 5th ed. World Scientific Publishing Co., Inc., Singapore

86. Goh GD, Yap YL, Agarwala S, Yeong WY (2018) Recent progress in additive manufacturing of fiber reinforced polymer composite. Adv Mater Technol 4:1800271
87. Hu C, Sun Z, Xiao Y, Qin Q (2019) Recent patents in additive manufacturing of continuous fiber reinforced composites. Recent Pat Mech Eng 12:25–36
88. Parandoush P, Tucker L, Zhou C, Lin D (2017) Laser assisted additive manufacturing of continuous fiber reinforced thermoplastic composites. Mater Des 31:186–195
89. Gascons M, Blanco N, Matthys K (2012) Evolution of manufacturing processes for fiber-reinforced thermoset tanks, vessels, and silos: a review. IIE Trans 44:476–489
90. Holmes M (2017) High volume composites for the automotive challenge. Reinf Plast 61:294–298

Evaluation of Mechanical Properties of Composites Involving Wear, Hardness

Mechanical Properties of Aluminium Metal Matrix Composites: Advancements, Opportunities and Perspective

Adil Wazeer, Arnab Mukherjee, Apurba Das, Barnali Sengupta, Gurudas Mandal, and Arijit Sinha

Abstract Metal Matrix Composites (MMCs) have rapidly gained prominence for prospective deployments in the aerospace and automotive sectors owing to its greater strength-to-weight ratio and greater temperature tolerance. MMCs are formed by incorporating a reinforcing element into a metal matrix. Due to its exceptional strength, stiffness, wear resistance, thermal stability, and a variety of other characteristics that vary depending on the type and quantity of reinforcements used, aluminium-based metal matrix composites are considered as one of the best engineering structural elements. The chapter examines the mechanical properties of Aluminium Metal Matrix Composites as well as the progress accomplished in this area. Opportunities and perspectives are also provided in the chapter.

Keywords MMC · Composites · Mechanical properties · Reinforcements · Composition

---

A. Wazeer (✉)
School of Laser Science and Engineering, Jadavpur University, Kolkata 700 032, India
e-mail: wazeeradil@gmail.com

A. Mukherjee
Department of Mechanical Engineering, Netaji Subhash Engineering College, Kolkata 700 152, India

A. Das
Titragarh Wagons Ltd., Anandapur, Kolkata 700107, India

Titagarh Firema, Loc. Ponteselice, Via Appia Antica, 81100 Caserta, Caserta, Italy

B. Sengupta · G. Mandal · A. Sinha
Department of Metallurgical Engineering, Kazi Nazrul University, Asansol 713 340, India

S. B. Boppana et al. (eds.), *Structural Composite Materials*, Composites Science and Technology, https://doi.org/10.1007/978-981-99-5982-2_9

1 Introduction

The correct choice of correct material for a particular application is an arduous task in engineering. There are various types of materials to choose from but selecting the correct appropriate material is crucial in achieving a successful product. The properties of material are largely affected by geographical conditions like temperature, precipitation, and pressure. Light-weight materials is not seen to have enough strength whereas brittle materials are subjected to constant failure due to low toughness resistance. Aluminum, being the most abundant metallic element found in Earth has been used in various applications since its advent in fifteenth century. Nineteenth century marked the first commercial use of aluminum during which the uses were immense [1]. With the introduction of alloying aluminum, it could be widely used for more various purposes and in industrial sectors making it at par with iron and steel [2].

The performance (strength, stiffness, etc.) of the metal and its alloys can be further improved by combining it with another material mainly other metal, ceramic or organic compounds resulting in the formation of metal matrix composites (MMC). Incorporating particles like whiskers, fibres or hollow micro balloons of different material to meet specific design needs [3]. Properties for both the matrix and reinforcement can be achieved from a MMC by combining the reinforcement strength and the ductile properties of the matrix [4]. Aluminum has been a prime choice in the composition of metal matrix composites [5]. It is mainly due to its light weight, abundancy, economically viable, high corrosion, and wear resistance. Due to the aforesaid properties aluminum MMCs are mostly used in aerospace, automobile and in railway industries [6]. SiC, graphite, Al_2O_3, ZrO_2 and Zr are some of the most common reinforcements used in the manufacture of aluminum metal matrix composites [7]. The reinforced composites are generally of three types as (i) long-continuous-unbroken fibres reinforcement of high strength material, (ii) whiskers reinforcement, (iii) equiaxed particles reinforcement of high strength material [SiC reinforced]. Reinforcement with silicon carbide (SiC) increases the strength (tensile), hardness, density, and the wear properties [8]. The material's hardness can be increased by adding Boron Carbide (B_4C) reinforcement to the Aluminium matrix [9]. Zircon as a reinforcement improves the wear resistance by a significant amount [10]. Adding fly ash as reinforcement improves the wear resistance but decreases the corrosion resistance [11]. Using metallic glass as reinforcements improves the density and the overall strength of the metal matrix while reducing the ductility [12]. Stir casting method from liquid phase processing technique and powder metallurgy from solid phase fabrication techniques are the most used processing methods for the Al MMC [13, 14]. In stir casting technique reinforcement materials are mixed to the molten aluminum by mechanical stir process which are shaped thereafter [14, 15]. Fine, powdered particles are pushed together to form the appropriate shape in the powder metallurgy process, and then the materials are heated to join them [15].

2 Mechanical Properties from Mechanical and Tribological Behavior of Particulate

2.1 *Tensile Strength*

Direct and indirect strengthening are the two types of strength to be predicted in a matrix composite. The direct stress is due to the changes in the microstructure in the metal matrix. Whereas indirect strengthening is due to the microstructural changes and alterations that occurs during the process of fabrication and processing. Due to this reinforcement strengthening the overall mechanical characteristics of the matrix are changed significantly.

Reinforcements has a key part in governing the mechanical characteristics of the composite materials. Al6061/SiC has proved to have a greater tensile strength then Al6061/graphite composites [16]. Similarly, aluminum alloys like Al7075 reinforced with silicon carbide and red mud provides more tensile strength than that with silicon carbide and fly ash [17]. Reinforcements added to the aluminum matrices helps in increasing the modulus of elasticity and the overall strength of the composite by distributing the load bearing capacity to the reinforcements as well [18] instead of monolithic alloys [19–22]. Although addition of reinforcements beyond the needed limit led to decrease in tensile strength [23], dipping the reinforcement size is found to be an improvement for the overall strength in the matrix composites [24]. Sulaiman et al. investigated on heat treatment method to enhance the tensile properties of the composites due to weak interface between the matrix and the reinforcements [25]. For ease of fabrication and improvement in the mechanical characteristics, they are pre heated to a peak aged condition [26]. After heat treatment of the composites, an increase in strength and precipitation hardening is seen to improve with the reduction in the cracking tendency [27]. Pre-heating the reinforcements can increase the material's tensile strength [28]. Addition of commonly used ceramic reinforcements like SiC and Al_2O_3 in AMMCs results in an increase in elastic modules, strength as well as hardness of the composites. It has been seen that up to a 10% increase in SiC wt% reinforcement leads to an improved strength beyond which non uniform distribution causes cluster formation [16]. Increasing the wt% of Al_2O_3 reinforcement in Al 1100 alloy up to 6 wt% was seen to increase the tensile strength to 188 MPa, above which the strength value decreased [29]. Stir casting method helped in the fabrication of the A356/15%SiC composite. The mechanical properties of the composite were represented by the size of the impeller, temperature, and stirring rate [30, 31]. Altering the conditions resulted in dissimilar tensile properties up to 309.83 MPa. Hot rolling of A356/3%B_4C improved the maximum strength of the composites by reaching a maximum tensile strength of 310 MPa [32]. Fabrication of AA430/7.5% SiC MgO of reinforcement size 15 μm increased tensile properties with reinforcement size [33]. Centrifugal casting of Al/SiC composite increased the tensile strength, the maximum being 144 MPa. Using some other reinforcement particles, not conventionally used like Gr, WC, B_4C, ZrO_2, TiB_2 and others on various aluminium alloys has been studied and experimented by past researchers [34–39]. Addition Bio reinforcements like rice husk on A356.2/SiC composite improved the damping capacity by 0.06 at a frequency of 16 Hz [40]. Strength was found at to be 150 MPa with 10 wt% SiC reinforcement and 0% GSA (Ground nutshell) in Al–Mg–Si alloy [41].

2.2 Hardness

Adding TiC particles in Al6063 alloys increased the 20% hardness, i.e., hardness value is proportional to the amount of TiC particles added [42]. There was an improvement in the Brinell hardness value with 63.7% for single reinforced composite and 81.1% for hybrid composites [43]. Alumina particles enhances the hardness of the Al7075 alloy, the hardness is seen to decrease with the addition of graphite [44]. Reheating the aluminium MMC with reinforcements like TiC, MoC, WC and Fe_3C increases the hardness [45, 46]. For SiC reinforced aluminium alloys, the hardness increases with the increase in SiC particles [16, 47].

2.3 Impact Strength

When a load is suddenly applied on a body, the amount of load resisted by the impacted material is called impact strength or toughness. Among the mechanical properties studied on AMMCs, impact strength holds a relatively limited information. Studies made by Hasson et al. on the thermal treatment of Al6061/SiC composite showed that orientation of the grains depicted on the toughness value [48]. Highest toughness value was found for the case with least SiC whiskers in a crack plane. Table 1 shows us the toughness values for some aluminium composites [49]. Increasing in diameter of Boron reinforcement in Al/B composite, the toughness was seen to increase [50]. Fibre orientation was also seen to play a part on the toughness value [50]. High ductile matrices also improved the toughness value for boron reinforced aluminum alloys [50]. Nardone et al. studies on the Charpy impact test on SiC/6061 composite and concluded that a low tensile strength value was found for toughened composites by impacts [51]. The impact energy value was seen to decline within a few hours for thermal ageing of alumina reinforced AMMCs conducted by Unsworth et al. [52]. Impact behavior Al2014/Al_2O_3 and Al6061/Al_2O_3 AMMCs was studied in the higher temperature range by Bonollo et al. [53]. It was seen that for both the composites, the impact energy generally increased with the increase in temperature from 25 to 200 °C. An increase in temperature led to the value of fracture energy for Al2040/Al_2O_3 composite to decrease while the unreinforced aluminium alloy showed an increase in the fracture energy value [54]. Ozden et al. performed Charpy impact test on SiC reinforced Al alloys and concluded that the particle agglomeration, clustering, and low matrix bonding affected the impact behaviour of the composite [55].

Table 1 Al/B composite toughness with fibre diameter [49]

Matrix	Fibre diameter (μm)	Fracture energy (kJ/m^2)
Al1100/B	100	90
	140	150
	200	200–300

2.4 Compressive Strength

Increase in compressive strength of Al356 alloy reinforced with fly ash particles was observed to enhance the compressive strength to a maximum of 738.21 MPa [56]. For Al5083 alloy with 10 wt% SiC as reinforcement also increased the compressive strength to 350 MPa. Compressive strength was found to increase from 221.1 to 351.6 MPa for Al6063/6% Al_2O_3/1%Gr with hardening [34]. A 40% increase in compressive strength was observed by Kurtyka et al. for A339 alloy with SiC as reinforcement fabricated through FSP process [57].

3 Tribological Properties

The term tribology defines as the study of surfaces interacting under relative motion under the application of surface friction, wear, lubrication, and other aspects [58]. Due to sliding of the solid materials, there is a loss of material on the interacting surfaces [59, 60] and so the strengthening of solid surfaces against the effect of friction and wear is required [61]. In case of aluminum metal matrix composites, the strengthening can be achieved via two methods, the direct and indirect mechanisms. For the direct mechanism, the load bearing capacity is transferred to the reinforcement using shear lag theory [62–65] and homogenization method [62, 66–70]. For the indirect mechanism, the load bearing capacity gets transferred to the matrix from the reinforcement [71]. In case of the later mechanism, refinement of the microstructure predicts improved tribological and mechanical properties.

3.1 Wear Resistance

Relative motion between two mating surfaces results wear leading to loss of materials. Past research on wear property of MMCs have shown that MMCs tend to have better performance than the unreinforced ones [70, 72–74]. The wear resistance is seen to rise accordingly with the enhancement of reinforcement content in an MMC. This is due to the property of high strength and hardness of the MMC. Addition of reinforcements like alumina [75–77], silicon carbide [78–82], granite [83], glass [84], boron carbide [85, 86], nickel aluminide [87, 88], aluminum diboride [89] and others have shown an improvement in wear property. Carbon nanotubes were found to be a primary material in the improvement in the wear resistance of the aluminium MMCs [90–92]. Adhesive, fatigue, abrasive and corrosive wear are some of the common wear mechanisms for MMCs [93]. Adhesive wear is due to the formation of strong forces due to the relative contact motion of the surfaces leading to shearing of the adhered junctions. The rubbing of a hard surface on a softer surface or a hard particle getting trapped between the sliding surfaces lead to abrasive wear. Because of cyclic loading conditions, fatigue wear is predominant on all surfaces. Corrosive wear occurs due the reaction of surfaces with the environment leading to removal of surface material in the process.

Of all the reinforcements, silicon carbide (SiC) is the most widely used AMCs because of its good mechanical properties, availability and economically viable. Effect of wear reduces in AMCs with an increase in wt% of the SiC matrix reinforcements [94, 95]. Reinforcement size of 120 μm proved to be better in performance than those of 47 μm sized particles [96]. For a same volume fraction, dual sized particles also were found out to have better performance than the smaller ones. Apart from reinforcement types, the selection of the type of aluminium in the matrix also determines the wear resistance [97]. AA7010 alloy shows maximum wear resistance with the addition of 25 wt% SiC as reinforcement [98]. For Al6061 alloy, addition of SiC by 6 wt% showed better tribological properties than addition of 2 or 4 wt% SiCP [94]. Adding copper by 5 wt% in Al–Mg–Cu alloys, the wear resistance was seen to improve [99]. Improved wear resistance of SiC reinforcement are observed compared to Al_2O_3 for Al–Cu based alloys [79]. The wear constant is found out by the Eq. 1 as shown.

$$K = \frac{W}{L} H_v \quad (1)$$

where

K = wear constant,

W = wear rate,

L = load applied,

H_v = Hardness of material.

Effects of SiC emery paper and Al_2O_3 emery paper as counter face material on SiC reinforced AMCs is shown in Eqs. 2 and 3 respectively [79] to calculate the wear rate (WR).

$$\begin{aligned} \mathrm{WR(SiC)} = {} & 0.0206 + (0.0034 * \mathrm{D}) + (0.0046 * \mathrm{P}) + (0.0059 * \mathrm{R}) \\ & + (0.0012 * \mathrm{D} * \mathrm{P}) + (0.0007 * \mathrm{P} * \mathrm{R}) + (0.0014 * \mathrm{R} * \mathrm{D}) \\ & - (0.0003 * \mathrm{D} * \mathrm{P} * \mathrm{R}) \end{aligned} \quad (2)$$

$$\begin{aligned}\mathrm{WR}(\mathrm{Al_2O_3}) = {} & 0.0122 - (0.0014 * \mathrm{D}) + (0.0023 * \mathrm{P}) + (0.0054 * \mathrm{R}) \\ & - (0.0001 * \mathrm{D} * \mathrm{P}) + (0.0007 * \mathrm{P} * \mathrm{R}) - (0.0012 * \mathrm{D} * \mathrm{R}) \\ & - (0.0008 * \mathrm{D} * \mathrm{P} * \mathrm{R}) \end{aligned} \tag{3}$$

where

P = applied load,

R = size of abrasives,

D = sliding distance.

Considering two more parameters of sliding speed (N) and volume percentage (V), wear rate of Al7075/SiC can be determined by Eq. 4 [100].

$$\begin{aligned}\mathrm{WR} = {} & 24.604 + (1.9585 \times \mathrm{R}) - (3.375 \times \mathrm{V}) + (3.2086 * \mathrm{P}) + (4.042 * \mathrm{N}) \\ & - (2.209 * \mathrm{H}) + \left(0.9979 * \mathrm{R}^2\right) + \left(0.4979 * \mathrm{V}^2\right) + \left(0.4979 * \mathrm{H}^2\right) \\ & - (0.3130 * \mathrm{R} * \mathrm{V}) - (0.938 * \mathrm{R} * \mathrm{P}) + (0.3125 * \mathrm{R} * \mathrm{N}) \\ & + (0.56250 * \mathrm{V} * \mathrm{P}) - (0.938 * \mathrm{V} * \mathrm{N}) - (0.3130 * \mathrm{P} * \mathrm{N}) \end{aligned} \tag{4}$$

3D response surface curves for SiC reinforced AMC based on SiC volume fraction, on wear loss with respect to Al matrix size and SiC particle size was calculated out [101]. It was found out that with decreasing SiC size and increasing the aluminum matrix size, the wear loss increased. With a higher value of counter-surface temperature, a more ductile matrix was formed with the reinforced particles to enter the matrix leading to wear loss [102]. Improvement in wear performance was seen with the heat treatment of the aluminium metal composites due to the removal of dendritic structure on heat treatment [103]. Advanced composites like functionally graded composite materials (FGCM) can be controlled in the mechanical and tribological aspects by varying the microstructure and composition [104–106]. A356/SiC with 10 and 20 wt% SiC was investigated by Karun et al. [107] in which the author found out that a better wear resistance was achieved by centrifugal casting method for higher SiC reinforcement in the outer most ring of FGCM.

3.2 Friction

The resistance faced by a material when having relative motion between another surface or object is friction. Friction is calculated as $\mu = F/P$, where μ is the coefficient of friction, F is the frictional force acting and P is the force normally acting between the two bodies. The friction acting between surfaces lead to wear on the material surfaces leading to fatigue failure. So, researchers have been studying various methods to reduce this friction acting. One of them being the addition of DLC as a surface coating where aluminium proved to show a lesser value of coefficient of friction [108]. With the inclusion of reinforcements such SiC, Al_2O_3, and ceramics, the physical and mechanical characteristics of the MMC could be changed, which improves wear loss and lowers friction coefficient [109, 110]. Some solid lubricants like graphite also help in decreasing the frictional coefficient value to some extent [111]. So, the addition of reinforcement has a natural tendency to decrease the frictional coefficient and improve wear loss [70, 73, 74]. AMC has a lesser frictional coefficient value than those of unreinforced aluminium alloys [112].

4 Recent Developments

Manufacture of aerospace and automobile components are done by AMMCs owing to its outstanding properties like decreased wear resistance, low friction coefficient and some other mechanical properties [113, 114]. The ever-evolving properties of the AMMCs are replacing some of the conventional metal alloys used in aviation and in automotive field like the fuselage, the wings, the piston, the drums, and others [115, 116]. Two manufacturing process for aluminium composites are mainly solid state and liquid state processing. The non-uniform distribution and thorough mixing of the reinforcing particles in the liquified metal present a challenge during the ex-situ procedure [117]. However, the chemical reaction of the particles reinforced in the melt matrix in the in situ process led to the resolution of this issue.

Aluminium matrix metals are generally processed by three different processes namely solid-state processing friction stir processing and powdered metallurgy. Among this processes friction stir processing is a new method which provides us with a refined grain structure. In this process the metals attain plastic flow and ultimately grain refinement due to the heat generated from the friction. This process showed better tensile strength and yield strength then stir casting method while maintaining the ductile property as well [118, 119]. For powder metallurgy process, a good mechanical strength with minimal porosity was achieved using vacuum sintering process [120]. In case of microwave sintering, AA2024 alloy reinforced with SiC provided excellent strength while Al2900 reinforced with alumina proved to be better in ductility and formability properties [121]. Ultimately microwaved sintering process proved to be better in exhibiting mechanical properties than the other sintering process. In liquid processing process, the vapour infiltration method proved to have a good thermal conductivity in which reinforcement is done on infiltrated molten metal [122]. Casting method is the most popular of the liquid dispensation methods in the manufacture MMC. With the increase in stir time and temperature in stir casting method, a better incorporation of SiC particles is achieved [123].

Reinforcement plays the vital role in the characterization of the matrix composites in terms of strength, hardness, wear loss and other defects. Among the reinforcements used silicon carbide is the most extensively used reinforcement and has been thoroughly explained in the above sections. There are numerous recent developments on this matrix composite with SiC reinforcement. Minimal defect and homogenous distribution of the matrix particles were observed for Al/SiC and Al/TiB_2

[124]. Also, similar homogenous distribution of SiC reinforcement was observed using AA7075 matrix processed via the squeeze casting process [125]. For A359/SiC matrix composites, failure or fracture occurred in the particle due to the stress concentration effect of the sharp edges [125]. The load bearing capacity of the TiC reinforced AA6082 composite was increased due to homogenous distribution of the TiC particles without any segregation [126]. Hot forging method results in the grain refinement which breaks down the TiC clusters easily and thereby reduces the size of the particles [117]. This hot forging fabrication method was studied for the Al6061/TiC composite using salt reaction synthesis [127]. Nano sized hexagonal and rectangular ZrB_2 particles was examined by the in-situ reinforcement of the AA5052/ZrB_2 composites [98]. For in-situ ZrB_2 particles reinforced with aluminium alloys, grain refinement was achieved due to nucleation sites formed by the reinforced particles thereby resisting the grown of the α-Al during the solidification process [128]. Nano alumina acts as nucleation sites when reinforced with A356 alloy and thus a smaller grain size when manufactured using the compo casting method [129]. This inclusion of nanoparticles is responsible for grain refinement but a slower solidification time. A finer grain structure can also be attained by restricting the grain growth according to the need [130]. A composite material free of porosities lead to an improved tensile strength. This can be achieved by the reinforcement of alumina oxide in the Al-Si alloy fabricated by a high speed sir casting method and so getting a homogenous composite [131]. The high elastic modulus value of carbon nanotubes (CNT) makes it a suitable and effective reinforcement for the AMMCs. Addition of $MgZn_2$ to aluminium alloys mainly Al7075 reinforced with CNT improves the interfacial bonding of the composite where the CNT helps in quick precipitation hardening [132]. Plasma-Sprayed ZrO_2, Al_2TiO_5, and Cr_2O_3 experimentation was done on Al–Si (LM13) materials as ceramic coatings to study on the microstructure and the hardness [133].

5 Opportunities and Perspective

AMMCs finds its usage in various sectors and domains like automobile, aerospace, railways, biomedical, marines, construction and much more. Using combinations of various aluminum alloys and reinforcements enable researchers to alter the properties (mechanical and physical) of the composite according to the purpose related to the different sectors. It is used mainly for its wear resistance, light weight, and good fatigue life in the automotive and aircraft industries. Using the composite material for railway cars also offers in better strength and durability while keeping the overall weight to a minimum. The aluminium composites are also used in construction and biomedical purposes for its overall performance in terms of strength and rigidity. Friction stir casting process for the manufacture of the composites are much utilized in the recent times due to the improvement in the ductile property and the wear resistance of the composite fabricated. The FSP casting method is also a prime choice in the fabrication of the aluminium composites due to cost reduction and is a time saving process. Though hybrid aluminium metal matrix composites are not different from the conventional aluminium matrix, it is used in mechanical and electronic sectors for its improvement in the ductility and malleability of its properties.

As useful as it sounds, aluminium metal matrix composites face many hurdles in the steps of processing and manufacturing. Choosing exact materials and composite particles for specific purposes requires numerous trial and errors as the materials tend to show properties with a wide range of deviation. Carbon nanotubes as a reinforcement has not been thoroughly investigated and very few literatures have been published in this area. Finding of other reinforcement sources also holds a great potential in improving the composites and needs to be studied in the future. Utilization of bio wastes or bio products as reinforcement needs more attention to detail. Bio reinforced composites could play a major role in the medical sector in developing ultra-light weight components. Porosity is formed due to much nucleation sites when nano reinforced particles are used, and few studies have been conducted on this topic to eliminate or minimize porosity. Studies related to controlled grain growth to get a uniform homogenous composite particle also needs further research studies.

6 Conclusion

This work discusses on the importance of composite structures with AMMCs with the focus on the mechanical properties with different types of reinforcements. Addition of reinforcements helps to enhance tensile strength of the AMMCs. Heat treatment prior to manufacture of composite can enhance tensile properties along with improving other mechanical properties. Better results with the tensile strength of the composites were obtained by adding SiC reinforcement. Altering the weight proportions of SiC and Al_2O_3 reinforcements lead to gain the required hardness. Researchers have explained that a uniform distribution of reinforced particles led to an increase in the toughness value to an extent. Among the various manufacturing techniques of the composites, FSP technique is mostly a prime choice. Tribological properties show a significant amount of improvement with implementation of the indirect method of strengthening in which the load carrying capacity is transferred from reinforcement to the matrix. Hybrid AMMCs show better results than those of single reinforced aluminium composites by possessing better mechanical and microstructural properties when fabricated with proper reinforcements as per need.

References

1. Nappi C (2013) The global aluminium industry 40 years from 1972. In: World aluminium, pp 1–27

2. Nturanabo F, Masu L, Kirabira JB (2019) Novel applications of aluminium metal matrix composites. In: Aluminium alloys and composites
3. Macke A, Schultz BF, Rohatgi PK, Gupta N (2013) Metal matrix composites for automotive applications. In: Advanced composite materials for automotive applications: structural integrity and crashworthiness, 24, pp 311–344
4. Kurumlu D, Payton EJ, Young ML, Schöbel M, Requena G, Eggeler G (2012) High-temperature strength and damage evolution in short fiber reinforced aluminum alloys studied by miniature creep testing and synchrotron microtomography. Acta Mater 60(1):67–78
5. McDanels DL (1985) Analysis of stress-strain, fracture, and ductility behavior of aluminum matrix composites containing discontinuous silicon carbide reinforcement. Metall Trans A 16(6):1105–1115
6. Ralph B, Yuen HC, Lee WB (1997) The processing of metal matrix composites—an overview. J Mater Process Technol 63(1–3):339–353
7. Rajan TPD, Pillai RM, Pai BC (1998) Reinforcement coatings and interfaces in aluminium metal matrix composites. J Mater Sci 33(14):3491–3503
8. Murty SN, Rao BN, Kashyap BP (2003) On the hot working characteristics of 6061Al–SiC and 6061–Al_2O_3 particulate reinforced metal matrix composites. Compos Sci Technol 63(1):119–135
9. Previtali B, Pocci D, Taccardo C (2008) Application of traditional investment casting process to aluminium matrix composites. Compos A Appl Sci Manuf 39(10):1606–1617
10. Das S, Das S, Das K (2007) RETRACTED: abrasive wear of zircon sand and alumina reinforced Al–4.5 wt% Cu alloy matrix composites–a comparative study
11. Ramnath BV, Elanchezhian C, Annamalai RM, Aravind S, Atreya TSA, Vignesh V, Subramanian C (2014) Aluminium metal matrix composites–a review. Rev Adv Mater Sci 38(5):55–60
12. Xie MS, Wang Z, Zhang GQ, Yang C, Zhang WW, Prashanth KG (2020) Microstructure and mechanical property of bimodal-size metallic glass particle-reinforced Al alloy matrix composites. J Alloy Compd 814:152317
13. Panwar N, Chauhan A (2018) Fabrication methods of particulate reinforced Aluminium metal matrix composite-a review. Mater Today: Proc 5(2):5933–5939
14. Sharma R, Jha SP, Kakkar K, Kamboj K, Sharma PA (2017) A review of the aluminium metal matrix composite and its properties. Int Res J Eng Technol 4(2):832–842
15. Sharma P, Khanduja D, Sharma S (2014) Tribological and mechanical behavior of particulate aluminum matrix composites. J Reinf Plast Compos 33(23):2192–2202
16. Rahman MH, Al Rashed HM (2014) Characterization of silicon carbide reinforced aluminum matrix composites. Proc Eng 90:103–109
17. Vinitha BS (2014) Motgi, evaluation of mechanical properties of Al 7075 alloy, flyash, SiC and red mud reinforced metal matrix composites. Inter J Sci Res Develop 2:190–193
18. Chen R, Iwabuchi A, Shimizu T, Shin HS, Mifune H (1997) The sliding wear resistance behavior of NiAl and SiC particles reinforced aluminium alloy matrix composites. Wear 213(1–2):175–184
19. Martin A, Martinez MA, Llorca J (1996) Wear of SiC-reinforced Al-matrix composites in the temperature range 20–200 °C. Wear 193(2):169–179
20. Sekine H, Chent R (1995) A combined microstructure strengthening analysis of SiCp/Al metal matrix composites. Composites 26(3):183–188
21. Chen R, Zhang G (1993) Casting defects and properties of cast A356 aluminium alloy reinforced with SiC particles. Compos Sci Technol 47(1):51–56
22. Singh PM, Lewandowski JJ (1993) Effects of heat treatment and reinforcement size on reinforcement fracture during tension testing of a SiCp discontinuously reinforced aluminum alloy. Metall and Mater Trans A 24(11):2531–2543
23. Raturi A, Mer KKS, Pant PK (2017) Synthesis and characterization of mechanical, tribological and micro structural behaviour of Al 7075 matrix reinforced with nano Al_2O_3 particles. Mater Today: Proc 4(2):2645–2658

24. Ma ZY, Tjong SC (1997) In situ ceramic particle-reinforced aluminum matrix composites fabricated by reaction pressing in the TiO_2 (Ti)-Al-B (B_2O_3) systems. Metall and Mater Trans A 28(9):1931–1942
25. Sulaiman S, Marjom Z, Ismail MIS, Ariffin MKA, Ashrafi N (2017) Effect of modifier on mechanical properties of aluminium silicon carbide (Al-SiC) composites. Proc Eng 184:773–777
26. Llorca J (1993) Failure micromechanisms in particulate-reinforced metal-matrix composites. Le Journal de Physique IV 3(C7):C7-1793
27. Sawla S, Das S (2004) Combined effect of reinforcement and heat treatment on the two body abrasive wear of aluminum alloy and aluminum particle composites. Wear 257(5–6):555–561
28. Thakur SK, Dhindaw BK (2001) The influence of interfacial characteristics between SiCp and Mg/Al metal matrix on wear, coefficient of friction and microhardness. Wear 247(2):191–201
29. Ghanaraja S, Kumar KV, Ravikumar KS, Madhusudan BM (2017) Mechanical properties of Al_2O_3 reinforced cast and hot extruded Al based metal matrix composites. Mater Today: Proc 4(2):2771–2776
30. Hashim J, Looney L, Hashmi MSJ (1999) Metal matrix composites: production by the stir casting method. J Mater Process Technol 92:1–7
31. Dwivedi SP, Sharma S, Mishra RK (2014) RETRACTED: microstructure and mechanical properties of A356/SiC composites fabricated by electromagnetic stir casting
32. Khademian M, Alizadeh A, Abdollahi A (2017) Fabrication and characterization of hot rolled and hot extruded boron carbide (B4C) reinforced A356 aluminum alloy matrix composites produced by stir casting method. Trans Indian Inst Met 70(6):1635–1646
33. Kumar SM, Pramod R, Govindaraju HK (2017) Evaluation of mechanical and wear properties of aluminium AA430 reinforced with SiC and Mgo. Mater Today: Proc 4(2):509–518
34. Saravanakumar A, Sasikumar P, Sivasankaran S (2014) Synthesis and mechanical behavior of AA 6063–x wt% Al_2O_3-1% Gr (x = 3, 6, 9 and 12 wt%) hybrid composites. Proc Eng 97:951–960
35. Mohanavel V, Rajan K, Senthil PV, Arul S (2017) Mechanical behaviour of hybrid composite (AA6351+ Al_2O_3+ Gr) fabricated by stir casting method. Mater Today: Proc 4(2):3093–3101
36. Ravikumar K, Kiran K, Sreebalaji VS (2017) Characterization of mechanical properties of aluminium/tungsten carbide composites. Measurement 102:142–149
37. Sharma P, Sharma S, Khanduja D (2015) Production and some properties of Si_3N_4 reinforced aluminium alloy composites. J Asian Ceramic Soc 3(3):352–359
38. Madhusudhan M, Naveen GJ, Mahesha K (2017) Mechanical characterization of AA7068-ZrO_2 reinforced metal matrix composites. Mater Today: Proc 4(2):3122–3130
39. Radhika N, Raghu R (2016) Development of functionally graded aluminium composites using centrifugal casting and influence of reinforcements on mechanical and wear properties. Trans Nonferr Metals Soc China 26(4):905–916
40. Prasad DS, Shoba C (2016) Experimental evaluation onto the damping behavior of Al/SiC/ RHA hybrid composites. J Market Res 5(2):123–130
41. Alaneme KK, Bodunrin MO, Awe AA (2018) Microstructure, mechanical and fracture properties of groundnut shell ash and silicon carbide dispersion strengthened aluminium matrix composites. J King Saud Univ Eng Sci 30(1):96–103
42. Kumar KR, Kiran K, Sreebalaji VS (2017) Micro structural characteristics and mechanical behaviour of aluminium matrix composites reinforced with titanium carbide. J Alloy Compd 723:795–801
43. Kannan C, Ramanujam R (2017) Comparative study on the mechanical and microstructural characterisation of AA 7075 nano and hybrid nanocomposites produced by stir and squeeze casting. J Adv Res 8(4):309–319
44. Kannan C, Ramanujam R, Balan ASS (2018) Machinability studies on Al 7075/BN/Al_2O_3 squeeze cast hybrid nanocomposite under different machining environments. Mater Manuf Processes 33(5):587–595
45. Narayan S, Rajeshkannan A (2017) Hardness, tensile and impact behaviour of hot forged aluminium metal matrix composites. J Market Res 6(3):213–219

46. Veeravalli RR, Nallu R, Mohiuddin SMM (2016) Mechanical and tribological properties of AA7075–TiC metal matrix composites under heat treated (T6) and cast conditions. J Market Res 5(4):377–383
47. Ramnath BV, Elanchezhian C, Jaivignesh M, Rajesh S, Parswajinan C, Ghias ASA (2014) Evaluation of mechanical properties of aluminium alloy–alumina–boron carbide metal matrix composites. Mater Des 58:332–338
48. Hasson DF, Hoover SM, Crowe CR (1985) Effect of thermal treatment on the mechanical and toughness properties of extruded SiCw/aluminium 6061 metal matrix composite. J Mater Sci 20(11):4147–4154
49. Friend CM (1989) Toughness in metal matrix composites. Mater Sci Technol 5(1):1–7
50. McDanels DL, Signorelli RA (1976) Effect of fiber diameter and matrix alloys on impact-resistant boron/aluminum composites (No. E-8648)
51. Nardone VC, Strife JR, Prewo KM (1991) Microstructurally toughened particulate-reinforced aluminum matrix composites. Metall Trans A 22(1):171–182
52. Unsworth JP, Bandyopadhyay S (1994) Effect of thermal ageing on hardness, tensile and impact properties of an alumina microsphere-reinforced aluminium metal-matrix composite. J Mater Sci 29(17):4645–4650
53. Bonollo F, Ceschini L, Garagnani GL (1997) Mechanical and impact behaviour of (Al_2O_3) p/2014 and (Al_2O_3) p/6061 Al metal matrix composites in the 25–200 °C range. Appl Compos Mater 4(3):173–185
54. Surappa MK, Sivakumar P (1993) Fracture toughness evaluation of 2040-Al/Al_2O_3 particulate composites by instrumented impact. Compos Sci Technol 46(3):287–292
55. Ozden S, Ekici R, Nair F (2007) Investigation of impact behaviour of aluminium based SiC particle reinforced metal–matrix composites. Compos A Appl Sci Manuf 38(2):484–494
56. Kulkarni SG, Meghnani JV, Lal A (2014) Effect of fly ash hybrid reinforcement on mechanical property and density of aluminium 356 alloy. Procedia Mater Sci 5:746–754
57. Kurtyka P, Rylko N, Tokarski T, Wójcicka A, Pietras A (2015) Cast aluminium matrix composites modified with using FSP process–changing of the structure and mechanical properties. Compos Struct 133:959–967
58. Bhushan B (2013) Introduction to tribology. John Wiley & Sons
59. Davim JP (2013) Wear of advanced materials. John Wiley & Sons
60. Almen JO (1943) Shot blasting to increase fatigue resistance. SAE Trans:248–268
61. Davim JP (Ed) (2012) Statistical and computational techniques in manufacturing. Springer Science & Business Media
62. Yang LJ (2003) Wear coefficient equation for aluminium-based matrix composites against steel disc. Wear 255(1–6):579–592
63. Riahi AR, Alpas AT (2001) The role of tribo-layers on the sliding wear behavior of graphitic aluminum matrix composites. Wear 251(1–12):1396–1407
64. Yu SY, Ishii H, Tohgo K, Cho YT, Diao D (1997) Temperature dependence of sliding wear behavior in SiC whisker or SiC particulate reinforced 6061 aluminum alloy composite. Wear 213(1–2):21–28
65. Wilson S, Alpas AT (1997) Wear mechanism maps for metal matrix composites. Wear 212(1):41–49
66. How HC, Baker TN (1997) Dry sliding wear behaviour of Saffil-reinforced AA6061 composites. Wear 210(1–2):263–272
67. Wilson S, Alpas AT (1996) Effect of temperature on the sliding wear performance of Al alloys and Al matrix composites. Wear 196(1–2):270–278
68. Gurcan AB, Baker TN (1995) Wear behaviour of AA6061 aluminium alloy and its composites. Wear 188(1–2):185–191
69. Garcia-Cordovilla C, Narciso J, Louis E (1996) Abrasive wear resistance of aluminium alloy/ceramic particulate composites. Wear 192(1–2):170–177
70. Sharma SC (2001) The sliding wear behavior of Al6061–garnet particulate composites. Wear 249(12):1036–1045
71. Hutchings IM (1987) Wear by particulates. Chem Eng Sci 42(4):869–878

72. Naplocha K, Granat K (2008) Dry sliding wear of Al/Saffil/C hybrid metal matrix composites. Wear 265(11–12):1734–1740
73. Shafiei-Zarghani A, Kashani-Bozorg SF, Zarei-Hanzaki A (2009) Microstructures and mechanical properties of Al/Al_2O_3 surface nano-composite layer produced by friction stir processing. Mater Sci Eng, A 500(1–2):84–91
74. Ghasemi-Kahrizsangi A, Kashani-Bozorg SF (2012) Microstructure and mechanical properties of steel/TiC nano-composite surface layer produced by friction stir processing. Surf Coat Technol 209:15–22
75. Yang LJ (2003) The transient and steady wear coefficients of A6061 aluminium alloy reinforced with alumina particles. Compos Sci Technol 63(3–4):575–583
76. Al-Qutub AM, Allam IM, Qureshi TW (2006) Effect of sub-micron Al_2O_3 concentration on dry wear properties of 6061 aluminum based composite. J Mater Process Technol 172(3):327–331
77. Al-Qutub AM, Allam IM, Samad A (2008) Wear and friction of Al–Al_2O_3 composites at various sliding speeds. J Mater Sci 43(17):5797–5803
78. İzciler MAHMUT, Muratoglu M (2003) Wear behaviour of SiC reinforced 2124 Al alloy composite in RWAT system. J Mater Process Technol 132(1–3):67–72
79. Sahin Y (2003) Wear behaviour of aluminium alloy and its composites reinforced by SiC particles using statistical analysis. Mater Des 24(2):95–103
80. De Salazar JG, Barrena MI (2004) Influence of heat treatments on the wear behaviour of an AA6092/SiC25p composite. Wear 256(3–4):286–293
81. Muratoğlu M, Aksoy M (2006) Abrasive wear of 2124Al–SiC composites in the temperature range 20–200 C. J Mater Process Technol 174(1–3):272–276
82. Ghosh SK, Saha P (2011) Crack and wear behavior of SiC particulate reinforced aluminium based metal matrix composite fabricated by direct metal laser sintering process. Mater Des 32(1):139–145
83. Singh M, Prasad BK, Mondal DP, Jha AK (2001) Dry sliding wear behaviour of an aluminium alloy–granite particle composite. Tribol Int 34(8):557–567
84. Hemanth J (2002) Wear behavior of chilled (metallic and non-metallic) aluminum alloy–glass particulate composite. Mater Des 23(5):479–487
85. Tang F, Wu X, Ge S, Ye J, Zhu H, Hagiwara M, Schoenung JM (2008) Dry sliding friction and wear properties of B4C particulate-reinforced Al-5083 matrix composites. Wear 264(7–8):555–561
86. Shorowordi KM, Haseeb ASMA, Celis JP (2004) Velocity effects on the wear, friction and tribochemistry of aluminum MMC sliding against phenolic brake pad. Wear 256(11–12):1176–1181
87. Wang Y, Rainforth WM, Jones H, Lieblich M (2001) Dry wear behaviour and its relation to microstructure of novel 6092 aluminium alloy–Ni_3Al powder metallurgy composite. Wear 251(1–12):1421–1432
88. Jiménez AE, Bermúdez MD, Cintas J, Herrera EJ (2009) Dry wear of $NiAl_3$-reinforced mechanically alloyed aluminium with different microstructure. Wear 266(1–2):255–265
89. Melgarejo ZH, Suárez OM, Sridharan K (2006) Wear resistance of a functionally-graded aluminum matrix composite. Scripta Mater 55(1):95–98
90. Zhou SM, Zhang XB, Ding ZP, Min CY, Xu GL, Zhu WM (2007) Fabrication and tribological properties of carbon nanotubes reinforced Al composites prepared by pressureless infiltration technique. Compos A Appl Sci Manuf 38(2):301–306
91. Kim IY, Lee JH, Lee GS, Baik SH, Kim YJ, Lee YZ (2009) Friction and wear characteristics of the carbon nanotube–aluminum composites with different manufacturing conditions. Wear 267(1–4):593–598
92. Choi HJ, Lee SM, Bae DH (2010) Wear characteristic of aluminum-based composites containing multi-walled carbon nanotubes. Wear 270(1–2):12–18
93. Rohatgi PK, Tabandeh-Khorshid M, Omrani E, Lovell MR, Menezes PL (2013) Tribology of metal matrix composites. In: Tribology for scientists and engineers. Springer, New York, NY, pp 233–268

94. Kumar GV, Rao CSP, Selvaraj N (2012) Studies on mechanical and dry sliding wear of Al6061–SiC composites. Compos B Eng 43(3):1185–1191
95. Sahin Y, Acılar M (2003) Production and properties of SiCp-reinforced aluminium alloy composites. Compos A Appl Sci Manuf 34(8):709–718
96. Bindumadhavan PN, Wah HK, Prabhakar O (2001) Dual particle size (DPS) composites: effect on wear and mechanical properties of particulate metal matrix composites. Wear 248(1–2):112–120
97. Mistry JM, Gohil PP (2017) An overview of diversified reinforcement on aluminum metal matrix composites: tribological aspects. Proc Inst Mech Eng Part J: J Eng Tribol 231(3):399–421
98. Rao RN, Das S (2010) Effect of matrix alloy and influence of SiC particle on the sliding wear characteristics of aluminium alloy composites. Mater Des 31(3):1200–1207
99. Hassan AM, Alrashdan A, Hayajneh MT, Mayyas AT (2009) Wear behavior of Al–Mg–Cu–based composites containing SiC particles. Tribol Int 42(8):1230–1238
100. Kumar S, Balasubramanian V (2008) Developing a mathematical model to evaluate wear rate of AA7075/SiCp powder metallurgy composites. Wear 264(11–12):1026–1034
101. Diler EA, Ipek R (2013) Main and interaction effects of matrix particle size, reinforcement particle size and volume fraction on wear characteristics of Al–SiCp composites using central composite design. Compos B Eng 50:371–380
102. Dvivedi A, Rajeev VR, Kumar P, Singh I (2012) Tribological characteristics of Al 6063–SiCp metal–matrix composite under reciprocating and wet conditions. Proc Inst Mech Eng Part J: J Eng Tribol 226(2):138–149
103. Aksöz S, Bican O, Çalın R, Bostan B (2014) Effect of T7 heat treatment on the dry sliding friction and wear properties of the SiC-reinforced AA 2014 aluminium matrix composites produced by vacuum infiltration. Proc Inst Mech Eng Part J: J Eng Tribol 228(3):312–319
104. Vieira AC, Sequeira PD, Gomes JR, Rocha LA (2009) Dry sliding wear of Al alloy/SiCp functionally graded composites: influence of processing conditions. Wear 267(1–4):585–592
105. Duque NB, Melgarejo ZH, Suhrez OM (2005) Functionally graded aluminum matrix composites produced by centrifugal casting. Mater Charact 55(2):167–171
106. Rajan TPD, Pai BC (2009) Formation of solidification microstructures in centrifugal cast functionally graded aluminium composites. Trans Indian Inst Met 62(4):383–389
107. Karun AS, Rajan TPD, Pillai UTS, Pai BC, Rajeev VR, Farook A (2016) Enhancement in tribological behaviour of functionally graded SiC reinforced aluminium composites by centrifugal casting. J Compos Mater 50(16):2255–2269
108. Yang W, Wang AY, Ke PL, Jiang BL (2011) Characterizations of dlc/mao composite coatings on az80 magnesium alloy. Acta Metall Sin 47(12):1535–1540
109. Natarajan N, Vijayarangan S, Rajendran I (2006) Wear behaviour of A356/25SiCp aluminium matrix composites sliding against automobile friction material. Wear 261(7–8):812–822
110. Kumar GH, Bapu BR, Sagar R, Mohit H (2010) The abrasive wear behaviour of Al-SiC p composites for automotive parts. In: Frontiers in automobile and mechanical engineering. IEEE, pp 54–59
111. Valiei M (2020) Improving tribological and mechanical properties of copper-based friction materials for brake pad applications
112. Sannino AP, Rack HJ (1995) Dry sliding wear of discontinuously reinforced aluminum composites: review and discussion. Wear 189(1–2):1–19
113. Make A, Schultz BF, Rohatgi P (2012) Metal matrix composites. Adv Mater Processes 170(3):19–23
114. Dursun T, Soutis C (2014) Recent developments in advanced aircraft aluminium alloys. Mater Des 1980–2015(56):862–871
115. Rohatgi P (1991) Cast aluminum-matrix composites for automotive applications. Jom 43(4):10–15
116. Rambabu PPNKV, Eswara Prasad N, Kutumbarao VV, Wanhill RJH (2017) Aluminium alloys for aerospace applications. In: Aerospace materials and material technologies, pp 29–52

117. Samal P, Vundavilli PR, Meher A, Mahapatra MM (2020) Recent progress in aluminum metal matrix composites: a review on processing, mechanical and wear properties. J Manuf Process 59:131–152
118. Bajakke PA, Malik VR, Deshpande AS (2019) Particulate metal matrix composites and their fabrication via friction stir processing–a review. Mater Manuf Processes 34(8):833–881
119. Ma ZY (2008) Friction stir processing technology: a review. Metall and Mater Trans A 39(3):642–658
120. Zhang X, Liang S, Li H, Yang J (2017) Mechanical and optical properties of transparent alumina obtained by rapid vacuum sintering. Ceram Int 43(1):420–426
121. Ashwath P, Xavior MA (2018) Effect of ceramic reinforcements on microwave sintered metal matrix composites. Mater Manuf Processes 33(1):7–12
122. Tao P, Wang Y (2019) Improved thermal conductivity of silicon carbide fibers-reinforced silicon carbide matrix composites by chemical vapor infiltration method. Ceram Int 45(2):2207–2212
123. Soltani S, Azari Khosroshahi R, Taherzadeh Mousavian R, Jiang ZY, Fadavi Boostani A, Brabazon D (2017) Stir casting process for manufacture of Al–SiC composites. Rare Met 36(7):581–590
124. Zhao M, Wu G, Jiang L, Dou Z (2006) Friction and wear properties of TiB2P/Al composite. Compos A Appl Sci Manuf 37(11):1916–1921
125. Kalkanlı A, Yılmaz S (2008) Synthesis and characterization of aluminum alloy 7075 reinforced with silicon carbide particulates. Mater Des 29(4):775–780
126. Thangarasu A, Murugan N, Dinaharan I, Vijay SJ (2015) Synthesis and characterization of titanium carbide particulate reinforced AA6082 aluminium alloy composites via friction stir processing. Arch Civ Mech Eng 15(2):324–334
127. Kumar GSP, Koppad PG, Keshavamurthy R, Alipour M (2017) Microstructure and mechanical behaviour of in situ fabricated AA6061-TiC metal matrix composites. Arch Civ Mech Eng 17(3):535–544
128. Kumar N, Gautam RK, Mohan S (2015) In-situ development of ZrB2 particles and their effect on microstructure and mechanical properties of AA5052 metal-matrix composites. Mater Des 80:129–136
129. Sajjadi SA, Ezatpour HR, Parizi MT (2012) Comparison of microstructure and mechanical properties of A356 aluminum alloy/Al_2O_3 composites fabricated by stir and compo-casting processes. Mater Des 34:106–111
130. Valibeygloo N, Azari Khosroshahi R, Taherzadeh Mousavian R (2013) Microstructural and mechanical properties of Al-4.5 wt% Cu reinforced with alumina nanoparticles by stir casting method. Int J Min Metall Mater 20(10):978–985
131. Zhang P, Zhang W, Du Y, Wang Y (2020) High-performance Al-1.5 wt% Si-Al_2O_3 composite by vortex-free high-speed stir casting. J Manuf Process 56:1126–1135
132. Zhang HB, Wang B, Zhang YT, Li Y, He JL, Zhang YF (2020) Influence of aging treatment on the microstructure and mechanical properties of CNTs/7075 Al composites. J Alloy Compd 814:152357
133. Wazeer A, Mondal V, Kennedy S (2021) Comparative studies on microstructure and hardness of plasma-sprayed Al_2 TiO_5, ZrO_2 and Cr_2O_3 ceramic coatings on Al–Silicon (LM13). In: Recent advances in mechanical engineering 2021. Springer, Singapore, pp 907–916

Mechanical Properties of Light Weight Particulate Metal Matrix Composites

D. Shivalingappa and N. Raghavendra

Abstract The knowledge of mechanical properties is essential for designing the mechanical and structural components used in all engineering applications. As lighter metallic structural material, aluminium and magnesium alloys play a vital role in aerospace, automotive and defence sectors because of their low density, higher strength and stiffness combined with high wear resistance. Desired specific properties of these alloys can be enhanced or altered, by using reinforcements while making composite materials, based on the applications. Metal matrix composites are well recognized for their combination of light weight and superior mechanical behaviour. The hardness and tensile properties are essentially required to control the dry sliding wear characteristics of the materials. Many researchers have investigated the hardness behaviour of particulate metal matrix composites (PMMCs) and reported that presence of particulate reinforcements have led to a considerable increase in hardness of a matrix material. The tensile properties, except the ductility, show improved values by reinforcing light weight structural metals with particles. Further, researchers have reported better mechanical properties in PMMCs fabricated by generating the reinforcing particles in the matrix material during processing, over conventional or ex-situ technique, in which reinforcement particles are gradually poured from outside to the matrix material during processing.

Keywords Light weight metal · MMCs · Particulate reinforcement · Mechanical characteristics

1 Introduction

The mechanical characteristics, especially the hardness and tensile, are the important characteristics which control the dry sliding wear characteristics of the materials. The hardness and tensile characteristics of the PMMCs depend upon the type, size,

D. Shivalingappa (✉) · N. Raghavendra
Department of Mechanical Engineering, B. N. M. Institute of Technology, Bengaluru 560 070, India
e-mail: dsivadvg@gmail.com

S. B. Boppana et al. (eds.), *Structural Composite Materials*, Composites Science and Technology, https://doi.org/10.1007/978-981-99-5982-2_10

shape, amount of reinforcement and the particles distribution inside the composites. Normally, mechanical properties of MMCs are depending on the particle–matrix interface characteristics, which are governed by the wettability of reinforced particles in a melt and affinity of the particles to be attached to the gas bubbles either sucked into the melt during mixing by stirring or being available on release of dissolved gases by the molten metal during its solidification from a high processing temperature. Many researchers have tried to enhance the hardness and tensile properties of light weight structural metals (aluminium, magnesium and titanium) by particulate reinforcement through stir casting or powder metallurgy technique and some of them are discussed in the following sections.

2 Hardness Behaviour

The hardness of a material is its ability to resist permanent deformation by indentation. Investigation of hardness for a material is very important especially, when it is subjected to wear. Many researchers have investigated hardness behaviour of number of particulate metal-matrix composites.

(a) **Powder Metallurgy Technique**

Micro-hardness and macro-hardness investigations conducted out by Hassan and Gupta [1] on nano-Al_2O_3 reinforced magnesium composites produced using powder metallurgy process (PM) and by hot extrusion are as reported in Table 1. There is an increment in hardness of magnesium (matrix) with increased amount of nano-Al_2O_3. The main causes of this behaviour are (a) the addition of substantially tougher particles to the matrix, (b) the fact that localised matrix deformation is restricted as a result of their existence, and (c) smaller grain size. The hardness behaviour of cast magnesium and magnesium/2.5 wt% Al_2O_3 nano-composites created using the disintegrating melt deposition process (DMD) has also been studied by Hassan and Gupta [2]. They reported that nano-Al_2O_3 reinforcement has improved macro-hardness of Mg in the case of DMD processed material (65 15HRT) compared with PM processed material (60 15HRT). Due to rather uneven distribution and minimal impact of grain refinement of reinforcement, increase in hardness of PM-processed material (40%) as compared to DMD-processed material (76%) is not very significant.

Wang and co-workers [4] investigated the hardness values of TiB_2/Mg particulate composites fabricated through PM technique with varied volume fractions of TiB_2 particulates. According to their findings, hardness of composites supplemented with 10, 20, and 30 vol% TiB_2 particles rose by 41% (45 HB), 106% (66 HB), and 181% (90 HB), respectively, as compared to as cast pure Mg (32 HB). This is because (a) the magnesium matrix comprises tougher TiB_2 particulates and (b) the presence of TiB_2 prevents localised matrix deformation during indentation.

Wang et al. [5] investigated the hardness characteristics of magnesium alloy composites strengthened with 2, 5 and 7.5 wt% fine TiB_2. In the fabrication of

Table 1 Hardness values of Mg and alumina strengthened Mg composites (Hassan and Gupta [1]; Srikanth et al. [3])

Materials	Mg	Mg/ 0.22 vol% alumina	Mg/ 0.66 vol% alumina	Mg/ 1.11 vol% alumina	Mg/ 1.00 vol% alumina
Macro-hardness (15HRT)	43 ± 0	51 ± 0	56 ± 0	60 ± 1	–
Micro-hardness (HV)	37 ± 0	44 ± 0	50 ± 1	70 ± 0	52 ± 3

composites, TiB_2–Al compacted powder was prepared and added to magnesium in semi-solid state. They have mentioned that, hardness of Mg alloy composite is higher and increases with increasing TiB_2 content compared to as-cast AZ91 (Fig. 1).

Comparison of hardness values of typical in-situ particulate AZ91 composites and their matrix alloys (Table 2). Jiang et al. [6] have reported the highest hardness value (83 HB) in their in-situ 10 vol% TiC_p/AZ91 composite. They have used powder metallurgy route in the preparation of master alloy as well as in-situ composite. Wang et al. [7] and Ma et al. [8] have also used similar technique in the preparation of master alloy/green perform for their in-situ composites. By dispersing copper particulates (between 5 and 15 wt%) in an aluminium matrix using a stir-cast technique, the effect of particulate composition was explored [9]. It was found that while hardness increases with particulate contents, strength and strain decrease by 13% and 15%, respectively, compared to matrix alloy. Susila et al. [10] synthesised Al–Cu–Mg alloy reinforced with 6 wt% and 7 wt% SiC_p through powder metallurgy technique and

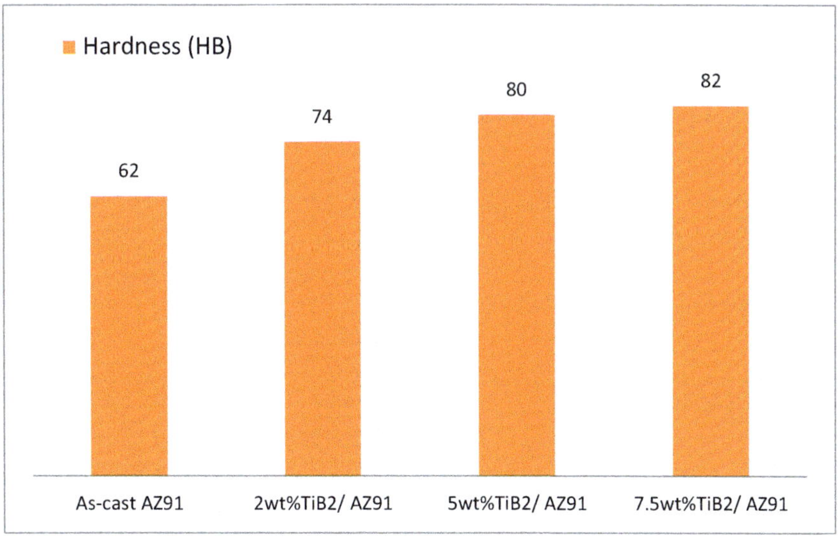

Fig. 1 Brinell hardness values of Mg composites strengthened with TiB_2

16 h of aging and concluded that 7 wt% SiC_p content Al–Cu–Mg alloy composite gives better hardness values.

(b) **Stir-casting Technique**

Hardness behaviour of cast in-situ Mg-9wt% Al composite resulted by reinforcing Al_3Ti–Al_2O_3 particles studied by Shivalingappa et al. [11] have shown the improvement by a factor of 1.41 from 607 ± 32 to 794 ± 65 MPa. Their studies are showing negligible effect of the processing temperature on the hardness and enhanced with increased amount reinforcing particles. Saravanan and Surappa [12] have measured hardness values of pure Mg-30 vol% SiC_p composite fabricated through stir casting method and reported higher hardness value for the composite (55 VHN) compared to pure magnesium (45 VHN).

Abdulhaqq and co-workers [13] have investigated hardness characteristics of Al (Mg, Mn)–Al_2O_3 (MnO_2) in-situ composite with different processing variables (processing temperature 670–850 °C and stirring time 3–15 min). They found that when the processing temperature was raised up to 780 °C for in-situ composite, the hardness increased, but dropped once it was raised over 780 °C. However, they found negligible influence of stirring time on hardness of composite.

They [14] have also investigated hardness behaviour of commercially pure Aluminium strengthened with in-situ formed Al_2O_3 particles by addition of TiO_2 particles and small amount of Mg along the height of the casting synthesized at 780 °C. Mg was put to the Al melt prior to addition of TiO_2 to improve wettability of generated Al_2O_3 particles during reaction between Al and TiO_2. Authors have concluded that, the hardness of the composite was relatively constant from the bottom to the middle of casting, but drastically decreased at the top, may be due to increased porosity.

Jenix et al. [15] studied the mechanical behaviour of Al6063 strengthened with $ZrSiO_4$ and Al_2O_3 particles of total 8 wt% hybrid reinforcement in the combinations of (0 + 8)%, (2 + 6)%, (4 + 4)%, (6 + 2)%, (8 + 0)% using stir casting technique

Table 2 Hardness values of AZ91 and in-situ particulate AZ91 composites

Sl. No.	Primary processing	Materials	Hardness (HB)	Investigators
1	Semisolid slurry stirring technique	As-cast AZ91	60	Jiang et al. [6]
		In-situ 10 vol% TiC_p/AZ91	83	
2	Stir casting technique	AZ91D alloy	60	Wang et al. [7]
		In-situ 5 wt% TiC_p/AZ91D	80	
3	Remelting and dilution stir casting technique	AZ91 alloy	56	Ma et al. [8]
		In-situ 5 wt% $(TiB_2\text{-}TiC)_p$/AZ91	79	

and reported (4 + 4)% combination is giving optimum mechanical (hardness and tensile) properties.

3 Tensile Properties

Tensile properties, except the ductility, show higher values when reinforced with magnesium, aluminium and their alloys with particulates. Following four different mechanisms are thought to be responsible for increasing strength and elastic modulus [16].

- Orowan mechanism: The particles and dislocations are interdependent.
- Strengthening by reducing grain size or stabilizing.
- Dislocation around the particles Increases and the internal stress are developed due to misfit during thermal expansion.
- Different strain properties of both matrix and particles will be strengthening the matrix.

Due to their high stiffness, strength, and chemical stability, SiC particles are frequently utilised to create particulate reinforced metal matrix composites. For example, Laurent et al. [17] explored the tensile properties of extruded AZ91D-15 vol% SiC_p composites processed in the semi-solid temperature range between 581 and 587 °C prior to extrusion. The tensile properties were compared with the room-temperature tensile test results published by the Dow chemical company for AZ91D alloys. They found that in composites containing 15 vol% SiC_p of size 54 μm, the room-temperature yield strength (YS) was 257 MPa, a value 20% higher than the yield strength of 215 MPa observed in unreinforced AZ91D alloy. UTS in the same composite was 289 MPa, a value close to 296 MPa, observed in the unreinforced alloys. Improvement in modulus of elasticity (E) of up to 46% observed in the AZ91D alloy but the total elongation of the composite decreased drastically to 0.7% compared to 10.2% measured in the unreinforced alloy for die cast AZ91-15 vol% SiC_p reinforced composites.

Lloyd [18] reported typical properties of some of the commercially available magnesium alloy composites and unreinforced base alloys as provided by Dow co. The YS for AZ91 alloy strengthened with 9.4 vol% and 15.1 vol% SiC_p were found to be 191 and 208 MPa, respectively, although the UTS for both of these composites were 236 MPa. In these composites, the elastic moduli were 47.5 and 54 GPa in the composites containing 9.4 vol% and 15.1 vol% SiC_p respectively but the elongation observed were 2 and 1% respectively. When the base alloy has been changed to AZ61 containing lower amount of aluminium but reinforced with 20 vol% of SiC_p, it was observed that the YS of the unreinforced alloy, 157 MPa, increased to 260 MPa in the composite. The UTS of the unreinforced alloy, 198 MPa, also increased to 328 MPa in the composite. There was very little reduction of percentage elongation with reinforcement from 3.0 to 2.5%. The elastic modulus increased considerably from 38 to 80 GPa on reinforcement. Thus, it appears that one may get relatively better tensile strength, elastic modulus and percentage elongation in lower aluminium alloys.

Lee and co-workers [19] have investigated the tensile properties of AZ91 and AZ91/SiC_p composites produced by powder metallurgy technique, using 8, 30, 50 μm size SiC particles. According to their findings, extruded composites have YS and UTS than Mg alloy, and the YS of magnesium alloy/$SiC_{p(8\,\mu m)}$ composites is higher compared to Mg alloy/$SiC_{p(50\,\mu m)}$ composites. This improvement is attributable to the matrix's finer grain structure.

Chua et al. [20] investigated influence of size (15, 20, 25, 38, 50 μm) of SiC_p reinforcement on mechanical characteristics of Mg/SiC_p composites. Authors have observed small differences in YS and UTS for composites reinforced with 15–25 μm particles and much lower strengths with 50 μm particles. However, they reported an increased modulus and decreased ductility with increased particle size. Tensile properties of some of the SiC_p reinforced magnesium composites are summarized and presented in Table 3.

After SiC particulate reinforcement, extensive work has been taken place in Al_2O_3 particulate reinforced aluminium based composites, however, very few researchers have tried to synthesize Al_2O_3 particulate reinforced magnesium based composites. The tensile characteristics of elemental magnesium reinforced with nano-Al_2O_3 particulates made using PM technique and hot extrusion were examined by Hassan and Gupta [1]. The gradual addition of nano-sized Al_2O_3 particles, according to their research, significantly improved the YS, UTS, and ductility of Mg in tensile tests carried out at room temperature (Table 3). Improvement in tensile properties (YS, UTS and ductility) of composite compared to pure Mg may be attributed to the combined effect of applied load transfer to the properly bonded strong nano-alumina and the creation of dislocation at the matrix-reinforcement interface.

Table 3 Tensile properties of Mg/SiC_p and Mg alloy/SiC_p composites

Sl. No.	Material	0.2% YS (MPa)	UTS (MPa)	E (GPa)	Ductility (%)	Investigators
1	AZ91/15 vol% SiC_p	257	289	_	0.7	Laurent et al. [17]
2	Mg/10 vol% SiC_p	120	160	45	2.0	Krishnadev et al
3	AZ91/16.1 wt% SiC_p	191	236	47.5	2.0	Lloyd [18]
4	Mg/10 vol% SiC_p	135	152	44.7	0.8	Luo [21]
5	Mg/15.1 vol% SiC_p	100	150	_	1.0	Vijaymohan et al. [22]
6	Mg/30 vol% SiC_p	229	258	59	2	Saravanan et al. [12]
7	Mg/9.3 vol% SiC_p	120 ± 5	181 ± 6	44 ± 2	4.7 ± 1.3	Gupta et al. [23]

Hassan and Gupta [2] have also studied tensile behaviour of magnesium/2.5 wt% Al_2O_3 nano-composites and magnesium processed through DMD technique and hot extrusion.

They reported a substantial improvement in UTS, YS, and ductility of Mg due to the inclusion of Al_2O_3 nano particles processed sing DMD technique compared to PM technique as shown in Table 3. They have mentioned better tensile properties for alumina reinforced magnesium nano-composites compared to AZ91Mg alloy strengthened with higher quantity of micron size SiC particles and reinforcement of pure Mg with nano-Al_2O_3 particles had changed its brittle fracture to ductile.

Shivalingappa et al. [11] have studied tensile behaviour of cast magnesium, Mg-9 wt% Al alloy and Mg-9 wt% Al/Al_3Ti–Al_2O_3 particles composite. Great improvement in modulus of the composite compared to cast Mg and Mg-9 wt% Al alloy is reported (Table 4). However, reinforcing particles have negative impact on tensile strength and ductility. Increasing trend in UTS with increasing processing temperature and negligible effect on ductility have been reported. Decreasing tensile strength and ductility with increasing porosity content in the composite is also highlighted.

Even though SiC_p reinforcement has been extensively used to manufacture particulate metal matrix composites, one can understand from the available literature that, the inclusion of SiC_p in Mg, in spite of increasing its Young's modulus and YS, often reduces the ductility of magnesium, and therefore its use in composite fabrication is not so favourable. The reduced ductility of MMCs reinforced with SiC_p might be due to coefficient of thermal expansion mismatch, modulus of elasticity and crystal structure between SiC and magnesium.

Table 4 Tensile characteristics of Mg and Mg/Al_2O_3 MMCs

Sl. No.	Material	0.2% YS (MPa)	UTS (MPa)	E (GPa)	Ductility (%)
1	Pure Mg [24]	69–105	165–205	40–44	5–8
2	Mg (Cast) [12]	135	196	38	12
3	Mg (PM) [1]	132 ± 7	193 ± 2	41.2	4.2 ± 0.1
4	Mg (DMD) [2]	97 ± 2	173 ± 1	42.8	7.4 ± 0.2
5	Cast Mg-9 wt% Al [2]	–	203.7 ± 9.2	38.6 ± 1.6	9.3 ± 0.9
6	Mg/0.22 vol%Al_2O_3 (PM) [1]	169 ± 4	232 ± 4	42.5	6.5 ± 2.0
7	Mg/0.66 vol%Al_2O_3 (PM) [1]	191 ± 2	247 ± 2	43.4	8.8 ± 1.6
8	Mg/1.11 vol%Al_2O_3 (PM) [1]	194 ± 5	250 ± 3	44.5	6.9 ± 1.0
9	Mg/1.11 vol%Al_2O_3 (DMD) [2]	175 ± 3	246 ± 3	52.7	14.0 ± 2.4
10	Cast in-situ Mg-9 wt% Al/ Al_3Ti–Al_2O_3 [12]	–	144.8 ± 1.5	47.8 ± 1.3	2.8 ± 0.1
11	AZ91/16.1 vol% SiC_p [18]	191	236	47.5	2

Alternatively, some of the metallic particulates and intermetallics with stronger and comparable physical properties to magnesium may be suitable for reinforcing magnesium and magnesium alloys. In this direction, few of the researchers have investigated the reinforcing effects of intermetallics, copper, nickel and titanium particulates.

Lu et al. [25] examined the tensile characteristics of in-situ formed Mg_2Si reinforced magnesium composite through powder metallurgy technique and reported improved yield and ultimate tensile strengths compared with magnesium alloy. Further, there is improvement in YS of the composite from 201 to 245 MPa as the amount of Mg_2Si reinforcement increased from 5 to 15 vol%. Formation and microstructure refinement of Mg_2Si might be attributed improvement in the mechanical characteristics of the composite. Addition of aluminium to magnesium silicon alloy will cause the solution strengthening effect by the Al in Mg leads to higher tensile strength than magnesium silicon alloy. Results of the investigations by Manjunath et al. [26] from the effect of bismuth on tensile characteristics of Mg–Al alloy strengthened with in-situ magnesium silicide particles using stir casting technique reveals that magnesium silicide (Mg_2Si) particles are highly brittle and appear to deteriorate the tensile characteristics. However, by refining the Mg_2Si particles using bismuth these properties can be improved for Mg–Al alloy.

Hassan and Gupta [27] investigated tensile properties of magnesium reinforced with fine elemental titanium particles and reported an improvement of YS and ductility, and marginal decrement in UTS values due the presence of titanium reinforcement (Fig. 2). Perez et al. [28] studied the mechanical characteristics of a Mg-10 vol% Ti_p composite in 25–300 °C temperature range. They reported 160 MPa tensile strength with 8% elongation, which is higher than most of magnesium composites strengthened with ceramic particles. At temperatures above 100 °C, there is a high ductility and a declining trend in strength, from 160 MPa at room temperature to 70 MPa at 100 °C. Ho et al. [29] showed that AZ91 reinforced with small copper particles had improved stiffness, 0.2% YS, and UTS with only minor ductility deterioration.

Xi and co-workers [30] studied the tensile characteristics of Mg strengthened withTi–6Al–4V particles reinforced composite and reported a marginal improvement in modulus of elastic, YS, UTS values and slight reduction in ductility of magnesium alloy (ZK51) by the inclusion of Ti–6Al–4V particles. According to the authors, incorporation of Ti–6Al–4V particles appears to be better choice for the ductility of Mg composites than the incorporation of SiC_p. Xiaoqing Liu et al. [31] examined mechanical characteristics of pure Al strengthened with CNT using PM technique. They have milled the CNT and Al powder mixtures for 2 to 12 h before compaction and sintering and observed highest yield strength (210 MPa) and the UTS (253 MPa) for the powder mixture milled for 12 h. Longer milling time may have helped with the enrichment of mechanical characteristics by facilitating the in-situ production of Al_4C_3, grain refining of the Al matrix, and uniform distribution of CNTs.

Lu et al. [32] examined mechanical characteristics of magnesium composites strengthened with different amount of SiC, TiC, TiB_2 and ZrB_2 particles processed using powder metallurgy route. They observed the strengths of 246, 245 and 235 MPa

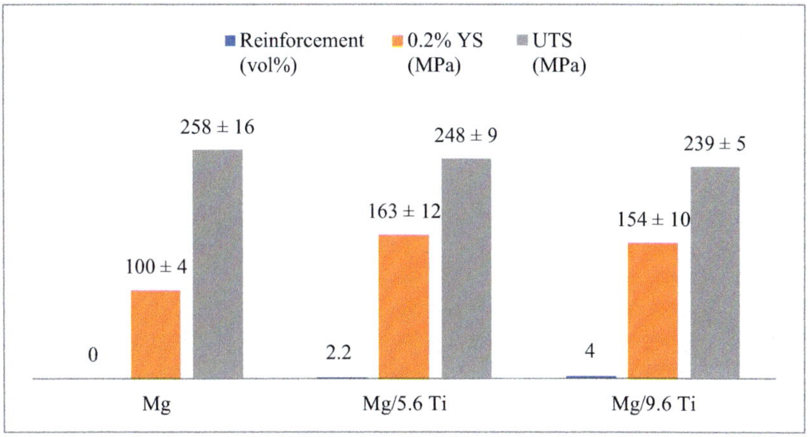

Fig. 2 Tensile characteristics of Mg and Mg/Ti_p MMCs

for the Mg composites strengthened with 5 vol% ZrB_2, TiC and SiC particles respectively. Authors reported maximum yield strength for composite reinforced with ZrB_2 particles and minimum for composite reinforced with TiB_2 particles. Decreased ductility with increased amount of particles has been reported for all the four composites and maximum ductility for SiC particle MMC while second maximum ductility for ZrB_2 particle MMC. Even though TiB_2 is a prominent reinforcement for aluminium MMCs, it is not a suitable reinforcement for magnesium MMCs.

The mechanical characteristics of in-situ particulate composites have been evaluated by several workers. Better mechanical and wear properties can be expected in in-situ composites compared to composites processed by conventional method due to better dispersion of finer in-situ formed strengthening particles. Westwood and Winzer [33] were the earliest researchers to report the superior mechanical characteristics of in-situ composites. TiB_2 particle reinforced Al in-situ composite produced by XD process, exhibit up to 40% more moduli than that of pure aluminium [33]. Kuruvilla and co-workers [34] compared tensile characteristics of in-situ TiB_2/Al composites synthesized by XD process with those of the composites processed by conventional methods containing similar amount of reinforcing phase (20 vol%). Their results have summarized and presented in Fig. 3. Although the conventionally processed composite exhibits significantly better qualities than those of unreinforced pure aluminium, it appears that the hardness, modulus and UTS of the in-situ composite are comparably better compared to the conventionally processed composite. Such better properties in in-situ composites are primarily due to uniform distribution of finer high-modulus TiB_2 particles and cleaner interfaces leading to strong bonding with the matrix.

Ma and co-workers [35, 36] studied the tensile behaviour of in-situ (Al_2O_3 + TiB_2)/Al particulate composites processed by reactive hot-pressing route. They demonstrated that as boron content in the Al–TiO_2–B system is increased, TiAl3

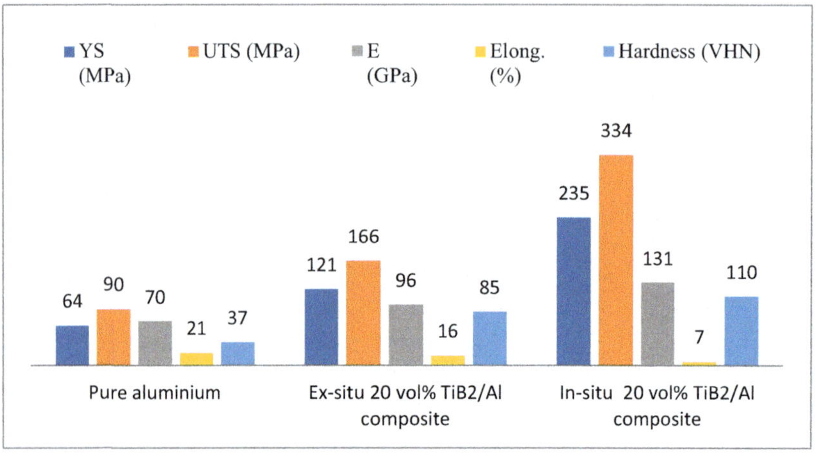

Fig. 3 Mechanical properties of pure aluminium and TiB_2/Al composites

intermetallic phase size and amount decrease, and more TiB_2 particles form in-situ, boosting the strength of the in-situ particulate composites significantly. They reported the YS and UTS values of 109.53 MPa and 144.76 MPa respectively in in-situ (10.5 vol%Al_2O_3)/Al composite, whereas those properties improved to 339.74 MPa and 381.16 MPa respectively in in-situ (10.5 vol%Al_2O_3 + 9.5 vol%TiB_2)/Al composite.

4 Conclusion

Hardness, yield strength, ultimate tensile strength and Young's modulus are the primary characteristics which control the wear behaviour of materials. In the above studies it is observed that the better mechanical properties, both at ambient temperature as well as at higher temperatures can be obtained by reinforcing magnesium and aluminium alloys with suitable ceramic or metallic particles. Al_2O_3, SiC, TiC, TiO_2 and TiB_2 ceramicparticles are compatible and stable particulate reinforcements with magnesium and aluminium matrix composite materials to achieve better mechanical properties. Reinforcement particle size may vary from few nm to 50 μm and quantity of reinforcement in the range as small as less than 1 vol% upto 20 vol%. Even few of the researchers have investigated the reinforcing effects of intermetallics, copper, nickel and titanium particulates, and the results found are not so attractive.

References

1. Hassan SF, Gupta M (2005) Development of high-performance magnesium nanocomposites using nano-Al_2O_3 as reinforcement. Mater Sci Eng, A 392:163–168
2. Hassan SF, Gupta M (2006) Effect of primary processing on the microstructure, CTE and mechanical properties of magnesium/alumina nanocomposites. Compos Struct 72:19–26
3. Srikanth N, Zhong XL, Gupta M (2005) Enhancing damping of pure magnesium using nano-size alumina particulates. Mater Lett 59:3851–3855
4. Wang HY, Jiang QC, Wang Y, Ma BX, Zhao F (2004) Fabrication of TiB_2 particulate reinforced magnesium matrix composites by powder metallurgy. Mater Lett 58:3509–3513
5. Wang Y, Wang H-Y, Xiu K, Wang H-Y, Jiang Q-C (2006) Fabrication of TiB_2 particulate reinforced magnesium matrix composites by two-step processing method. Mater Lett 60:1533–1537
6. Jiang QC, Wang HY, Wang JG, Guan QF, Xu CL (2003) Fabrication of TiC_p/Mg composites by the thermal explosion synthesis reaction in molten magnesium. Mater Lett 57:2580–2583
7. Wang HY, Jiang QC, Li XL, Zhao F (2004) Effect of Al content on the self-propagating high-temperature synthesis reaction of Al-Ti-C system in molten magnesium. J Alloy Compd 366:L9–L12
8. Ma BX, Wang HY, Wang Y, Jiang QC (2005) Fabrication of $(TiB_2\text{-}TiC)_p$/AZ91 magnesium matrix hybrid composite. J Mater Sci 40:4501–4504
9. Madhusudan S, Sarcar MMM, Rao NBRM (2016) Mechanical properties of aluminium-copper(p) composite metallic materials. J Appl Res Technol 14:293–299
10. Susila AB, Sugihartono I, Marpaung MA (2019) Study on mechanical properties of metal matrix composites (MMCs) Al-Cu-Mg/SiC_p with powder metallurgy. J Phys: Conf Ser 1402(4)
11. Shivalingappa D, Sundar Daniel BS, Ray S (2012) Effect of reinforcing phase inherited from another composite on the mechanical properties of cast magnesium base composite, Elsevier, Mater Sci Eng A 541:172–180
12. Saravanan RA, Surappa MK (2000) Fabrication and characterisation of pure magnesium-30 vol.% SiC_P particle composite. Mater Sci Eng, A 276:108–116
13. Abdulhaqq AH, Ghosh PK, Jain SC, Ray S (2005) Processing, microstructure, and mechanical properties of cast in-situ Al(Mg, Mn)-$Al_2O_3(MnO_2)$ composite. Metall Mater Trans 36A:2211
14. Abdulhaqq AH, Ghosh PK, Jain SC, Ray S (2006) Processing, microstructure, and mechanical properties of cast in-situ Al(Mg, Ti)-$Al_2O_3(TiO_2)$ composite. Metall Mater Trans 37A:469
15. Jenix Rino J, Shivalingappa D, Koti H (2013) Properties of Al6063 MMC reinforced with zircon sand and alumina. IOSR J Mech Civ Eng 5(5):72–77
16. Friedrich HE, Mordike BL (2006) Magnesium technology. Springer-Verlag, Berlin Heidelberg, Germany
17. Laurent VJ, Regazzoni G (1992) Processing, microstructure relationships in compocast magnesium/SiC. J Mater Sci 27:4447–4459
18. Lloyd DJ (1994) Particle reinforced aluminium and magnesium matrix composites. Int Mater Rev 39:1–23
19. Lee DM, Suh BK, Kim BG, Lee LS, Lee CH (1997) Fabrication, microstructure, and tensile properties of magnesium alloy AZ91/SiC_p composites produced by powder metallurgy. Mater Sci Technol 13:590
20. Chua BW, Lu L, Lai MO (1999) Influence of SiC particles on mechanical properties of Mg based composite. Compos Struct 47:595–601
21. Luo A (1995) Processing, microstructure and mechanical behaviour of cast magnesium metal matrix composites. Metall Mater Trans 26A:2445–2455
22. Vijayamohan V, Gopalakrishna V (1996) Technology for magnesium metal matrix composites. Indian Foundry J:5–9
23. Gupta M, Lai MO, Saravanaranganathan D (2000) Synthesis, microstructure and properties characterization of disintegrated melt deposited Mg/SiC composites. J Mater Sci 35:2155–2165
24. Handbook ASM (1990) Properties and selection: non-ferrous alloys and special-purpose materials 2:1134

25. Lu L, Thong KK, Gupta M (2003) Mg-based composite reinforced by Mg_2Si. Compos Sci Technol 63:627–632
26. Majunath G, Shivalingappa D, Kishore Kumar R, Umesha DE (2017) Effect of Mg_2Si along with bismuth on the microstructure and mechanical properties of Mg-Al/Mg_2Si composite. Int Journal of ChemTech Res 10(14):88–92
27. Hassan SF, Gupta M (2002) Development of ductile magnesium composite materials using titanium as reinforcement. J Alloy Compd 345:246–251
28. Perez P, Garces G, Adeva P (2004) Mechanical properties of a Mg 10 (vol.%)Ti composite. Compos Sci Technol 64:145–151
29. Ho KF, Gupta M, Srinivasan TS (2004) The mechanical behaviour of magnesium alloy AZ91 reinforced with fine copper particulates. Mater Sci Eng, A 369:302–308
30. Xi YL, Chai DL, Zhang WX, Zhou JE (2006) Titanium alloy reinforced magnesium matrix composite with improved mechanical properties. Scripta Mater 54:19–23
31. Liu X, Li C, Eckert J, Prashanth KG, Renk O, Teng L, Liu Y, Bao R, Tao J, Shen T, Yi J (2017) Microstructure evolution and mechanical properties of carbon nanotubes reinforced Al matrix composites. 133:122–132
32. Lu L, Lim CYH, Yeong WM (2004) Effect of reinforcements on strength of Mg-9%Al composites. Compos Struct 66:41–45
33. Westwood ARC, Winzer SR (1987) Advanced materials research. National Academy Press, Washington, DC, pp 225–226
34. Kuruvilla AK, Prasad KS, Bhanuprasad VV, Mahajan YR (1990) Microstructure-property correlation in Al/TiB_2 (XD) composites. Scr Metall Mater 24:879–883
35. Ma ZY, Li JH, Li SX, Ning XG, Lu YX, Bi J (1996) Property-microstructure correlation in situ formed Al_2O_3, TiB_2 and Al_3Ti mixture-reinforced aluminium composites. J Mater Sci 31:741–747
36. Ma ZY, Tjong SC (1997) In-situ ceramic particle-reinforced aluminum matrix composites fabricated by reaction pressing in the TiO_2 (Ti)-Al-B (B_2O_3) systems. Metall Mater Trans A 28A:1931–1942

Open Hole Tensile Test for Measuring Residual Tensile Strength and Delamination of Glass Fibre Metal Mesh Polymer Composites

M. Sakthivel, P. Raja, V. Parthiban, and A. Nagaraj

Abstract Glass Fibre Metal Mesh Polymer Composites (GFMMPC) are one of the alternative for engineering materials because their elevated mechanical properties. This work focuses on tensile test with open hole which is based on optimization of parameters. For the drilling experiment, Taguchi L8 two level orthogonal array a strategy was used to examine each parameter's influence and systematically evaluate the experimental parameters. Spindle speed, feed rate, and condition (dry and wet) with two levels are some of the drilling parameters that were chosen. Eight drilling tests were carried out, each in a different order. After the drilling trials are finished, hole properties are observed using Scanning Electron Microscope (SEM) this was done to assess the degree to which drilled holes are delaminated. The Corel draw was used for measuring the delamination factor. Also residual tensile properties measured using tensile test. Based on responses the optimized parameters are higher for high spindle speed (4000 rpm) and low feed rate (50 mm/min). Morphological study of the fractured samples are analysed using SEM.

Keywords Open hole · Residual tensile strength · Delamination factor · Drilling · Glass fibre

M. Sakthivel (✉)
Department of Mechanical Engineering, Adhiyamaan College of Engineering, Hosur 635130, India
e-mail: metalsakthi@gmail.com

P. Raja
Department of Mechanical Engineering, Prathyusha Engineering College, Chennai, India

V. Parthiban
Department of Mechanical Engineering, IIT-Madras, Chennai 600001, India

A. Nagaraj
Associate Professor, Department of Mechanical Engineering, Star Lion College of Engineering and Technology, Thanjavur 614206, India
e-mail: naga.nagaraj.raja@gmail.com

S. B. Boppana et al. (eds.), *Structural Composite Materials*, Composites Science and Technology, https://doi.org/10.1007/978-981-99-5982-2_11

1 Introduction

Most of the engineering disciplines, including the transportation, commercial aviation, shipbuilding, sporting events, and civil infrastructure industries the composite materials widely used. As a result, only the primary parts—not the secondary ones—are produced using composite materials. As is common knowledge, both automobiles and airplanes have a complex structural design. Numerous sorts of failure or damage result from the load being applied to composite materials because the joints between the components with holes used for joining the elements. In order to prevent failure, the industrial field relies heavily on the understanding of forces, stress, and strain [1–5].

Metal fibre laminate is a combination of metal and Fibre Reinforced Polymer (FRP) points of interest to produce greater mechanical qualities than traditional laminates. They have a few key points, like improved impact and fatigue fracture growth damage tolerance. They are typically employed in a variety of practical and crucial applications, including transportation, aviation, and other barrier applications [6–8].

These composite structures must be drilled in order to add fasteners for assembly. Numerous defects, including fibre pull-out, fuzzing and flaws akin to those seen in metal drilling, are created during the drilling of composite. The faults in the holes account for about 60% of the rejections of aeronautical parts. These flaws would result in a loss of fundamental stability, causing distinction in the dynamic execution of the complete structure. The use of suboptimal machining conditions and cutting tool outlines is the cause of the aforementioned problems [9–13].

The Taguchi design of the experiment is used to guide the execution of the experiments. The findings imply that spindle speed and tool shape were most susceptible to changes in feed rate, followed by residual strength and delamination factor [14].

Open-hole composites with three different layers were put through tensile strength tests. A probabilistic neural network method is projected to predict the tensile strength of composite panels based on sparse experimental data. The predictor uses the statistical parameters, layup characteristics and tensile stress of open-hole composites as inputs and uses the classification function of PNN to provide the protection status as an transitional result. The results are satisfactory, and the errors of the forecasts are analogous to the coefficient of variation of the investigational data. To show the model's effectiveness in predicting the experimentally recorded open-hole tensile strength of composite plates data from earlier investigations, additional assessments are conducted [15].

Long kenaf composites with drilled holes and long kenaf/woven glass reinforced polyester composites' tensile properties. The measurements were made using the residual tensile strength of open-hole and impact-damaged specimens. The damaged area of the composites might then be projected from there. It was found that the hybrid kenaf/glass composite was fewer notch sensitive than the kenaf composite. The hybrid composite outperformed kenaf composite in terms of strength [16].

Due to their increased strength, glass fibre with metal wire mesh reinforced in polymer which effectively to form a barrier [17]. According to the authors, the created replica was capable to precisely forecast the investigational values of tensile strength [18, 19].

There have only been a few studies on GFMMPC open hole tensile strength. The study emphasises OHT strength. The hole on the specimen is drilled using the Taguchi method, and the GFMML composites are constructed. The responses, such as the delamination factor and tensile strength, are then optimised. Using a scanning electron microscope, the fragmented surface is examined (SEM).

2 Experimental Methods

A Glass fibre woven material is used as reinforcement. The use of the epoxy resin and hardener improves the composite's interfacial bonding. Based on their literature research, a 10:1 resin to hardener ratio was chosen. Stainless Steel Wire Mesh (SSWM) is another reinforcing material used in composites. The mesh was washed with alkaline soap for 30 min prior to usage, and then rinsed with deionized water. The layer of glass fibre was detached before metal mesh was applied to pile in order to retain the same laminate thickness.

Vacuum bag moulding is used to create the GFMMP composites. The centre layer of the composite has a 10% weight fraction of SSWM remaining, with the top and bottom layers of the composite containing 50% weight fractions of woven glass fibre [17]. A releasing agent is used after cleaning the mould surface. The composite is then employed under pressure for two hours, followed by a 24-h pre-curing period in the mould. The finished laminate is then post-cured for up to three hours at 100 °C in the hot air oven. The layers are organised in this fashion to create the six possible combination composites, each with a 2 mm thickness. The fabrication process is shown in Fig. 1.

The drilling experiment used the Taguchi method to methodically examine the experimental parameters and determine the relative contributions of each component. One of the crucial elements in the Taguchi method's parametric analysis is the choice of appropriately controlled parameters and their levels are shown in Table 1 with two levels. For experimentation, a carbide tool with a 6 mm diameter [9] is employed. The drilling experiments are conducted using machining centre which is shown in Fig. 2.

Figure 3 standard tensile specimen in accordance with ASTM 3039. This was accomplished using UTM, as shown in Fig. 4, which included a 10 kN load cell. For static tests in stress, the testing apparatus' crosshead speed was 2 mm/min. The specimens with the hole in the middle had nominal length, breadth, and thickness measurements of 250 mm, 25 mm, and 2 mm, respectively.

On a numerically controlled machine, holes were machined out of GFMMP composite. The centres of the test specimens the drill holes were placed. The selected parameter levels the investigational configuration is created by means of the Taguchi

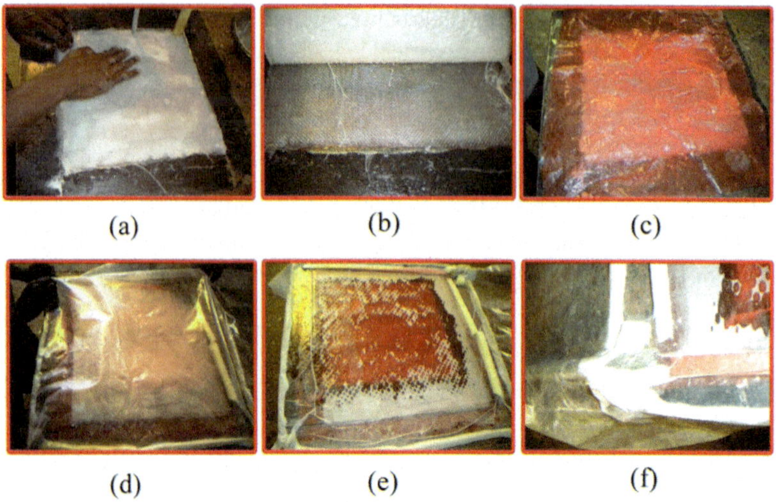

Fig. 1 GFMMPC fabrication process

Table 1 Selected process parameters and levels

Level	Spindle speed (rpm)	Feed rate (mm/min)	Condition
1	2000	50	Dry
2	4000	100	Wet

Fig. 2 Drilling experimental setup

method, as exposed in Table 2. The process are carried out, each in a different order. After the drilling experiments are finished, SEM is used to observe the characteristics of the holes. This was done to assess the degree to which drilled specimens delaminated.

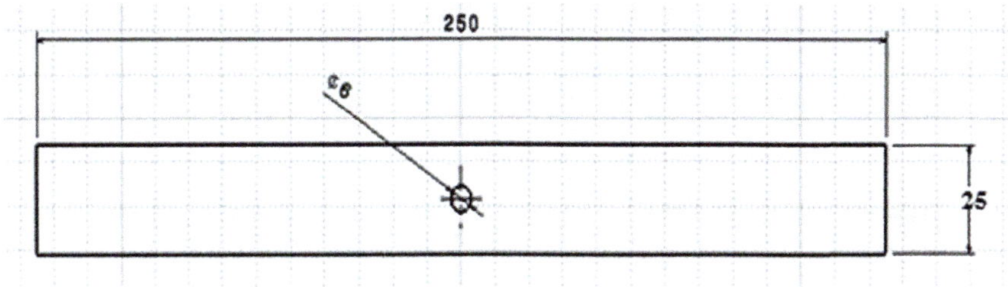

Fig. 3 Standard tensile specimen as per the ASTM 3039

Fig. 4 UTM for open hole tensile test

Table 2 Taguchi L8 orthogonal array with parameters and results

Sl. No.	Spindle speed v (rpm)	Feed rate f (mm/min)	Condition	Residual tensile strength (MPa)	Delamination factor
1	2000	50	Dry	132.25	1.115
2	2000	100	Dry	120.25	1.125
3	4000	50	Dry	135.11	1.105
4	4000	100	Dry	130.56	1.121
5	2000	50	Wet	133.93	1.109
6	2000	100	Wet	125.25	1.118
7	4000	50	Wet	135.68	1.095
8	4000	100	Wet	129.25	1.115

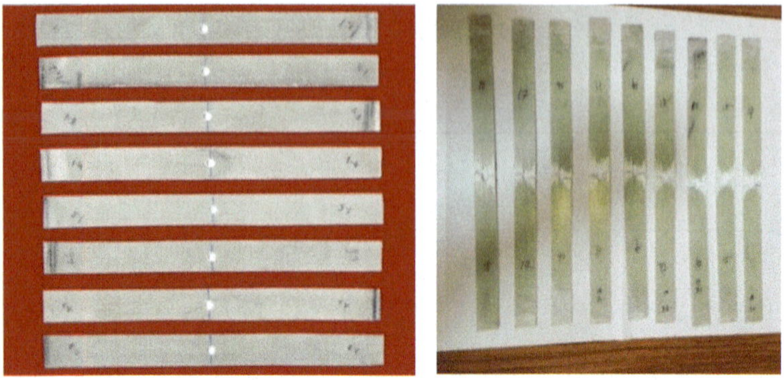

Fig. 5 Open hole tensile and fractured tensile specimens

The delamination factor is given by

$$F_d = \frac{D_{\max}}{D} \tag{1}$$

where $D_{\max}$ = Maximum diameter of delaminated area, D = Actual diameter.

GFMMP composites residual tensile characteristics were assessed in accordance with ASTM 3039. The open hole tensile specimen, UTM, and fractured tensile specimens are displayed in Fig. 5. In every experiment, at least three samples are examined. The residual tensile strength is calculated using Eq. 2 [20].

$$\sigma = \frac{F_u}{(W - D)t} \tag{2}$$

where σ = Residual Tensile strength (MPa) F_u = Ultimate load (N) t and W are Thickness and Width of the specimen (mm), D = Diameter of the drilled hole (mm).

3 Results and Discussion

Due to hole is in the specimen's centre, where stress concentration is more likely to occur, the open hole strength of composites is decreased. Based on these findings, it can be said that the GFMMPC composites created in this work have acceptable mechanical properties for structural applications.

On the GFMMPC specimen with a hole in the centre, the tensile stresses as a function of strain curves for eight trials are shown in Fig. 6. The experimental tensile strengths are, as is clear, between 120 and 140 MPa. These consequences of the GFMMPC composites strength was considerably reduced to just under half related to the non-drilled samples. Though, it's possible that the residual tensile strength is not accurately represented by direct measurement [14]. Therefore, it is recommended that the apt residual tensile strength of samples are calculated using Eq. 2 based on the narrated tensile force [20]. Figure 7 shows the GFMMPC composite specimens tensile strength under various experimental circumtances.

When it comes to delamination damage, it is frequently linked to the inter-ply failure phenomenon brought on by drilling. Equation 1 is used to compute the delamination factor. In summary, these data clearly show delamination factor with a highest value of 1.125 and a minimum of 1.095, which are evident that delamination damage has been minimised compared to non-drill composite specimen. The comparison between delamination factor with experiments is exposed in Fig. 8.

The response graphs are shown in Figs. 9 and 10, indicate the impacts of varying parameters specifically spindle speed, feed rate, and condition (dry or wet), on the responses. The linear trend of the slope on the graph illustrates how that factor influenced the experiment's outcomes. These numbers demonstrate that increasing residual tensile strength and decreasing delamination factor is most significantly affected by adjusting feed rate. The outputs are clearly adversely affected by adjusting

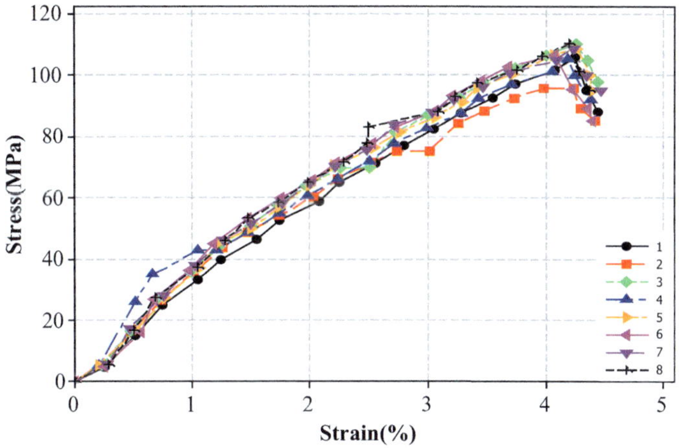

Fig. 6 Tensile stress verses tensile strain curves

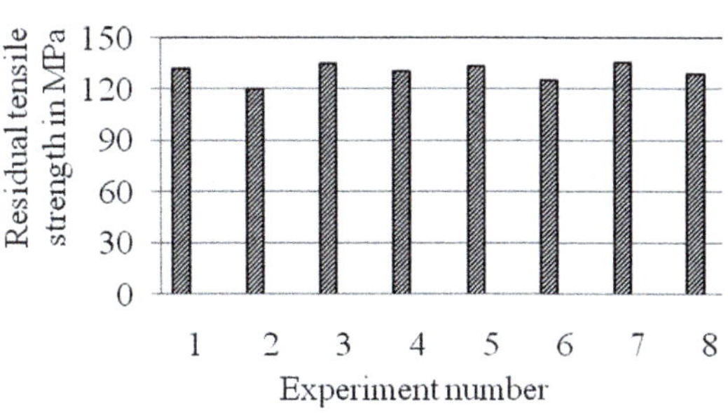

Fig. 7 Comparison of residual strength with experiments

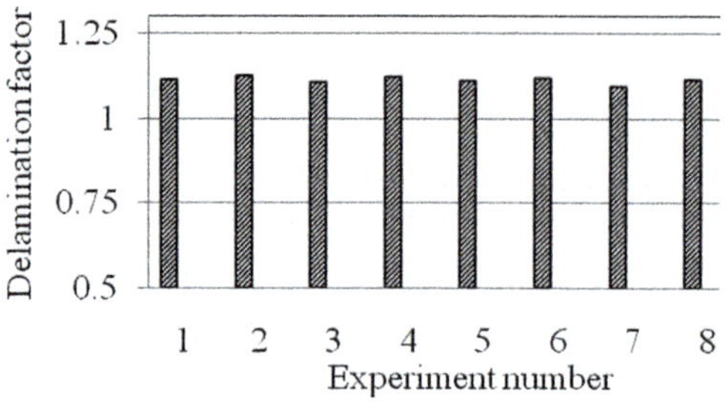

Fig. 8 Comparison of delamination factor with experiments

the feed rate, and a lesser feed rate is preferred to increase residual strength and reduce delamination. Furthermore, it is evident that tool geometry has little impact on these two outputs. To obtain elevated residual tensile strength and minimal delamination factor, v2f1c2 is the best drilling parameter combination.

The ANOVA Tables 3 and 4 shows how the total variations of the mean are used to calculate the specified factor and the errors. These findings are frequently used to assess the relative contributions of each element. In addition, the estimated F-values for each component are compared to the crucial F-value at the 95% confidence level to establish their relative significance. It is significant to keep in mind that a factor's impact on performance characteristics depends on the size of its F-value. As is clear,

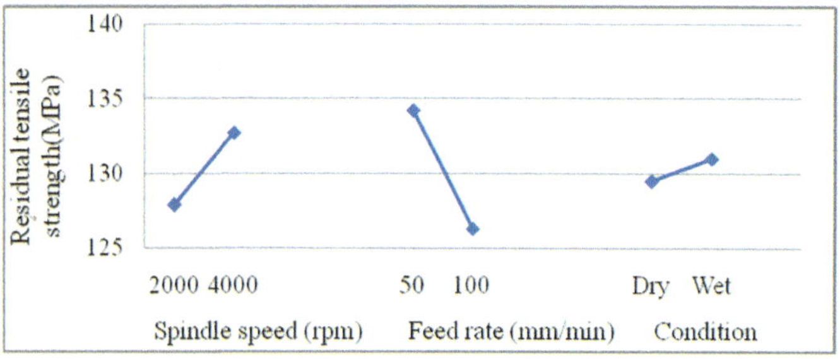

Fig. 9 Response graph for residual tensile strength

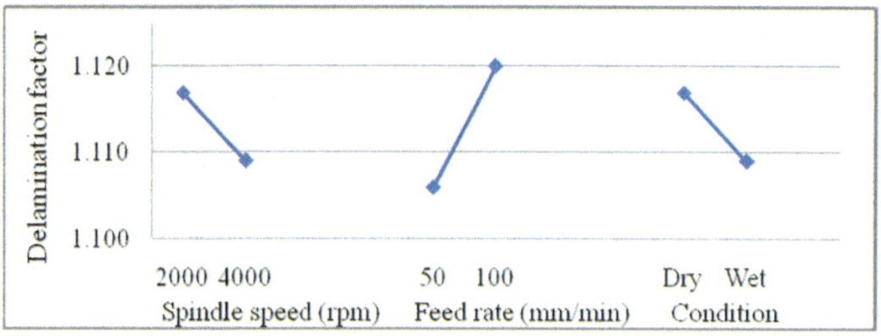

Fig. 10 Response graph for delamination factor

Table 3 ANOVA table for residual strength

Source	Sum of squares	DoF	Mean square	F value	% Contribution
v	44.78	1	44.78	13.24	22.76
f	**125.28**	**1**	**125.28**	**37.04**	**63.68**
c	4.40	1	4.40	1.30	2.24
vf	11.74	1	11.74	3.47	5.97
vc	6.88	1	6.88	2.03	3.5
fc	0.26	1	0.26	0.08	0.13
Residual	3.38	1	3.38		1.72
Cor total	196.73	7			100

Bold indicates feedrate is the most significance parameter compared with other parameters

Table 4 ANOVA table for delamination factor

Source	Sum of squares	DoF	Mean square	F value	% Contribution
V	1.20E−04	1	1.20E−04	38.44	18.61
F	**3.78E−04**	**1**	**3.78E−04**	**121**	**58.61**
C	1.05E−04	1	1.05E−04	33.64	16.28
Vf	3.60E−05	1	3.6E−05	11.56	5.68
Vc	1.00E−06	1	1.00E−06	0.36	0.155
Fc	1.00E−06	1	1.00E−06	0.36	0.155
Residual	3.00E−06	1	3.00E−06		0.465
Cor total	6.45E−04	7			100

Bold indicates feedrate is the most significance parameter compared with other parameters

the feed rate has a 63.68% impact on residual tensile strength, whereas the spindle speed has 22.74% impact. The impact of altering feed rate also makes up 58.61% of the equation. The impact of varying feed rate has a similar effect, accounting for 58.61% of the delamination damage. According to the ANOVA analysis, the condition had a negligible impact on the both responses, with percentage contributions 2.24% and 16.28%, respectively.

Figure 11 depicts the surface morphology of the GFMPC composites' cracked surface after a tensile test. In this instance, fibre ripping is seen along with matrix cracking and SSWM necking. The surface of the fibres that were pulled out reveals the tenuous interfacial bonds between the fibre and matrix. The formation and spread of a microcrack in the matrix, which is prejudiced by the shape, size and orientation of the reinforcement, determines the composite's strength. Figure 11 clearly shows the crack propagation through the matrix and glass layers debonding at the contact.

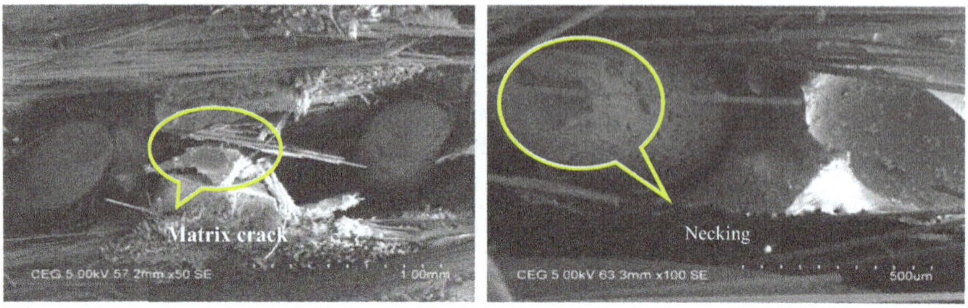

Fig. 11 Tensile fractured surface morphology

4 Conclusion

The following findings are drawn from the experimental work:

- Open hole strength analysis it was discovered that for high spindle speed and low feed rate, residual tensile strength is higher.
- It was noted that the samples damage behaviour happened in the matrix near the hole, which caused cracks to spread between the layers. The laminate cracked as a result of the stiffness reduction caused by propagation between layers.
- The outcome shows that the feed rate has a noteworthy impact on both responses.
- The spindle speed at 4000 rpm, feed rate at 50 mm/min, and wet condition are the best settings for both responses.
- The matrix crack, fibre failure as well as pull-out are the failure mechanisms identified by the SEM micrograph of the fractured specimen. In composites with high strengths, there was very little fibre pull-out due to the matrix and fibre wettability.

References

1. Wei L, Zhu W, Yu Z, Liu J, Wei X (2019) A new three-dimensional progressive damage model for fiber-reinforced polymer laminates and its applications to large open-hole panels. Compos Sci Technol 182:107757
2. Zhang D, Zheng X, Wu T (2019) Damage characteristics of open-hole laminated composites subjected to longitudinal loads. Compos Struct 230:111474
3. Rajkumar K, Ramraji K, Manojkumar M, Gnanavelbabu A (2020) Validation of tensile strength of notched hybrid polymer composite using Whitney-Nusimer point stress model. Mater Today: Proc 1(27):940–945
4. Azadi M, Sayar H, Ghasemi-Ghalebahman A, Jafari SM (2019) Tensile loading rate effect on mechanical properties and failure mechanisms in open-hole carbon fiber reinforced polymer composites by acoustic emission approach. Compos Part B: Eng 1(158):448–458
5. Jalmeida JH Jr, Bittrich L, Spickenheuer A (2020) Improving the open-hole tension characteristics with variable-axial composite laminates: optimization, progressive damage modeling and experimental observations. Compos Sci Technol 5(185):107889

6. Hasselbruch H, Von Hehl A, Zoch HW (2015) Properties and failure behaviour of hybrid wire mesh/carbon fibre reinforced thermoplastic composites under quasi-static tensile load. Mater Des 66:429–436
7. Sinmazcelik T, Avcu E, Bora MO, Coban O (2011) A review: fibre metal laminates, background, bonding types and applied test method. Mater Des 32(7):3671–3685
8. Zitoune R, Krishnaraj V, Collombet F (2010) Study of drilling of the composite material and aluminium stack. Compos Struct 92(5):1246–1255
9. Sakthivel M, Vijayakumar S (2016) Optimization of drilling parameters for fiber metal mesh laminate composites using Grey relational analysis. Material Testing 58:7–8
10. Stone R, Krishnamurthy K (1996) A neural network thrust force controller to minimise delamination during drilling of graphite-epoxy laminate. Int J Mach Tools Manuf 36(9):985–1003
11. Sakthivel M, Vijayakumar S, Jenarthanan MP (2017) Grey-fuzzy logic to optimise process parameters in drilling of glass fibre-reinforced stainless steel mesh polymer composite. Pigment Resin Technol 46(4)
12. Budhe S, Banea MD, De Barros S (2018) Bonded repair of composite structures in aerospace application: a review on environmental issues. Appl Adhes Sci 6(1):1–27
13. Tong L, Soutis C (2003) Recent advances in structural joints and repairs for composite materials
14. Abdul Nasir AA, Azmi AI, Khalil ANM (2015) Measurement and optimisation of residual tensile strength and delamination damage of drilled flax fibre reinforced composites. Measurement 75:298–307
15. Fan H-T, Wang H (2014) Predicting the open-hole tensile strength of composite plates based on probabilistic neural network. Appl Compos Mater 21(6):827–840
16. Arifin AMT, Abdullah S, Rafiquzzaman M, Zulkifli R, Wahab DA (2014) Failure characterisation in polymer matrix composite for un-notched and notched (open-hole) specimens under tension condition. Fibers Polymers 15(8):1729–1738
17. Sakthivel M, Vijayakumar S, Vijaya Ramnath B (2018) Investigation on mechanical and thermal properties of stainless steel wire mesh-glass fibre reinforced polymer composite. Silicon 10:2643–2651
18. Dave HK, Prajapati AR, Rajpurohit SR, Patadiya NH, Raval HK (2020) Open hole tensile testing of 3D printed parts using in-house fabricated PLA filament. Rapid prototyping 26(1):21–31
19. Hassoon OH, Enzi A, Kashkool LH (2022) Experimental study of failure analysis in notched and repaired fiberglass epoxy composite. Adv Sci Technol Res J 16(2):99–108
20. Zarif Karimi N, Heidary H, Ahmadi M (2012) Residual tensile strength monitoring of drilled composite materials by acoustic emission. Mater Des 40:229–236

Prediction of Tribological Behaviour of AA5083/CSA-ZnO Hybrid Composites Using Machine Learning and Artificial Intelligence Techniques

A. Nagaraj, S. Gopalakrishnan, M. Sakthivel, and D. Shivalingappa

Abstract Aluminium Alloys AA5083 dispersed with varying fractions of reinforcement was fabricated through the stir casting method. In varying weight percentage combinations, zinc oxide (ZnO) and coconut shell ash (CSA) particles were combined to create hybrid reinforcement particles. Using a pin-on-disc tribometer, the wear characteristics of the developed AA5083 hybrid composites were estimated. The volumetric proportion of hybrid reinforcement particles CSA (3, 6, 9 and 3 ZnO wt%), load (20, 30, 40 N), sliding velocity (2, 3, and 4 m/s), Cumulative Time (4.16, 5.55, and 8.33 min), and sliding distance are some of the experimental parameters (1000 m). Wear analysis revealed effective bonding and homogeneous dispersion of hybrid reinforcement particles onto the AA5083. Analysis of Specific Wear Rate (SWR) results showed that Specific Wear Rate rose with load, sliding velocity, and sliding duration while decreasing with hybrid particle dispersion. This research proposes the use of several intelligent classification techniques using Machine Learning (ML) and Artificial Neural Network (ANN) to predict the wear rate of an AA 5083 hybrid composite. For estimating wear quantities, the algorithms Random Forest (RF), Neural Network (NN), and k-nearest neighbours (kNN) are utilized. Six inputs are utilized to train and evaluate the Machine Learning (ML) algorithms: the Applied Load (N), Sliding Velocity, Sliding Speed, Cumulative Time, Percentage of Reinforcements, and Sliding Distance. The output is the Specific Wear Rate (SWR). The

A. Nagaraj (✉)
Associate Professor, Department of Mechanical Engineering, Star Lion College of Engineering and Technology, Thanjavur 614206, India
e-mail: naga.nagaraj.raja@gmail.com

S. Gopalakrishnan
Department of Mechanical Engineering, K. S. Rangasamy College of Technology, Thiruchengode 637215, India

M. Sakthivel
Department of Mechanical Engineering, Adhiyamaan College of Engineering, Hosur 635109, India

D. Shivalingappa
Department of Mechanical Engineering, B. N. M. Institute of Technology, Bengaluru 560070, India

S. B. Boppana et al. (eds.), *Structural Composite Materials*, Composites Science and Technology, https://doi.org/10.1007/978-981-99-5982-2_12

RF, NN, and KNN algorithms all produced success rates of correlation between experimental to anticipated of 0.90, 0.84, and 0.90, respectively. The same model data was utilised to train and evaluate Artificial Neural Networks (ANN), with the Multilayer Perceptron (MLP) network having the lowest Mean Square Error (MSE) to improve machine learning prediction accuracy. Maximum estimate error range of 0.1%, training and cross-validation of 0.00000496 and 0.0261, respectively, with linear correlation coefficient in testing of 0.9999 or 99.9% better prediction accuracy rate. The AA 5083 composites were designed and implemented using this machine learning and artificial neural network model for forecasting specific wear rate.

Keywords AA 5083 · Stir casting · Hybrid composites · Wear · Machine learning (ML) · Artificial Neural Network (ANN)

1 Introduction

Aluminium Metal Matrix and Hybrid Composites (AMMHCs) are used in aerospace, automotive, ballistic, electrical, aviation, tribological, space and air vehicle, thermal, structure, defence industries, military, transportation, engineering, and mineral processing applications due to their high strength-to-weight ratio, good corrosion, oxidation, and wear resistance, and high thermal conductivity [1].

Due to extensive passivation, aluminium (Al) is one of the most frequently utilised metals in industries. It is a very strong, wear- and corrosion-resistant alloy that is lightweight. By changing their elemental compositions, aluminium alloys may have improved chemico-physical properties. To create varied concentrations of defect-free and evenly dispersed aluminium composites, stir casting technology is often used. AA5083 Due to its light weight, fabricability, physical characteristics, corrosion resistance, and affordability, aluminium–magnesium alloys are often utilised in the aerospace, automotive, shipbuilding, and construction sectors. The majority of aluminium–magnesium alloys, nevertheless, seem to have minimal wear resistance. There have been some recent findings on the tribological behaviour of different aluminium alloys [2].

Zhang and Li investigated the impact of yttria addition on aluminum's wear resistance in dry and corrosive environments. They discovered that the scattered yttria particles significantly improved the aluminium matrix composites resistance of aluminum to corrosion, corrosive and dry wear [3].

The use of agricultural waste (rice husk ash, coconut shell ash, bagasse ash, and corn cob ash), industrial waste (fly ash), or recycled materials has been a new breakthrough in composite materials. Agro-waste products that have been reinforced have appealing qualities including cheap cost, low density, and less environmental contamination. Coconut shell ash (CSA) particles are used extremely seldom, despite the fact that many studies have researched the improvement of mechanical characteristics of AMCs by the inclusion of agro-wastes. The authors' decision to concentrate on creating high-performance aluminium matrix composites with reinforcements made from coconut shell ash was driven by the paucity of existing research [4].

Zinc oxide (ZnO), an n-type semiconductor, is a very interesting material because it can be used to make solar cells, sensors, displays, gas sensors, varistors, piezoelectric devices, electro-acoustic transducers, photodiodes, UV light emitting devices, and antibacterial materials [5]. Due to its distinctive mechanical, electrical, and optical characteristics as well as its many uses, zinc oxide (ZnO) is a significant substance in the metal oxide family [6]. The mechanical characteristics of the extruded Mg materials are improved by the addition of modest volume fractions of nano-particulate reinforcements such as nano-Al_2O_3, nano-ZnO, to pure Mg/alloys [7].

This study looked at how ZnO particles affected the stir-cast aluminium metallic matrix composite's compressive strength, hardness, and wear properties. The reinforcement included various weight percentages of ZnO (0, 2, 4, 6, 8 and 10) with wear characteristics that improve as the weight percentage of ZnO rises. For specimens of aluminium reinforced with 2, 4, 6, 8, or 10 weight percent ZnO particles, the increases in Brinell hardness are (15%), (25%), (35%), (40%), and (50%) accordingly. It was discovered that by increasing the zinc oxide content and going above the composite minimum quantity, which reflects the overall percentage of zinc oxide, the volume loss was significantly decreased (10 percent) [8].

Aluminium matrix composite with graphite and Coconut Shell Ash (CSA). Modified stir-casting creates Al-1100 composites. Three more Aluminium with Aluminium oxide, Aluminium–Aluminium oxide Graphite, and Aluminium-Coconut Shell Ash composites were developed. The Al-CSA composite outperforms the other three composites in terms of mechanical and tribological characteristics including tensile strength and hardness. Gr incorporation aids in the hybrid Al-CSA-Gr composite's improved tribological characteristics while allowing for a somewhat lower specific strength [9].

Composites A and B with Al-5083 matrices reinforced with 5 and 10% wt% B4C particles were manufactured by cryomilling and consolidating. Composite pins were tested for dry sliding wear using a pinon-disc tribometer. Composite B (10 wt% B4C) wore 40% less than composite A (5 wt% B4C) under the identical circumstances. This experiment suggests that B4C particles improve composites wear resistance [10].

After ECAE, the alloy's wear resistance rose with the least amount of mass loss and friction coefficient. As a result of ECAE processing, the alloy is now resistant to scratch deformation. The mechanical and wear characteristics of aluminium 5083 alloy were improved via ECAE processing. The alloy may now be used in a variety of technical applications that call for high strength because to improvements in its mechanical and wear qualities [11].

According to wear experiments, the Al 5083/SiCp nanocomposite has a much lower specific wear rate than the nanostructured Al 5083 alloy. In contrast to Al5083/SiCp, which showed a wear mechanism of adhesive wear to abrasive wear, nanostructured Al5083 alloys had a mix of abrasive and delamination wear processes [12].

There have not been many research specifically looking at coconut shells in metal matrix composite. A low-weight metal matrix composite with high thermal and wear resistance has been produced using CSAp in certain studies [13].

The Al 6063 alloy/coconut shell composites mechanical and corrosion characteristics. They observed an improvement in hardness and tensile strength as coconut shell % weight increased and corrosion resistance decreased [14]. Similar results were obtained when coconut shell was used to enhance recycled scrap aluminium's mechanical and wear qualities [15].

Machine-Learning (ML) techniques have recently been used to anticipate the characteristics of metallic materials. ML models, for instance, may be used to look for shape memory alloys with certain transition temperatures [16]. A method for designing materials that combine tests with ML models to create high-entropy alloys with high toughness. A deep neural network was used by Feng et al. to forecast the flaws in stainless steel. ML models to forecast the capacity of binary metallic alloys to make glass. The design of tailored metallic glasses was confirmed using commercially feasible manufacturing techniques using a machine learning framework for speeding design. High-throughput experimentation and machine learning-based iteration to quickly find novel glass-forming systems ML techniques have been used in many research to determine the relationships between an alloy's capacity to produce glass and its empirically observed characteristics [17]. Thus, these experiments provided conclusive evidence that ML techniques were effective and dependable for identifying novel metal matrix composites and predicting their properties.

Random Forests are trained using bootstrapped datasets of the same size as the training set (RF). Randomly resampling the training set produces these datasets. Once a tree is generated, out-of-bag (OOB) samples are used as the test set. OOB generalization error estimate is all test sets' categorization error rate. Bagged classifiers have the same OOB error as using a training-set-sized test set. OOB estimation no longer requires a separate test set. Each CART tree votes for one class, and the forest predicts the class with the most votes to categorize input data [18].

A mathematical or computer model called a neural network replicates the structure and functionality of a biological neural network. It uses artificial neurons to process information in a connectionist manner. Modern neural networks are non-linear statistical modelling tools that are often used for simulating intricate relationships between input and output while looking for patterns in data. The neural network approach is based on the same assumptions as how the human brain functions. The human brain has a vast network of neurons that link sensory and motor nerves. The majority of scientists thought that neurons in the brain communicate with one another by firing electrical impulses across synapses [19].

A broad family of algorithms used in classification, regression, and density estimation is called artificial neural networks (ANN-s). A function known as a Multilayer Perceptron (MLP) may be imagined as a network made up of multiple layers of neurons coupled in a feedforward fashion. Input neurons are the neurons that represent input variables in the first layer. The output neurons in the top layer are those that deliver the function result value. Hidden layers are those layers that exist between the first and final levels. Every neuron in the network acts as a perceptron, accepting input values x1, x2, … xk and using the formula to calculate output value o;

$$O = \emptyset \sum_{i=1}^{k} (\mathrm{wixi} + \mathrm{b}) \tag{1}$$

where wi, b are the neuron's weights and bias, and is a nonlinear function. (x) is 1 / 1 + e ax or tanh (x). The multilayer perceptron is trained by finding the weights and biases of all the neurons that will result in the network having the least amount of error on the training set. A nonlinear decision border between classes is modelled by the multilayer perceptron, a nonlinear classifier. Since the training data we employed was linearly separable, as was discussed in the preceding section, using a nonlinear decision boundary had little chance of enhancing generalization performance. As a consequence, the basic perceptron's output is the best one we could hope for. Another issue in our situation is that a network with 20,000 input neurons makes it difficult to perform effective backpropagation learning. Therefore, reducing the number of features to a manageable number would be the only practical approach to deploying multilayer perceptron [20].

The k-NN method is the most fundamental ML technique, and it may be used for both classification and regression. To put it simply, k-NN uses the average of the object's k-NNs as the property value for regression purposes. It is common practise to utilise K-NN to predict forest attributes from data collected across disciplines. The distance measure and closest neighbour weighting used in k-NN implementations are also factors. The accuracy and efficiency of computations depend on accurately determining k. To determine the best value for k, we employed a combination of the leave-one-out (LOO) cross-validation technique and the v-fold (dividing training samples into two v-fold halves, v − 1 for prototype and one fold for validation) (1–20). For the k-NN estimator, the LOO is a recognized technique for producing unbiased estimates of predicted classification or estimation error. In this method, the algorithm selects k with the lowest RMSE after computing RMSE in validation sample sets for each value of k [21].

To assess the wear resistance of abrasion-resistant tribological materials to be employed under various operating circumstances, several research are needed. These tests may take a long period. Thus, there has been a growing need to create machine learning algorithms, such as the ANN-MLP algorithm, which have recently been able to utilize experimental data to anticipate wear behaviours of materials in order to decrease the number of tests and lower the cost of experimental investigations.

2 Materials and Methods

2.1 Preparation of Coconut Shell Ash

The surface skins of the coconut shells were cleaned and smoothed after they were purchased from a nearby market. The material was then crushed and ground in a ball mill and a jaw crusher, respectively. The ground material was separated using 100 mesh-size screens. The powder was burned for three hours at 1200 °C (1473 K). Once again, a ball mill was used to grind the resulting ash. The material was screened to a 240 mesh size (50 μm).

2.2 Preparation of AA5083-Based Hybrid Composites

The matrix alloy chosen for the development of composite material is Al–Mg–Mn alloy and designated by the aluminium association as AA 5083. The chemical compositions of the matrix material are given in Table 1.

Percentage of Zinc oxide and the chemical composition of coconut shell ash burned at 700 degrees Celsius were also identified by XRF, and the results are shown in Table 2.

Aluminium Alloy (grade AA5083) was cut from an ingot and preheated at 300 °C for 1 h. The material is in a 1073 K bottom-pouring furnace (800 °C). Inert Ar prevented oxidation during melting. The furnace's stirrer was controlled by a PID rheostat. Slowly lowering the rotating spindle formed a vortex in the pool. Reinforcement particles were put into the liquid melt vortex at 1223 K (950 °C). The spindle was moved from top to bottom with such a 2-mm clearance. This ensured a smooth melt. Slowly decreasing the melt's temperature between 1123 K (850 °C) and 1023 K (750 °C) increases its viscosity. This retained fine particles in the liquid melt. The liquid that contained the particles was put into a steel mould that had been warmed to 673 K (400 °C). The melting furnace is shown in Fig. 1. Four Aluminium alloy composites, namely AA5083, AA5083-3%CSA-3%ZnO, AA5083-6%CSA-3%ZnO and AA5083-9%CSA-3%ZnO reinforcement were cast under different conditions were added separately with different casting in the form of Table 3.

Table 1 Chemical composition of aluminium (5083) alloy in weight percentage

Constituent	Al	Mg	Cr	Cu	Mn	Fe	Si	Ti	Zn	Others
Percentage	92.6	4.9	0.25	0.1	0.8	0.4	0.4	0.15	0.25	Max 0.20

Table 2 The wt% of CSA and ZnO in the chemical composition

Elements	CSA	ZnO
SiO_2	45.05	–
Cuo	–	7.44
TiO_2	–	0.26
Al_2O_3	15.6	–
Fe_2O_3	12.4	5.68
CaO	0.57	–
MgO	16.2	0.50
SO_3	–	–
K_2O	0.52	0.07
Na_2O	0.45	0.22
ZnO	0.3	84.6
MnO	0.22	0.33
Others	Balance	Balance
LOI	8.69	0.94

3 Experimental Work

From the castings, ASTM test specimens were machined. Three of each test type were performed. Figure 2 demonstrates dry slide wear testing using DUCOM's pin-on-disc apparatus [22].

The pin (workpiece) was against the disc's 105 mm wear track. The disc's pin was deadweight-loaded. Many samples were examined with 20, 30, and 40 N loads at 2, 3, and 4 m/s. Similar wear testing reveals a 1000-m sliding distance. 6 mm pins were 35 mm. Before the test, the pin worm surfaces were slid with an emery sheet to touch the steel disc. The sample and worn track were cleaned with acetone and weighed to 0.0001 g before and after each test [22].

3.1 Microstructure Analysis

The microstructure of cast samples and wear debris surface morphology were examined using a LEICA S440i SEM equipped with an Oxford INCATM EDS system. Before placing the samples in the sample chamber, they were taped on using double-sided carbon tape. The SEM accelerated at 5 to 20 kV.

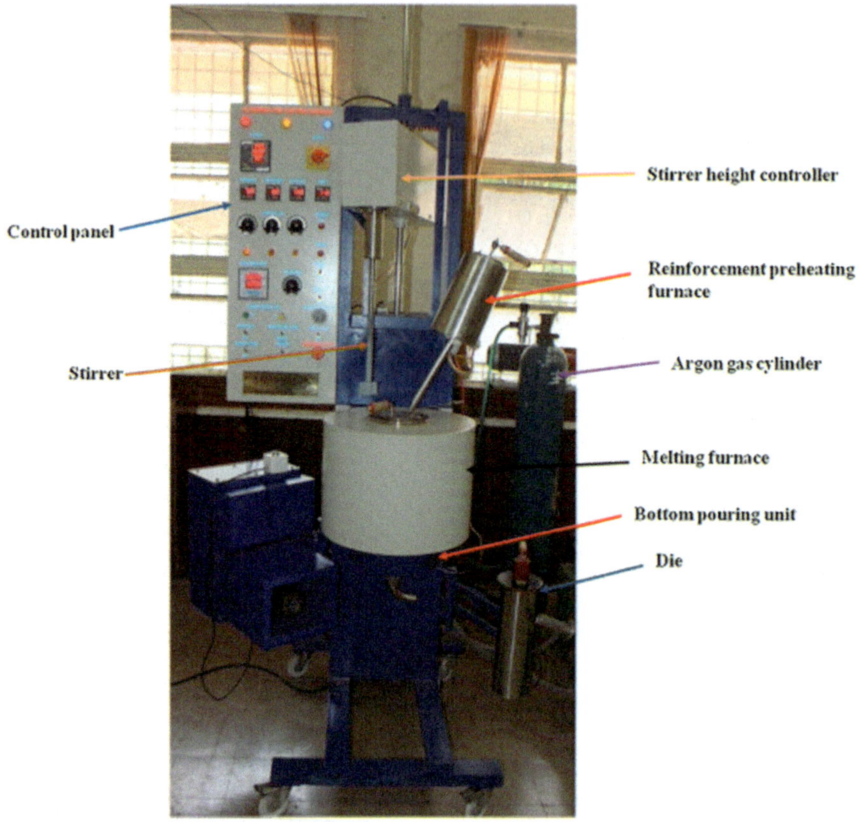

Fig. 1 Melting furnace-stir casting

Table 3 Percentage of AA 5083 hybrid composites

Samples	Aluminium 5083%	CSA %	ZnO %
1	100	0	3
2	94	3	3
3	91	6	3
4	88	9	3

3.2 *Architecture of Artificial Neural Networks*

An effective data modelling technique that can capture and depict complex input and output interactions is an artificial neural network. Identification of the network architecture, including the number of input and output neurons, hidden layers, and neurons in each hidden layer, as well as the network parameters, is necessary for the construction of ANN models (Activation Function and Learning Rate). Using both

Fig. 2 Pin-on disc wear testing machine

supervised and unsupervised learning techniques, artificial neural networks include at least three layers, an input layer, several hidden layers, and an output layer.

3.3 Inputs and Outputs for ML and MLP-ANN

Six input neurons (Ni = 6) represent the variables of loads (L; N), sliding velocities (v; m/s), sliding speeds (N; RPM), reinforcement percentages (Reinforcement;%), cumulative time (t; mints), co-efficient of friction (μ) for AA 5083 CSA and ZnO hybrid composites. One neuron represents the value of the corresponding Specific Wear Rate (mm^3/N.M) in the output layer. For predicting the Specific Wear Rate of AA 5083 composites, ANN (MLP) was tested. The experimental data sets for Specific Wear Rate are 36 samples.

4 Result and Discussion

4.1 Tribological Behaviour

The tribological behaviour of the hybrid composite with the AA 5083 matrix was followed through several phases. The tests were performed in conditions without lubrication on samples with the best structural, mechanical, and anti-corrosion characteristics. The wear loss was measured during the testing. The wear loss, one of the major parameters for wear monitoring, was estimated based on the volume of worn material, sliding velocity, sliding time, and a constant1000 m sliding distance.

Tribological Sample, Table 4 provides the specific wear rate values for the tested materials based on the loads, sliding velocity, and sliding speed. Due to the extensiveness of the obtained results, a partial number of experimental values of wear loss is shown in Table 4. First, the testing of the base material aluminium alloy AA 5083, was performed, and then 3 wt% CSA, 6 wt% CSA 9 wt% CSA with 3 wt% ZnO was added in the base. The Specific Wear Rate decreased by increasing CSA with 3% percentage reinforcements and increasing load and sliding velocity to the base material of the hybrid composites. The coefficient of friction of the AA 5083 hybrid composites was significantly reduced by adding only 6 wt% CSA and 3% ZnO. With the addition of 3 and 9% CSA, the co-efficient of friction increased while increasing loads and Sliding velocity at certain intervals, while in others it decreased.

4.2 Effect of Applied Loads and Sliding Velocities on Specific Wear Rate and Co-efficient of Friction

The applied load is one of the most important determinants of the specific wear rate of the composites. The unreinforced aluminium 5083 alloy is shown to have a higher specific wear rate than hybrid composites. This is mainly because the hard dispersoids on the surface of the composites function as protrusions and protect the matrix from hard interaction with the counter surfaces, causing hybrid composites to wear less gradually than alloys under all loads. Figure 3 shows that the specific wear rate of the composites decreases with increasing load at a constant sliding distance (1000 m).

At 20 and 40 N loads, the composites containing 6% CSA and 3% ZnO had the lowest specific wear rate. At 30 and 40 N, composites containing 6% CSA and 3% ZnO hybrids exhibited an essentially same specific wear rate. When the load is increased from 20 to 40 N, composites with percentage increases reveal a lower specific wear rate than all the hybrid composites.

From Fig. 4, composites with more reinforcements have stronger wear resistance at 3 m/s sliding velocity, constant sliding distance, and constant load 20 N. This might be because CSA and ZnO particles are easily ploughed away from the matrix's surface, increasing wear at 4 m/s sliding velocity. In hybrid composites containing zinc oxide particles, the particles fragment into small pieces and continue to inhibit particle removal, decreasing wear.

Figures 5 and 6 By incorporating weight percentages of CSA and ZnO, the composites coefficient of friction was significantly lowered. The co-efficient of friction increased in certain periods and decreased in others with the addition of weight percent of reinforcements, increasing load, and sliding velocity.

Table 4 Experimental wear analysis of AA 5083 hybrid composites

Sl. No.	Composition of AA 5083	Load (N)	Sliding velocity (m/s)	Sliding speed (rpm)	Cumulative Time (sec)	Initial mass (m1)	Final mass (m2)	Wear Loss (Δm)	Coefficient of friction (Ff/N) (μ)	Wear rate $\Delta m/L \times 10^{-8}$ (m/L) (N/m)	Volumetric wear rate (wv × 10) [$\Delta m/\rho t$] (mm^3/s)	Specific wear rate (ws × 10^{-1}) [wv/vs FN] (mm^3/N-m)
1	Pure 5083	20	2	364	8.33	3.4649	3.4566	0.0083	0.6738	8.3	6.2	1.56
2	Pure 5083	30	2	364	8.33	3.4579	3.4487	0.0098	0.6375	9.8	7.4	1.23
3	Pure 5083	40	2	364	8.33	3.4487	3.4389	0.011	0.6388	11	8.3	1.03
4	Pure 5083	20	3	546	5.55	3.4699	3.4649	0.005	0.5825	5	5.6	0.94
5	Pure 5083	30	3	546	5.55	3.4389	3.4317	0.0072	0.645	7.2	8.1	0.9
6	Pure 5083	40	3	546	5.55	3.4317	3.4239	0.0078	0.6625	7.8	8.8	0.73
7	Pure 5083	20	4	728	4.16	3.4848	3.4744	0.0104	0.6075	10.4	15.6	1.95
8	Pure 5083	30	4	728	4.16	3.4744	3.463	0.0114	0.82	11.4	17.1	1.43
9	Pure 5083	40	4	728	4.16	3.4712	3.4588	0.0124	0.6931	12.4	18.6	1.55
10	3% CSA 3% ZnO	20	2	364	8.33	3.1619	3.1546	0.0073	0.7588	7.3	6	1.5
11	3% CSA 3% ZnO	30	2	364	8.33	3.1546	3.1457	0.0089	0.6083	8.9	7.3	1.22
12	3% CSA 3% ZnO	40	2	364	8.33	3.1457	3.1364	0.0098	0.6338	9.8	8.1	1.01
13	3% CSA 3% ZnO	20	3	546	5.55	3.1663	3.1619	0.0044	0.675	4.4	5.4	0.91
14	3% CSA 3% ZnO	30	3	546	5.55	3.1364	3.1311	0.0053	0.6058	5.3	6.5	0.73
15	3% CSA 3% ZnO	40	3	546	5.55	3.1311	3.1249	0.0062	0.6106	6.2	7.7	0.64

(continued)

Table 4 (continued)

Sl. No.	Composition of AA 5083	Load (N)	Sliding velocity (m/s)	Sliding speed (rpm)	Cumulative Time (sec)	Initial mass (m1)	Final mass (m2)	Wear Loss (Δm)	Coefficient of friction (Ff/N) (μ)	Wear rate $\Delta m/L \times 10^{-8}$ (m/L) (N/m)	Volumetric wear rate (wv × 10) [$\Delta m/\rho t$] (mm^3/s)	Specific wear rate (ws × 10^{-1}) [wv/vs FN] (mm^3/N-m)
16	3% CSA 3% ZnO	20	4	728	4.16	3.1789	3.1707	0.0082	0.6413	8.2	13.5	1.69
17	3% CSA 3% ZnO	30	4	728	4.16	3.1707	3.1667	0.004	0.7175	4	6.6	0.55
18	3% CSA 3% ZnO	40	4	728	4.16	3.1667	3.1652	0.0015	0.8213	1.5	2.5	0.15
19	6% CSA 3% ZnO	20	2	364	8.33	3.2208	3.2134	0.0074	0.8	7.4	5.7	1.43
20	6% CSA 3% ZnO	30	2	364	8.33	3.2124	3.2035	0.0089	0.5767	8.9	6.9	1.15
21	6% CSA 3% ZnO	40	2	364	8.33	3.2026	3.1916	0.0093	0.5113	9.3	7.2	0.9
22	6% CSA 3% ZnO	20	3	546	5.55	3.225	3.2208	0.0042	0.6363	4.2	4.9	0.81
23	6% CSA 3% ZnO	30	3	546	5.55	3.1916	3.1862	0.0054	0.6092	5.4	6.3	0.7
24	6% CSA 3% ZnO	40	3	546	5.55	3.1862	3.1791	0.0071	0.5344	7.1	8.2	0.69
25	6% CSA 3% ZnO	20	4	728	4.16	3.2551	3.2305	0.0078	0.49	7.8	12	1.51
26	6% CSA 3% ZnO	30	4	728	4.16	3.2305	3.2265	0.004	0.7242	4	6.2	0.51

(continued)

Table 4 (continued)

Sl. No.	Composition of AA 5083	Load (N)	Sliding velocity (m/s)	Sliding speed (rpm)	Cumulative Time (sec)	Initial mass (m1)	Final mass (m2)	Wear Loss (Δm)	Coefficient of friction (Ff/N) (μ)	Wear rate Δm/L × 10^{-8} (m/L) (N/m)	Volumetric wear rate (wv × 10) [Δm/ρt] (mm^3/s)	Specific wear rate (ws × 10^{-1}) [wv/vs FN] (mm^3/N-m)
27	6% CSA 3% ZnO	40	4	728	4.16	3.2265	3.2259	0.0006	0.6656	0.6	0.9	0.06
28	9% CSA 3% ZnO	20	2	364	8.33	3.0084	3.0017	0.0067	0.79	6.7	5.6	1.39
29	9% CSA 3% ZnO	30	2	364	8.33	3.0009	2.9927	0.0082	0.5717	8.2	6.8	1.14
30	9% CSA 3% ZnO	40	2	364	8.33	3.9927	3.9835	0.0085	0.605	8.5	7.1	0.88
31	9% CSA 3% ZnO	20	3	546	5.55	3.012	3.0084	0.0036	0.7188	3.6	4.5	0.75
32	9% CSA 3% ZnO	30	3	546	5.55	3.9835	3.9787	0.0048	0.6392	4.8	6	0.67
33	9% CSA 3% ZnO	40	3	546	5.55	3.9783	3.9722	0.0061	0.5856	6.1	7.6	0.63
34	9% CSA 3% ZnO	20	4	728	4.16	3.0216	3.0146	0.007	0.6125	7.0	11.6	1.46
35	9% CSA 3% ZnO	30	4	728	4.16	3.0146	3.0126	0.002	0.7092	2.0	3.3	0.28
36	9% CSA 3% ZnO	40	4	728	4.16	3.0126	3.0122	0.0004	0.7769	0.4	0.7	0.04

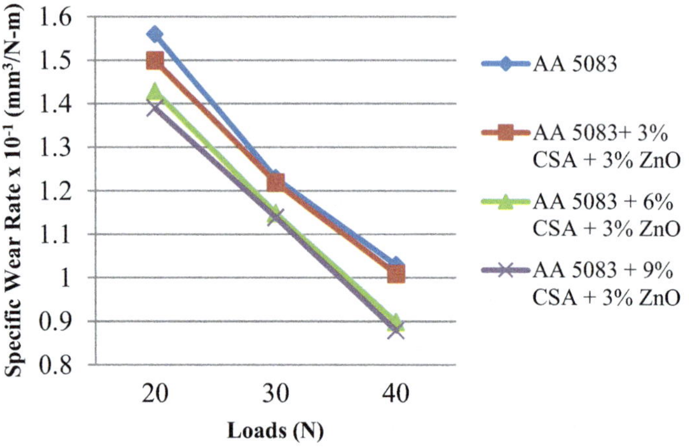

Fig. 3 Variation of aluminium alloy 5083 based composite specific wear rate with load

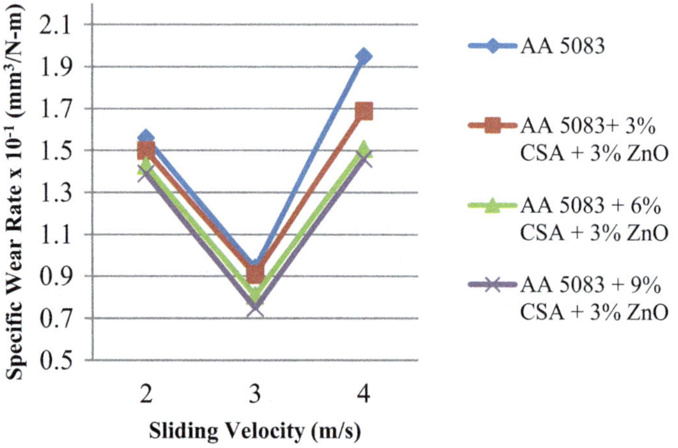

Fig. 4 Variation of aluminium alloy 5083 based composite specific wear rate with sliding velocity

4.3 Wear Mechanism

Multiple factors contribute to the specimens effective wear. Increased load causes hard asperities of the counter surface to penetrate the softer pin surface, micro cracking of the subsurface, and deformation and fracture of softer asperities. Beyond each composite critical load, the wear rate increases dramatically. The transition load is when a specific wear rate suddenly increases [23].

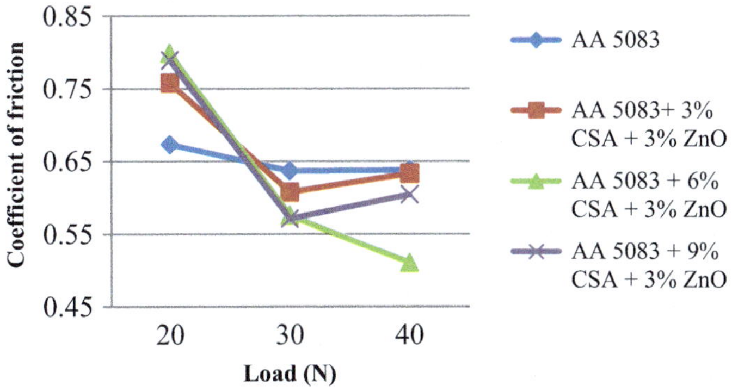

Fig. 5 Variation of aluminium alloy 5083 based composite coefficient of friction with Load

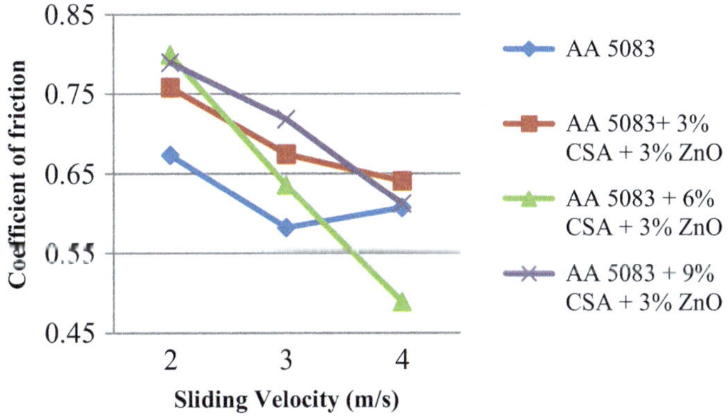

Fig. 6 Variation of aluminium alloy 5083 based composite coefficient of friction with sliding velocity

4.4 SEM Worn-Out Sample Images of AA 5083 Hybrid Composites

After Wear test worn-out samples were tested are shown in Fig. 7a AA 5083, Fig. 7b AA 5083 with 3% CSA + 3%ZnO, Fig. 7c AA 5083 with 6% CSA + 3%ZnO, Fig. 7d AA 5083 with 9% CSA + 3%ZnO.

AA 5083 HBMMCs are susceptible to delamination and adhesive wear as wear mechanisms. Here, it is described how surface morphology relates to each of the processes. Analyzed is a comparative research of hybrid composites made of 3%, 6%, and 9% CSA and 3% ZnO and 5083 aluminium. Unreinforced 5083 alloy and AA5083/CSA/ZnO hybrid composites are evaluated under constant load (40 N), sliding velocity (4 m/s), sliding time (4.16 min), and sliding distance conditions.

(a)

(b)

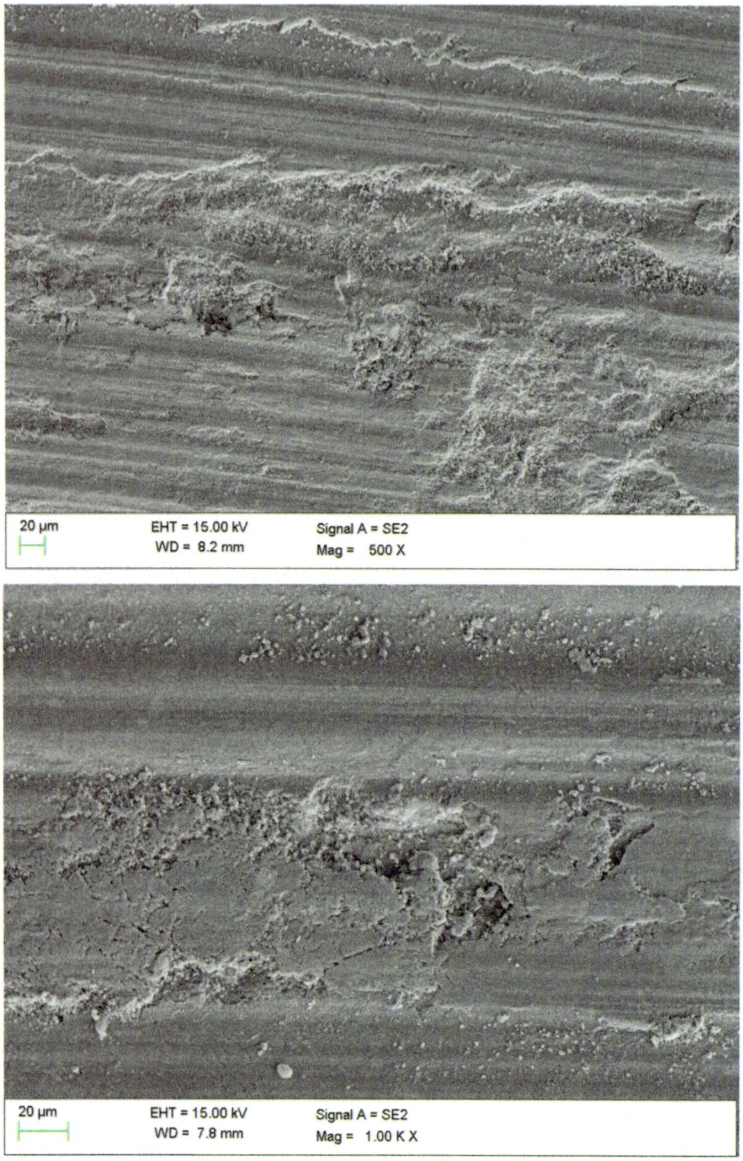

Fig. 7 SEM Images of AA 5083 and its hybrid composites

The results of these tests are photographed in scanning electron microscope (SEM) images (1000 m).

Figure 7a–d shows SEM micrographs and enlarged morphologies of AA 5083 matrix, CSA, and ZnO/AA 5083 hybrid composites at 40 N. AA5083/CSA/ZnO hybrid composites scratch less than AA5083. At 40 N, composites wear with large

(c)

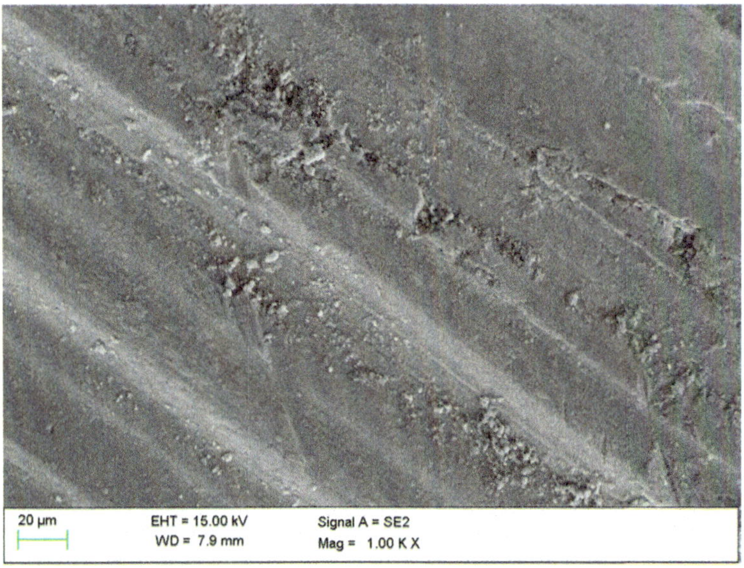

(d)

Fig. 7 (continued)

grooves and debris. The magnified morphology shows the AA5083/CSA/ZnO hybrid composites low applied load wear mechanism. Figure 7b–d reveal that micro-cutting and abrasive wear are the primary wear mechanisms Fig. 7d. The rather minor delamination layer occurrence of AA 5083/9% CSA/3% ZnO hybrid composite demonstrated significantly increased wear resistance, which is comparable to Fig. 7d [24].

When the increased load is 40 N, layers of delamination attach to worn surfaces, as seen in Fig. 7a–d. At 40 N and 4 m/s sliding velocity, AA5083/CSA/ZnO hybrid composites show substantial to wear. Figure 7b–d show the enlarged morphology of the worn surfaces of AA5083/CSA/ZnO hybrid composites under 40 N loads.

CSA hard reinforcement changed pin and disc contact characteristics. The AA 5083 matrix was worn out first due to the hardness difference between the reinforcement and matrix. Grooves are formed as a result of debris being removed and pushed into ridges along the direction of sliding during the wear process. The deficiency of AA 5083 matrix increased the load-bearing function of CSA and ZnO particles and desquamation. Desquamated CSA/ZnO and AA 5083 matrix altered wear behaviours and generated abrasive wear. As sliding speed increased, AA5083/CSA/major ZnO wear mechanism changed from abrasion to adhesion. In the pin and disc counter body wear system, shear stress desquamated the AA5083 matrix, causing periodic plastic deformation of the AA5083/CSA/ZnO hybrid composite. Adhesion wear and a delamination layer were generated as a result of stress concentration between the pin and the disc. These features provided to shield the composite from further friction and to increase its wear resistance [25].

Figure 7 AA5083/CSA/ZnO hybrid composite is economical, efficient, and high wear resistant. Figure 7b–d Co-efficient of friction and Fig. 7b wear loss of AA 5083 matrix and AA5083/CSA/ZnO hybrid composite at various loads and sliding velocities. CSA and ZnO changed the pin-disc interaction properties. Due to the hardness discrepancy between the reinforcement, The adhesion wear delamination layer improved the composites wear resistance. Considering its economy, efficiency, and good wear resistance, the hot-press sintered AA5083/CSA/ZnO hybrid composite can be widely utilized in wear resistance applications.

4.5 Machine Learning (ML)

Open source ML and data visualisation system evaluated data. These metrics represent ANNs and provided outputs in terms of Specific Wear Rate (SWR). Per sample, the algorithm estimated Specific Wear Rate (SWR) using RF, NN, and kNN. Table 5 displays ML training parameters.

Random Forest (RF) is the most suitable assessment method, the estimate that was created by (just) the NN methodology was provided in that figure. This is because the Standard Deviation percentage ($\sigma\%$) of Table 6 indicates that the RF is the most appropriate evaluation method. The NN demonstrates a good connection between the experimental dataset and the predicted Specific Wear Rate with a value of 32% and 37% for AA 5083 hybrid composites. However, this correlation is not perfect (SWR).

Table 5 Parameters for machine learning methods

Random Forest	
Number of trees	15
Fixed seed for random generator	32
Do not split subset smaller than	5
Neural network	
Learning speed	0.6
Inertial coefficient	0.5
Test mass tolerance	0.02
Tolerance of the learning set	0.03
Number of layers	8
k-Nearest Neighbours (kNN)	
Metric	Chebyshev
Number of neighbor	2
Weight	Uniform

Table 6 displays the Mean value (μ), the Standard Deviation (σ), and the Relative Standard Deviation (σ %) to illustrate the variability of the overall value and compare the various techniques.

This prediction ensures a substantial coincidence in SWR averages (0.98 vs. 0.96) and variability.

Additionally, it seems that any Random Forest (RF) approach under examination may provide a reliable estimate. Figure 8 illustrates this feature by showing values from the various approaches (RF, NN, and kNN) in the context of, for instance, AA5083 hybrid composites.

Even though the specimens were taken from identical tribological conditions, this result may be seen as being more than suitable since the experimental results were subject to some inherent variability ($\sigma = 0.96$). Even while it is also clear that there is a trend toward a decrease overall, this variability was translated via the ML procedure. Neither as a structure nor during training, ML algorithms have not been optimized. Without going into specifics of AI Methods, this decision is tied to an investigative technique that aims to demonstrate their universal applicability.

The correlation between measurements and estimates as predicted by RF, NN, and KNN is shown point by point in Fig. 9. It demonstrates the ability of the Machine Learning technique to identify correlation ($r = 0.92$ on RF, $r = 0.84$ on NN, and $r = 0.90$ on kNN). This excellent match is shown by the clustering of data around the diagonal. They also demonstrate that there are values that significantly vary from this linearity. The distribution of points above and below the line suggests no systemic errors in the estimate.

Table 7 Showed Random Forest (RF) R-square value is more than 0.92 and 0.90 on KNN and 0.84 on NN algorithms in machine learning. Compared with the least values of error MSE, MAE and RMSE, RF model is less and a good prediction algorithm in Machine Learning.

Table 6 Parameters in machine learning

S. No.	SWR	RF	NN	KNN
1	1.46	1.41	1.53	1.56
2	1.20	1.27	1.23	1.23
3	0.99	1.02	1.02	1.03
4	1.02	1.22	0.93	0.94
5	0.99	1.05	0.81	0.90
6	0.95	0.61	0.69	0.73
7	1.48	1.61	1.82	1.95
8	1.04	0.92	0.99	1.43
9	1.05	0.81	0.85	1.55
10	1.35	1.57	1.53	1.50
11	1.16	1.34	1.23	1.22
12	0.98	1.03	1.02	1.01
13	0.97	1.14	0.93	0.91
14	0.75	1.02	0.81	0.73
15	0.72	0.63	0.69	0.64
16	1.19	1.40	1.82	1.69
17	0.44	0.95	0.99	0.55
18	0.47	0.47	0.85	0.15
19	1.37	1.45	1.46	1.43
20	1.16	1.18	1.19	1.15
21	0.94	0.95	0.96	0.90
22	0.96	1.02	0.86	0.81
23	0.73	0.92	0.71	0.70
24	0.71	0.63	0.67	0.69
25	1.40	1.38	1.60	1.51
26	0.43	0.70	0.53	0.51
27	0.66	0.43	0.10	0.06
28	1.33	1.14	1.41	1.39
29	1.16	0.83	1.15	1.14
30	0.98	0.67	0.89	0.88
31	0.81	0.71	0.78	0.75
32	0.68	0.75	0.69	0.67
33	0.70	0.53	0.66	0.63
34	1.31	1.05	1.48	1.46
35	0.43	0.53	0.40	0.28
36	0.44	0.04	0.05	0.04
Mean (μ)	0.96	0.96	0.98	0.96

(continued)

Table 6 (continued)

S. No.	SWR	RF	NN	KNN
St Dev (σ)	0.31	0.36	0.42	0.47
St Dev (σ%)	32	37	43	49

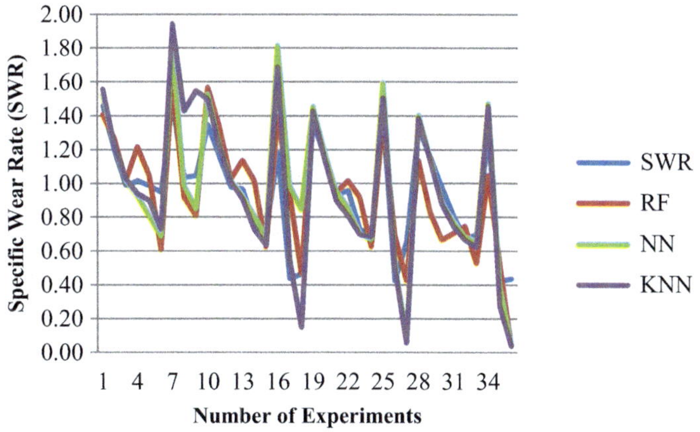

Fig. 8 Experimental SWR versus ML for AA 5083 and its hybrid composite

4.6 Artificial Neural Network (MLP Model)

The experimental data on AA 5083 composites was used to produce the data for training the ANN Multilayer Preceptron (MLP) model. A dataset consisting of a total of ($9 \times 4 = 36$) samples were collected, 3 for each combination of sample materials, specific wear rate and other parameters with six different inputs Load, Sliding velocity, Sliding speed, coefficient of friction, Cumulative time and % reinforcement as inputs and Specific Wear Rate as output (26 Samples) was used to train the ANN-MLP model. 15% (5 Samples) of total input data was used for test data and cross-validation (5 Samples). Cross-validation stops network training. This approach detects data error and stops training when it increases. The best generalization occurs here. Figure 12 specifies cross-validation and testing data sets. Figure 10 shows the MLP structure design, output, input, transfer functions, and hidden layers.

A wide range of composites were tested for Specific Wear Rate. Data has been statistically and ML-analyzed. ANN-MLP method improved specific wear rate prediction.

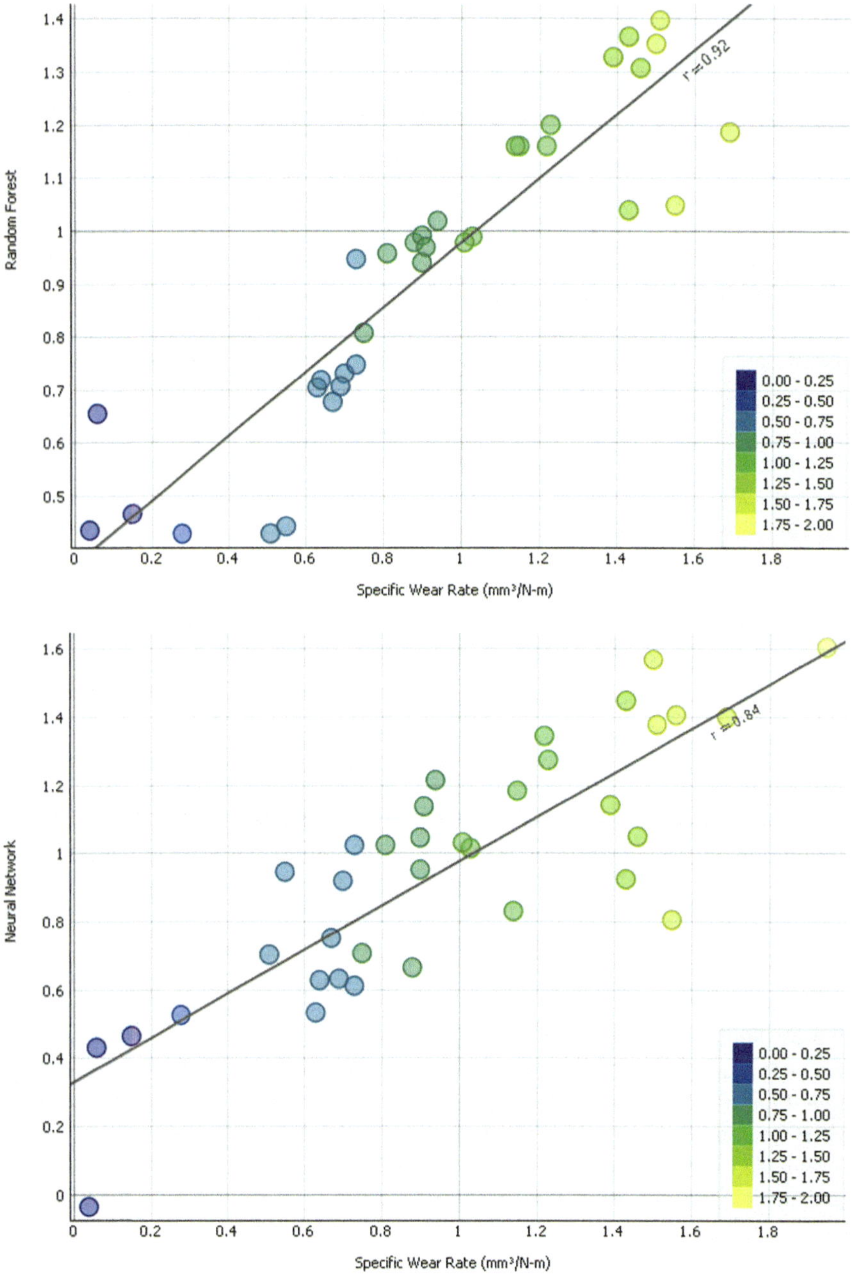

Fig. 9 The correlation between specific wear rate data and estimates is predicted by RF, NN, and kNN techniques

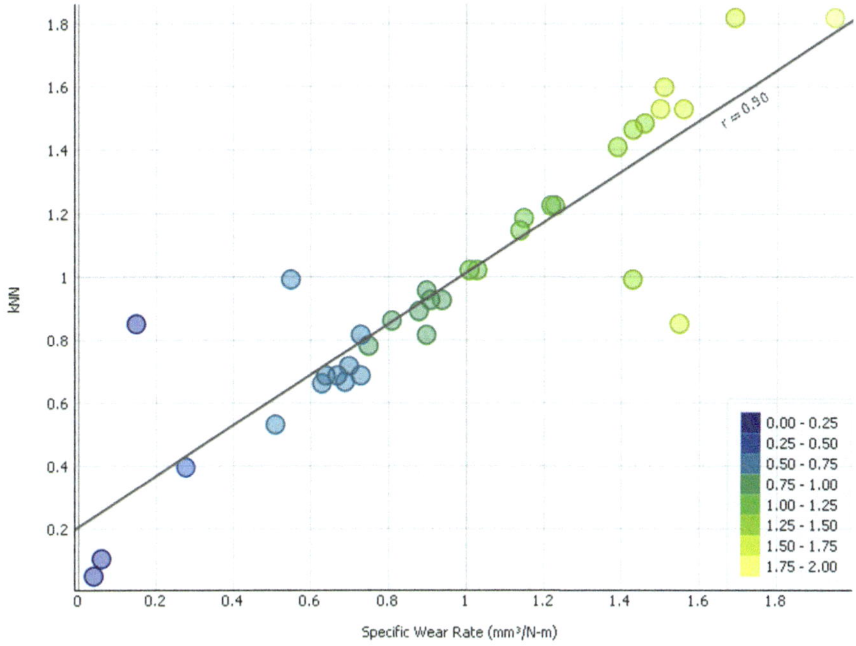

Fig. 9 (continued)

Table 7 Comparison of test results ML

Model	MSE	RMSE	MAE	R^2
Random Forest	0.040	0.201	0.099	0.92
Neural network	0.064	0.254	0.198	0.84
kNN	0.048	0.219	0.149	0.90

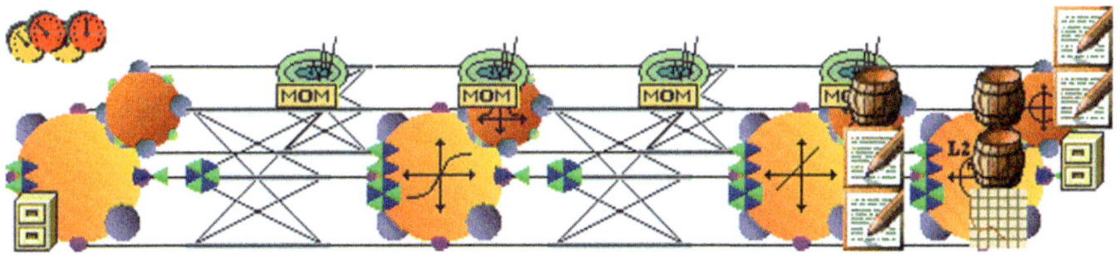

Fig. 10 Multilayer preceptron (MLP) network architecture design

4.7 *ANN-MLP Training and Testing*

The training results indicate good functionality and a low inaccuracy of 0.00000496 at 2991 epochs. 2991 epochs later, training is complete. Mean Square Error MSE curve

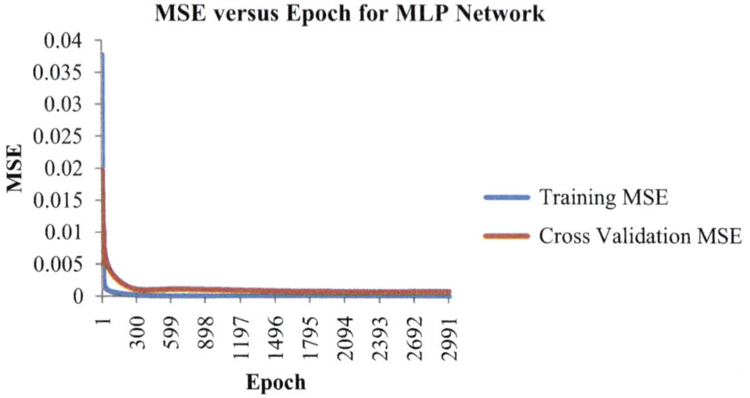

Fig. 11 Training performance of the multilayer preceptron (MLP) network

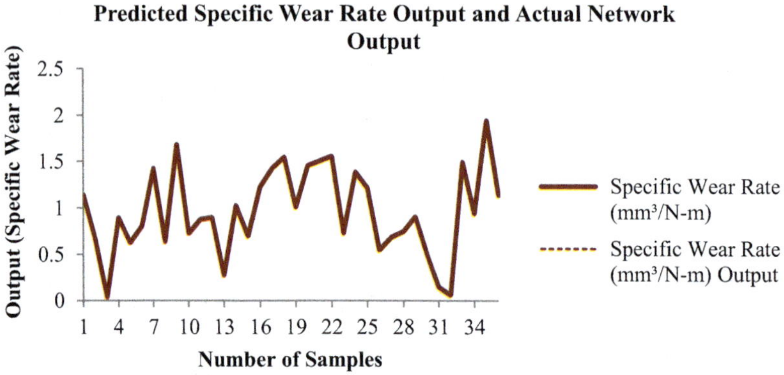

Fig. 12 Testing Performance of the multilayer preceptron (MLP) network

shows weight decay, indicating superior functioning. The network's functioning is further analyzed. Figure 11 shows training and cross-validation results.

The testing operation's MSE is 0.00001599 and MAE is 0.00276. This indicates good performance. NMSE is 0.0000749, while r is 0.99996. The reason that the components are so close to forming a straight line, as shown by the linear correlation value of 0.99996, suggests that there is a strong link between them. Figure 12 demonstrates the projected MLP Specific Wear Rate network output matches the actual rate.

5 Conclusion

The following are the findings of the present investigation:

Coconut Shell Ash (CSA) can be utilised to manufacture Aluminium Metal Matrix Hybrid Composite. It can replace aluminum-heavy materials. CSA can be used to make composites from agricultural waste. This helps with CSA storage and disposal.

Stir Casting successfully incorporates CSA and ZnO into aluminium alloy 5083. This approach produced hybrid composites with 3, 6, and 9 wt% CSA and 3 wt% ZnO. Stir Casting successfully incorporates CSA and ZnO into aluminium alloy 5083.

Specific wear rate reduces with increasing CSA weight%, and ZnO particles have outstanding tribological features, the least wear loss and coefficient of friction in all test settings. Increasing CSA and ZnO particles lowers Aluminium 5083 alloy friction. 6% CSA and 3% ZnO hybrid composites increase friction coefficient at 20 N and 30 N, 2 m/s sliding speed. Poor interfacial bonding between the reinforcement and matrix alloy increases friction.

At applied load 40 N, micro-cutting and abrasive wear were the major AA5083/CSA/ZnO wear mechanisms. Load 40 N supported micro-cutting and adhesion wear. Adhesion wear created a delamination layer that increased composite wear resistance.

Mchine learning methods predicted AA5083 hybrid composites wear rate. Using machine learning methods, the suggested model predicted wear.

Machine learning uses RF, NN, and KNN to predict SWR. NN had the lowest success rate, but RF and KNN were equal; R-squared was 0.92 in RF, 0.84 in NN, and 0.90 in KNN. The RF method had improved RMSE and MAE values. The RF method developed the most efficient model among the presented machine learning techniques for predicting wear rate.

To increase the prediction accuracy and R-square value ANN (MLP) Multilayer Perceptron neural network was used to train and test the network.

Accuracy and error percentage compared in both results for the regression coefficient of machine learning and artificial neural network with two models ML-NN design and ANN-MLP design Specific Wear Rate was adopted.

The regression R-square values for this network were 0.84–0.99. An Error of 0.16 to 0.01 and prediction accuracy of 84% to 99% and 15% improved by using this Multilayer Perceptron (MLP) network is more accurate for AA 5083 hybrid composites and any other types of materials. So, therefore the Artificial Neural Network can be used for predicting tribological parameters and showed good coincidence with the experimental results of these AA 5083 hybrid composites.

This results in the conclusion that both ML and ANN may be useful in preparing an AA 5083 hybrid composite with the optimal ratio of reinforcing elements.

References

1. Ikumapayi OM, Akinlabi ET, Pal SK, Majumdar JD (2019) A survey on reinforcements used in friction stir processing of aluminium metal matrix and hybrid composites. Proc Manuf 35:935–940

2. Sudherson DPS, Sunil J (2020) Dry sliding wear behaviour of novel AA5083-cadmium alloy prepared by stir casting process. Mater Today: Proc 21:142–147
3. Zhang T, Li DY (2001) Improvement in the resistance of aluminum with yttria particles to sliding wear in air and in a corrosive medium. Wear 251(1–12):1250–1256
4. Arulraj M, Palani PK (2018) Parametric optimization for improving impact strength of squeeze cast of hybrid metal matrix (LM24–SiC p–coconut shell ash) composite. J Braz Soc Mech Sci Eng 40(1):2
5. Ma X, Chang PR, Yang J, Yu J (2009) Preparation and properties of glycerol plasticized-pea starch/zinc oxide-starch bionanocomposites. Carbohyd Polym 75(3):472–478. https://doi.org/10.1016/j.carbpol.2008.08.007
6. Tun KS, Jayaramanavar P, Nguyen QB, Chan J, Kwok R, Gupta M (2012) Investigation into tensile and compressive responses of Mg–ZnO composites. Mater Sci Technol 28(5):582–588
7. Selvam B, Marimuthu P, Narayanasamy R, Anandakrishnan V, Tun KS, Gupta M, Kamaraj M (2014) Dry sliding wear behaviour of zinc oxide reinforced magnesium matrix nano-composites. Mater Des 58:475–481. https://doi.org/10.1016/j.matdes.2014.02.006
8. Jasim AH, Joudi WM, Radhi NS, Saud AN (2020) Mechanical properties and wear characteristic of (aluminum-zinc oxide) metal matrix composite prepared using stir casting process. Mater Sci Forum 1002:175–184. Trans Tech Publications Ltd
9. Raju RSS, Panigrahi MK, Ganguly RI, Rao GS (2017) Investigation of tribological behavior of a novel hybrid composite prepared with Al-coconut shell ash mixed with graphite. Metall Mater Trans A 48(8):3892–3903. https://doi.org/10.1007/s11661-017-4139-1
10. Tang F, Wu X, Ge S, Ye J, Zhu H, Hagiwara M, Schoenung JM (2008) Dry sliding friction and wear properties of B4C particulate-reinforced Al-5083 matrix composites. Wear 264(7–8):555–561. https://doi.org/10.1016/j.wear.2007.04.006
11. Thiyaneshwaran N, Sureshkumar P (2013) Microstructure, mechanical and wear properties of aluminum 5083 alloy processed by equal channel angular extrusion. Int J Eng Res Technol 2:17–24
12. Bathula S, Saravanan M, Dhar A (2012) Nanoindentation and wear characteristics of Al 5083/SiCp nanocomposites synthesized by high energy ball milling and spark plasma sintering. J Mater Sci Technol 28(11):969–975
13. Madakson PB, Yawas DS, Apasi A (2012) Characterization of coconut shell ash for potential utilization in metal matrix composites for automotive applications. Int J Eng Sci Technol 4(3):1190–1198
14. Daramola OO, Adediran AA, Fadumiye AT (2015) Evaluation of the mechanical properties and corrosion behaviour of coconut shell ash reinforced aluminium (6063) alloy composites. Leonardo Electron J Pract Technol 27:107–119
15. Agunsoyea JO, Talabib SI, Belloa SA, Awec IO (2014) The effects of Cocos Nucifera (coconut shell) on the mechanical and tribological properties of recycled waste aluminium can composites. Tribol Industry 36(2)
16. Xue D, Balachandran PV, Hogden J, Theiler J, Xue D, Lookman T (2016) Accelerated search for materials with targeted properties by adaptive design. Nat Commun 7(1):1–9. https://doi.org/10.1038/ncomms11241
17. Xiong J, Shi SQ, Zhang TY (2020) A machine-learning approach to predicting and understanding the properties of amorphous metallic alloys. Mater Des 187:108378. https://doi.org/10.1016/j.matdes.2019.108378
18. Sarica A, Cerasa A, Quattrone A (2017) Random Forest algorithm for the classification of neuroimaging data in Alzheimer's disease: a systematic review. Front Aging Neurosci 9:329. https://doi.org/10.3389/fnagi.2017.00329
19. Abd Jalil K, Kamarudin MH, Masrek MN (2010) Comparison of machine learning algorithms performance in detecting network intrusion. In: 2010 international conference on networking and information technology. IEEE, pp 221–226
20. Tretyakov K (2004) Machine learning techniques in spam filtering. In: Data mining problem-oriented seminar, MTAT, vol 3, No 177, pp 60–79. Citeseer

21. Shataee S, Kalbi S, Fallah A, Pelz D (2012) Forest attribute imputation using machine-learning methods and ASTER data: comparison of k-NN, SVR and random forest regression algorithms. Int J Remote Sens 33(19):6254–6280. https://doi.org/10.1080/01431161.2012.682661
22. Nagaraj A, Gopalakrishnan S (2021) A study on mechanical and tribological properties of aluminium 1100 alloys 6% of RHAp, BAp, CSAp, ZnOp and egg shellp composites by ANN. SILICON 13(10):3367–3376
23. Mazahery A, Shabani MO (2012) Study on microstructure and abrasive wear behavior of sintered Al matrix composites. Ceram Int 38(5):4263–4269
24. Alizadeh A, Abdollahi A, Biukani H (2015) Creep behavior and wear resistance of Al 5083 based hybrid composites reinforced with carbon nanotubes (CNTs) and boron carbide (B4C). J Alloy Compd 650:783–793
25. Zhao Q, Liang Y, Zhang Z, Li X, Ren L (2016) Microstructure and dry-Sliding wear behavior of B4C ceramic particulate reinforced Al 5083 matrix composite. Metals 6(9):227

Wear Behavior of Recycled Polyethylene Terephthalate Reinforced with Fly Ash Cenosphere

B. Krishna Prabhu, A. S. Saviraj, and Ajith G. Joshi

Abstract Polymers and their composites are emerging as viable alternative materials to metal and alloy-based ones in many general purpose and special purpose engineering applications. Further, the recycled engineering plastics like post-consumer Poly(Ethylene Terephthalate) (r-PET) can be viewed as an economical alternate feed systems for the production of parts which would have been otherwise made from non-recycled materials. In the current work, the tribological performance of Fly Ash Cenosphere (FAC) filled rPET composite is experimentally investigated to understand the influence of percentage of FAC and the Silane treatment of FAC (TFAC) using pin-on-disc apparatus. Also, wear behavior of TFAC filled rLDPE blended rPET (M-rPET/TFAC) composite is studied. Specific wear rate (SWR) of rPET/TFAC was found to be lower compared to rPET/FAC. Further, M-rPET/TFAC composite has exhibited better wear resistance yielding minimum SWR compared to other composite samples studied. Thus, it can be concluded that rLDPE blended rPET/ Silane coated FAC (M-rPET/TFAC) composite possess better tribological characteristic and reveal the fact that modification of both matrix and reinforcement enhances the wear resistance of rPET composite.

Keywords Recycled PET · Recycled LDPE · Fly ash cenospheres · Dry wear · Compression moulding

1 Introduction

A natural fiber composite material consists of natural fibers and polymeric resin which are glued together under optimum operating conditions. A proper knowledge on the properties of reinforcing fiber, polymeric matrix, the process of fabricating the composite material, and proper bonding at the interface is a crucial aspect which contributes a lion's share in determining the properties of the material [1]. In general,

B. K. Prabhu (✉) · A. S. Saviraj · A. G. Joshi
Department of Mechanical Engineering, Canara Engineering College, VTU Belagavi, Bantwal, Karnataka 574219, India
e-mail: kpbantwal@gmail.com

S. B. Boppana et al. (eds.), *Structural Composite Materials*, Composites Science and Technology, https://doi.org/10.1007/978-981-99-5982-2_13

a variety of natural fibers are available for the use of reinforcement. Among them, flax fiber proves to be the best in terms of.

There is an ever-growing interest in developing newer materials for engineering applications that suit the requirements such as high strength to weight ratio for general applications. Towards this end, polymers and their composites meet the requirements to a great extent. While most of these are tailored to meet the demands of high-tech applications, very few have been developed for low tech applications for instance gears in domestically used products. Manufacturing such products at low cost to meet the needs of customers is a big challenge [1]. In this context, use of post-consumer plastics can be an intelligent choice that not only reduces the cost involved with raw materials but also lessens the environment related problems. With increase in problems of dealing waste plastics, considerable attention is being paid in recent years for their recycling by blending them with suitable reinforcements/ inexpensive additives. Such approaches enhance their potential for manufacturing recycled products. As cost of the manufactured products is a major factor in a competitive market, post-consumer thermoplastics and inexpensive reinforcements offer a very good economic viability. One of the post-consumer plastics that have high recyclability is Poly(Ethylene Terephthalate) (PET) (resin code 1). PET is an engineering plastic with good stiffness. It is used in bottles and other packaging products in huge amounts and calls for recycling.

In the domain of plastic recycling, it is a well-established fact that multiple recycling cycles continuously degrades the properties of the plastic [2]. However, among various grades of plastics that are mechanically recycled, recycled Polyethylene Terephthalate (rPET) offers comparatively better retention of properties but with a small increase in brittleness [3, 4]. Retention of engineering properties post recycling combined with ease of availability makes rPET a front runner as an alternate material for several engineering applications. Reducing the brittleness of rPET and enhancing its retention capacity of properties respectively through blending and reinforcing have been a keen area of interest for several researchers [5], but still there is a wide scope for exploring different material combinations of rPET specifically focusing on the resulting wear properties. Wear resistance is an important consideration if low tech gears are intended application. Fly Ash Cenospheres could be one source of reinforcing material for rPET that have good mechanical and rheological properties and are environment hazardous [6].

Fly Ash Cenospheres (FAC) are hollow particles retrieved from waste fly ash, emitted in the vicinity of thermal power stations whose disposal has been a matter of concern throughout the globe [7]. FAC are the constituents of Fly Ash (FA) with least density. They are spherical hollow alumino-silicates with around 52.53% SiO_2 and 30% Al_2O_3 as main constituents. Micro structural investigation of fly ash filled composite for frictional properties reveals positive outcome towards incorporation of FA in the polymers [8]. The successful reinforcement of FAC in rPET matrix calls for the subsequent verification of mechanical and tribological behavior of the resultant composite [5, 9]. As the materials are procured from recycled arena, there would be a considerable cutoff in the cost of materials. The cost of processing, that decides final cost of the product, however can be reduced by selecting a suitable process for shaping the composite. The use of a low-cost moulding process such as compression moulding helps in reducing the cost due to processing. This moulding technique combined with materials from recycled sources could bring down the final cost of the product.

2 Materials and Methods

R-PET is the matrix system studied in this work. PET beverage bottles are cut into smaller pieces, treated with detergent to remove impurities and then washed with clean water. This is followed by extensive drying to remove moisture and then shredding into flakes with the help of an industrial granulator. Cenospheres used as reinforcements are extracted from Fly Ash (FA). FA is obtained from Raichur Thermal Power Station (RTPS), Raichur, India. A density separation is used for separating Cenospheres from the FA followed by drying. The Cenospheres used have an average particle size of about 50 microns. For making samples of composite, rPET is blended with recycled Low Density Poly-Ethylene (rLDPE) [10], 30% by wt and FAC are treated with 3 Amino trimethoxy silane (3APTMS—10% by weight) before introduction. Blended r-PET would be referred as M-rPET and the treated FAC as T-FAC hereafter. Blended r-PET matrix reinforced with treated FAC composite would be referred as M-rPET/TFAC.

Composite samples for wear testing are processed by compression moulding in the machine made by PLUS ONE MACHINEFABRIK (Belgaum, India) with a capacity of 32 tons, 250 $\times$ 250 mm platen, 200 mm ram stroke and 1.5 kW/platen heating. Samples are moulded in a male–female type of mould using pressure and heat. The dough prepared from r-PET and FAC is transferred to the lower half of pre heated mould. The upper mould half is then forced down to fill the cavity completely.

As per ASTM G 99-95 standards, bar specimens of the composites of size 8 mm $\times$ 8 mm $\times$ 4 ($\pm$0.4) mm were prepared by the above-mentioned process. A Pin-on-disc (POD) machine, TR-201CL, wear and friction monitor with facilities to monitor wear and friction is used in this study. Load on the specimen, sliding distance, speed of rotation and surface roughness of the disk are the four wear factors studied in this work. The former three factors are adjusted to the needs by the controls on the machine and the surface roughness is altered by fixing the SiC abrasive paper on the disk made of steel. Separate abrasive papers are used for each specimen in order to maintain required roughness. All the tests are performed with fixed wear track diameter (30 mm) in order to provide identical conditions for every set of experiments. Rotational speed (rpm) of the disk is, hence, considered as a test parameter instead of sliding velocity.

The pre-conditioning of the wear samples involves rubbing them over a 600 grade Si–C paper so as to ensure uniform contact with the counter surface. The samples are then thoroughly cleaned with soft paper soaked in acetone and dried to remove moisture. The initial weight of the samples is measured on a high precision Precisa XB120A. The counter face disc is covered with Si–C paper and the sample is made to touch the counter face using a dead weight. As soon as the sample runs the required sliding distance, the test is halted by the timer installed in the machine. The final weight of the sample is measured and weight loss is noted down. A minimum of five trials are carried out for each composition of the composite. From the weight loss details, specific wear rate K_o in m^3/Nm is calculated using (1):

$$K_o = W_l/(d \times D_s \times L) \quad (1)$$

where W_l is the weight loss in kg, d is the density of the composite in kg/m^3, D_s is the sliding distance in meters and L is the load in Newton.

Five specimens each for every combination of the materials are moulded. The samples are then tested on a POD machine. Subsequently, the weight loss data is collected and specific wear rate is determined [10–12].

3 Results and Discussions

The wear testing results of r-PET composites in terms of mean SWR (K_O) are computed for all experiments (Fig. 1).

In view that rPET/FAC composite is brittle in nature and that the reinforcing FAC is poorly bound to the matrix (Fig. 2), attempt is made to improve these properties of the composite treating FAC with 3-(Aminopropyl) Trimethoxy Silane (3APTMS), 10% by weight and rPET/TFAC composite is studied for its SWR. And then the matrix is modified by blending it with r-LDPE, 30% by weight and reinforced with T-FAC. The modified composite, M-rPET/TFAC, is also studied for its Specific Wear Rate. Figure 3 shows the wear behavior of rPET/TFAC and M-rPET/TFAC in comparison to that of r-PET/FAC composite.

It is evident from Fig. 3 that rPET/FAC composite shows an initial 34% increase in the wear rate which subsequently reduces to 4% with increase in FAC content from 5 to 15%. The initial increase in the wear rate could be the result of weakening of the composite owing to de-bonding of FAC as observed in Fig. 3. The drop in SWR at higher content of FAC could be due to some kind of lubricating action of

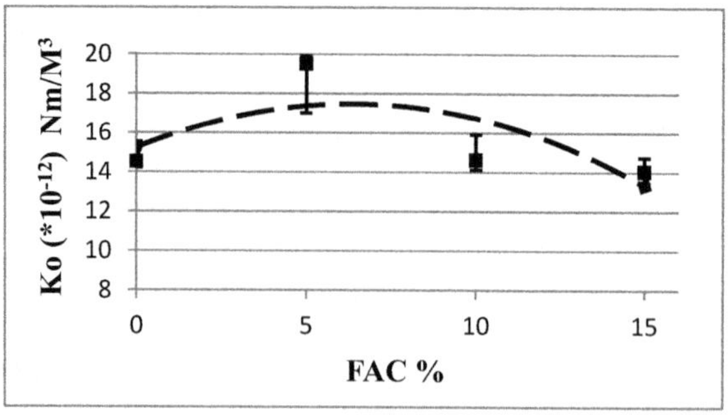

Fig. 1 Plot of specific wear rate of rPET/FAC composite for varying ratios of FAC

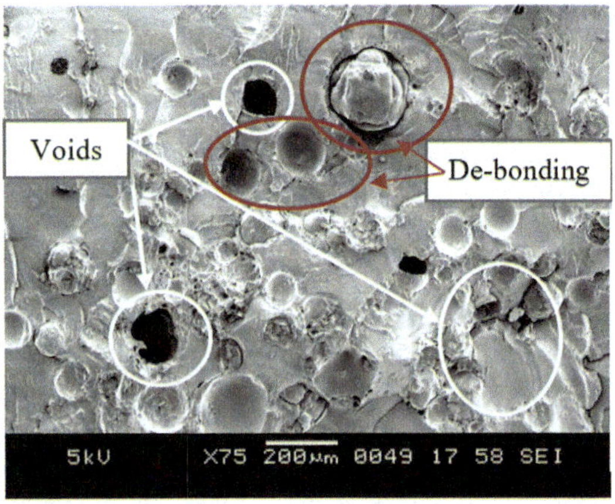

Fig. 2 SEM micrographs of failed surface showing brittle nature of the matrix and de-bonded FAC

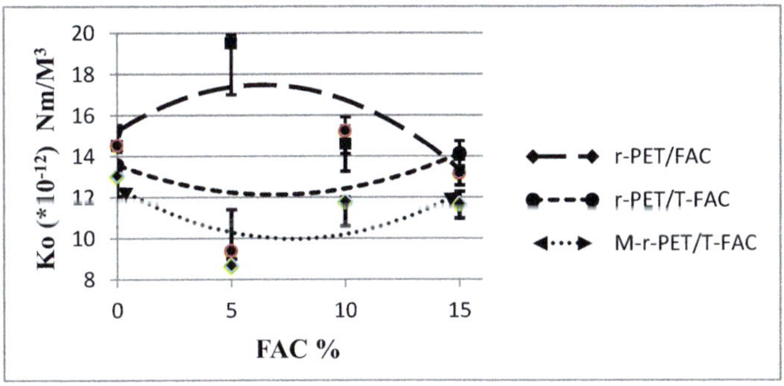

Fig. 3 Plot of specific wear rate of rPET/TFAC and M-rPET/TFAC composite in comparison with rPET/FAC composite for varying ratios of FAC

spherical, hard FAC particles. Surface treated FAC composite at 5% filler content display an initial drop of 35% in SWR. This could be due to the fact that there is improvement in binding of the filler to the matrix (Fig. 6b).

A marginal raise in SWR of these composites can be attributed to the agglomeration of fillers. FAC treated with 3-APTMS may increase the chances of electrostatic attraction between filler particles especially at higher concentrations of TFAC. Figure 4 exhibits the regions of agglomerations seen under SEM. Further, it is worth noting that, although there is raise in the wear rate of surface treated FAC composite, the Specific Wear Rate is less by at least 6% for higher FAC content when compared to rPET/FAC composite.

The wear behavior of M-rPET/TFAC composite displays a 52% drop in the SWR of the composite in comparison to rPET/FAC and 8% drop in comparison to r-PET/T-FAC composites at 5% FAC. This could be due to the surface treatment of FAC in matrix modified rPET composite. The rLDPE being a plastic with low glass transition

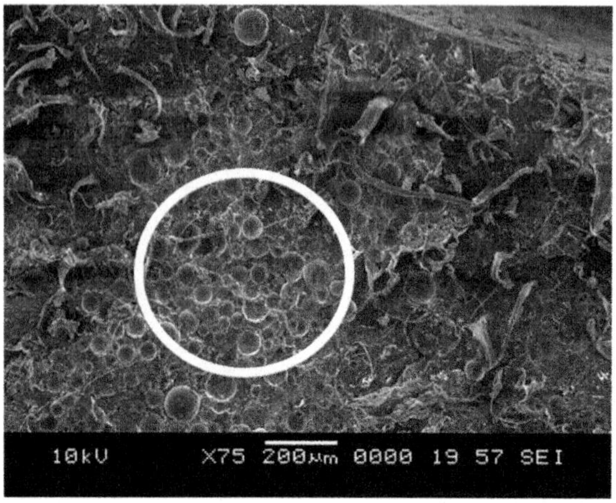

Fig. 4 r-PET/30% r-LDPE/15% FAC (10% APTMS) composite exhibiting regions of agglomeration of fillers on the failed surface

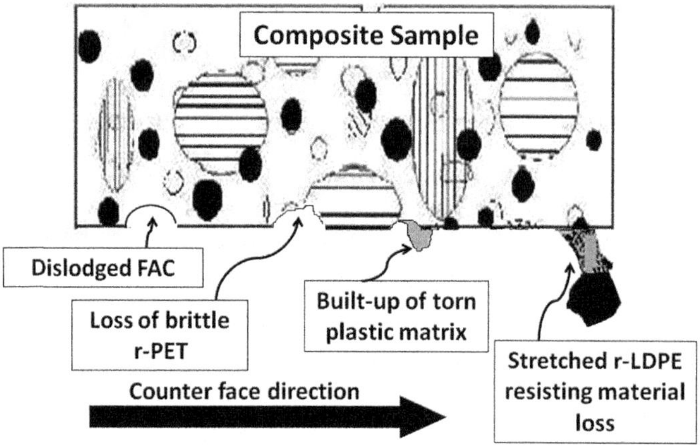

Fig. 5 Wear mechanism

temperature (T_g), improves the flexibility of the matrix. Under the action of tearing forces, the rLDPE component of the matrix modified composite can stretch and absorb energy from the moving counter face (Fig. 5). On releasing the contact, the stretched blend recoils and prevents the breakage of lump of the composite from the parent sample providing better resistance against wear to the composite. Such a pulling and recoiling of the composite could have led to the formation of hill like structure that can be seen in Fig. 6a. Further, better binding of FAC to the matrix improves resistance to the tearing forces and hence to the wear. Well bound FAC owing to surface modifications can be seen on wear surface in Fig. 6b.

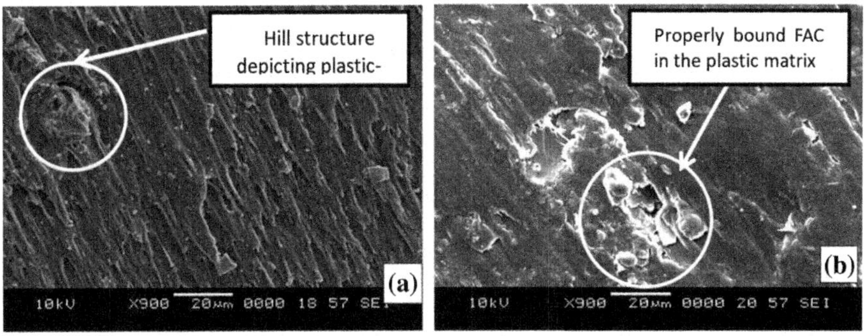

Fig. 6 SEM micrographs for M-rPET/TFAC composite

It could be deduced from the study as stated above that the overall wear loss of FAC reinforced rPET composites consists of weight loss due to dislodged FAC, weight loss due to brittle cracking of matrix and weight loss due to tearing of the matrix as depicted in (2), which is also observed in the SEM micrograms in Figs. 2 and 6.

As observed in Fig. 3, first and second components of (2) are predominant in r-PET/FAC composite. In contrast of this, M-rPET/TFAC could be reducing the brittleness of the matrix, wt loss due to tearing down and tendency of FAC to dislodge, as they are bound properly to the matrix by virtue of surface treatment. Such phenomena could be observed as hill like structures and proper binding in the micrograph in Fig. 6a, b.

$$\begin{aligned}\text{Overall wear loss} &= \text{Wt loss due to dislodged FAC particles}\\ &\quad + \text{Wt loss due to brittle cracking of the matrix}\\ &\quad + \text{Wt loss due to torn particles of the matrix} \qquad (2)\end{aligned}$$

4 Conclusions

The experimentation conducted to develop a rPET/FAC composite reveal an initial increase up to 34% in SWR followed by a drop of 4%. Moreover, the composite seems brittle with poor binding of FAC to the matrix.

R-PET/T-FAC composite displays 35% drop in SWR at 5% FAC content. A marginal raise in Specific Wear Rate owing to the agglomeration of TFAC is observed. M-rPET/T-FAC reveal the facts that both matrix and reinforcement modifications improve the wear property of rPET composite. Surface coated FAC binds the reinforcement better with the matrix reducing the loss of weight due to dislodging of FAC. Matrix modifications carried out by blending rPET with 30% rLDPE reduces the loss due to brittle cracking of the matrix. SWR of the order of 12×10^{-12} m^3/Nm is observed in rPET/TFAC composite which was further lowered to less than 10×10^{-12} m^3/Nm in M-rPET/TFAC composite at 5% FAC.

Acknowledgements The authors express their gratitude to the Department of Science and Technology (DST), Government of India, New Delhi and the management and the Principal of Canara Engineering College, Bantwal for providing support and assistance.

References

1. Tavčar J, Černe B, Duhovnik J, Zorko D (2021) A multicriteria function for polymer gear design optimization. J Comput Des Eng 8(2):581–599
2. Djapovic M, Milivojevic D, Ilic-Tomic T, Lješević M, Nikolaivits E, Topakas E, Nikodinovic-Runic J et al (2021) Synthesis and characterization of polyethylene terephthalate (PET) precursors and potential degradation products: toxicity study and application in discovery of novel PETases. Chemosphere 275:130005
3. Penumakala PK, Santo J, Thomas A (2020) A critical review on the fused deposition modeling of thermoplastic polymer composites. Compos B Eng 201:108336
4. Barba D, Arias A, Garcia-Gonzalez D (2020) Temperature and strain rate dependences on hardening and softening behaviours in semi-crystalline polymers: application to PEEK. Int J Solids Struct 182:205–217
5. Techawinyutham L, Tengsuthiwat J, Srisuk R, Techawinyutham W, Rangappa SM, Siengchin S (2021) Recycled LDPE/PETG blends and HDPE/PETG blends: mechanical, thermal, and rheological properties. J Market Res 15:2445–2458
6. Seena J, Bambola VA, Sherhtukade VV, Mahanwar PA (2010) Effect of flyash content, particle size of flyash, and type of silane coupling agents on the properties of recycled poly(ethylene terephthalate)/flyash composites. J Appl Polym Sci 119:201–208
7. Ahmaruzzaman M (2010) A review on the utilization of fly ash. Prog Energy Combust Sci 36:327–363
8. Tiwari A, Jaggi HS, Kachhap RK, Satapathy BK, Maiti SN, Tomar BS (2014) Comparative performance assessment of cenosphere and barium sulphate based friction composites. Wear 309(1–2):259–268
9. Deepthi MV, Sharma M, Sailaja RRN, Anantha P, Sampathkumaran P, Seetharamu S (2010) Mechanical and thermal characteristics of high density polyethylene-fly ash cenospheres composites. Mater Des 31:2051–2060
10. Krishna Prabhu B, Chomal GJ, Kulkarni SM (2012) Effect of recycled blends on the mechanical properties of recycled polyethylene terephthalate. ANTEC Mumbai (SPE, USA), pp 244–48
11. Akkurt S (1995) On the effect of surface roughness on wear of acetal-metal gear pair. Wear 184:107–109
12. Wright NA, Kukureka SN (2001) Wear testing and measurement techniques for polymer composite gears. Wear 251:1567–1578

Synthesis and Characterization of Al/MWCNT Composites Prepared Through Powder Metallurgy Technique

Production of Al/MWCNT Nanocomposite by Powder Metallurgy to Enhance Dry Sliding Wear Performance Aided by Design of Experiment

H. T. Shivaramu, U. Vignesh Nayak, V. Londe Neelakantha, and K. S. Umashankar

Abstract Aluminum (Al) reinforced with 0, 0.5, and 1.0 wt% MWCNT metal matrix nano-composites were fabricated via powder metallurgy (PM) route. Chemical stability between the matrix (Al) and reinforcement (MWCNT) was analyzed using X-ray diffractometry. Wear experimentations were investigated using cylindrical test pieces on a rotating disk wear apparatus. Compositions, applied loads, disk rotation speeds, and traversing distances were optimized using Taguchi's L9 orthogonal array. The 0.5 wt% MWCNT reinforced composite showed higher resistance to wear. The consequences of variables on rate of wear are presented. SEM micrographs of the investigated cracked surface were observed to judge the mechanism of fracture and MWCNTs effects.

Keywords Aluminium (Al) · Multi-walled carbon nano tubes (MWCNTs) · Wear · Powder metallurgy (PM) · Design of experiments (DOE)

H. T. Shivaramu (✉) · U. V. Nayak · V. L. Neelakantha
Department of Mechanical Engineering, Mangalore Institute of Technology & Engineering, VTU—Belagavi, Moodbidri, Karnataka 574225, India
e-mail: shivaramu@mite.ac.in

U. V. Nayak
e-mail: vignesh@mite.ac.in

V. L. Neelakantha
e-mail: neelakantha@mite.ac.in

K. S. Umashankar
Department of Mechanical Engineering, K V G College of Engineering, VTU—Belagavi, Sullia, Karnataka 574327, India

S. B. Boppana et al. (eds.), *Structural Composite Materials*, Composites Science and Technology, https://doi.org/10.1007/978-981-99-5982-2_14

1 Introduction

A sizable proportion of the materials utilized in engineering are aluminium alloys and composites. These materials indicate higher specific strength, excellent ductility, and anti-corrosion properties compared to steel. The properties of Al alloys cannot be tailored to the level of those provided by the composites of Al. Therefore, there is an increased research focus on Al composites. Typically, the AMMCs are produced using Al–Si alloys. These alloys have demonstrated greater wear resistance compared to pure Al and are the most popular aluminum alloy that is extensively utilized in the automobile sector. However, their wear resistance is limited and their wear performance is relatively unaffected at high fractions of Si additions [1, 2]. Indistinct findings have been published with respect to the consequence of parameters on the sliding performance of Al metal matrix composite (AMMC).

AMMC can be fabricated through various manufacturing techniques such as stir casting, gravity casting, centrifugal casting, squeeze casting, liquid metallurgy, forging, diffusion bonding, semi-solid, and powder metallurgy. Metal matrix composites (MMCs) are made using the stir casting process by mechanically stirring a molten metal with a reinforcing phase, such as ceramic particles or short fibers. The stirring action helps to distribute the reinforcement evenly throughout the matrix, resulting in a composite materials with improved strength, stiffness, and other properties. Gravity casting is a casting process in which molten metal is transferred into a mold under the force of gravity. This is in distinction to other casting methods, such as pressure casting, where the molten metal is forced into the mold under pressure. Centrifugal casting is a production procedure for creating hollow parts from composites. In this process, a mold is revolved at a higher speed while a mixture of resin and reinforcement is deposited inside. The centrifugal force forces the resin and reinforcement to spread evenly throughout the mold, resulting in a strong and uniform part. Squeeze casting is a production process that combines forging and casting. In this process, molten metal is poured into a heated die and then pressure was applied to force the metal to fill the mould cavity. This procedure is employable to produce high-quality metal matrix composites with good mechanical properties. The metal is shaped by compressive forces during the forging production process. This is a versatile process that can be adapted to create an extensive range of components, as well as those with intricate shapes and complex details. Forging can be utilized in addition to improving the strength and durability of metal parts. Diffusion bonding is a solid-state joining process in which materials were united by atomic diffusion at raised temperatures and pressures. It is a versatile process that can be applied to join a wide range of materials, including metals, plastics, and composites. Diffusion bonding is often used to produce composite parts because it can create strong, lightweight bonds without the requirement of adhesives or fillers. This can improve the strength, durability, and performance of composite parts. The semi-solid processing (SSP) is a production technique that uses a material in its semi-solid state to develop parts with intricate shape and high strength. The semi-solid state is achieved by heating the material to the temperature below its melt point but exceeding its solidus temperature. This causes the material to solidify into a mushy state with a high percentage of the liquid phase. SSP can be utilized to produce a broad variety of composite parts, including those made from metal matrix composites (MMCs), ceramic matrix composites (CMCs), and polymer matrix composites (PMCs). The method is particularly well-suited for production of parts with complicated geometry and internal features, such as turbine blades, gears, and brake rotors. Powder metallurgy is a production technique, which involves the fabrication of parts from powdered metals or metal-matrix composites. The process begins with the extraction of metal powders, which are then blended and compacted into the preferred shape. The next step is sintered, which employs heat to fuse the particles together after they have been compacted. The sintered parts are then heat treated to improve their properties. Powder metallurgy is a versatile process, which can be employed to produce a spacious variety of parts, as well as those with difficult shapes and intricate details. The process is also well-suited for producing parts from materials, which are hard to machine, such as titanium and tungsten. Of these techniques and their variants, powder metallurgy has emerged as the most suitable one due to its inherent advantages relative to other processing methods such as near-net shape casting, microstructure control, high production rate, and excellent dimensional accuracy.

Multiple reinforcement materials in standalone and in combination have been researched to study their effect on wear. Ceramics from the popularity of such elements. The materials also range in size from micro to millimeters to nanometers. All the reinforcements added to Al have shown increased wear resistance as they would inevitably enhance the hardness of the composites developed. Aluminium matrix composites reinforced with silicon carbide (SiC), silicon nitride (Si_3N_4), titanium carbide, titanium oxide, titanium boride, boron carbide (B_4C), aluminum oxide (Al_2O_3), and fly ash, are some of the composites materials that have been tried. Aluminium reinforced with different concentrations of SiC has also exhibited improved Young's modulus in addition to increased wear resistance. Al_2O_3 reinforcement has resulted in both wear properties and compression strength enhancement relative to the base alloys used. However, hard B_4C reinforced into Al, which had shown minimum wear resistance despite B_4C being one of the hardest substances. Fly-ash has been employed as reinforcement in Al-based composites for the past few years due to its inexpensive cost and ability to attain desired characteristics. Prasad and Ramachandra [3] evaluated the factors affecting the wear behaviour of squeeze-cast Al/fly ash (5–12.5 wt%) reinforced composites. Their findings demonstrated that the weight fraction of fly ash and casting pressure had an impact on the composite's wear resistance. Their hardness plots showed linear relation of hardness with the applied pressure during casting and the fly-ash content used. However, the investigation does not present any information regarding the dimension of the fly-ash particles utilized to form composites. Evaluation of wear behaviour with micrometer sized reinforcements has yielded the following results. Mishra et al. [4] examined the wear characteristics of 10.0 and 15.0 wt% SiC (150–160 μm) reinforced Al 6061 composite that was manufactured with the stir casting method. Their research exhibited that the application of load majorly affected the co-efficient of friction [COF]. Their research revealed that the major factor that leads to the increased rate of wear was the distance traversed for the 10.0 wt% SiC reinforcement whereas for the 15 wt% SiC composite it was applied load. Singla et al. [5] conducted a dry sliding wear investigation of Al–SiC composites with increasing reinforcement concentrations at a uniform speed. Their investigation confirmed that the rate of wear was the least and the COF was the highest with 20 wt% SiC reinforcement. Their reported plots showed a linear relationship between the rate of wear and COF with the applied load. The COF was observed to reduce and the rate of wear was seen to increase with increased load. The rate of wear of LM6 Al alloy reinforced with 37 μm sized SiC (5, 7.5, and 10 wt%) produced via the stir casting route was investigated by Ghosh et al. [6]. Experimentations were conducted with different SiC wt%, applied loads, sliding durations, and distances. ANOVA outcomes clearly disclosed that the rate of wear was majorly subjected by the SiC contents (41.50%) followed by speed (31.40%) and load (20.66%). Their findings clearly demonstrated that the loss of material reduced with an increase in the SiC content. Further, the wear rate is augmented with the increase of applied load and speed. Radhika et al. [7] investigated the tribological characteristics of the stir-cast Al–Si alloy reinforced with 9.0 wt% Al_2O_3 (average grain size 15–20 microns) and 3 wt% graphite (average grain size 50–70 microns) composites by varying load, velocity, and distance. Their findings specify that the influence of sliding length (around 47%) is more on the wear of the composite in comparison to the load applied and velocity. SEM images of the worn surfaces showed an increased amount of grooves and uncovered reinforcements with the increase in traversing velocity. They reported that at higher applied load, the pin materials smeared on the rotating disk. In their study, friction, and wear were found to decrease because of the self-lubricating nature of the graphite. The rate of wear tested in Al 6061T6/10 wt% SiC/10 wt% Al_2O_3/5 wt% graphite composite that was fabricated via stir casting. The developed composite materials were most predominantly affected by applied load, followed by traversing distance and speed [8]. These results imply that a high loading fraction with moderate to low external applied load, shorter distance traversed and moderate speed is the preferred condition to reduce wear.

The large reinforcement size of micro and millimeter reinforcement though beneficial in improving wear eventually does lead to excessive material loss owing to the 3 body friction. This limitation is overcome by using nanomaterials. Nanocomposites are a class of materials that are reinforced using nano-sized (1–100 nm) materials. Such materials have a higher surface-to-volume ratio. Another advantage is improved interaction between the nano reinforcements and matrix leading to better properties even at an elevated temperature. Metal matrix composite can be reinforced with nanomaterials such as nano oxides (Al_2O_3, WO_3, ZrO_2, B_2O_3 and Y_2O_3), nano-nitrides (Si_3N_4, WN, ZrN, BN and AlN), nano-carbides (TiC, WC, ZrC, B4C and SiC), nano-hydrates (TiH_2), nano-borides (TiB_2, WB and ZrB_2) and nano Ni–Al, Al_3Ti intermetallic compounds, nano-scaled carbon black, graphite (Gr), GNP, Single-walled and multi-walled carbon nanotubes [9–12]. An excellent review of the production techniques to incorporate nanomaterials to form composites was presented by Muley et al. [13]. The production techniques discussed were PM, mechanical alloying, high energy mechanical sphere milling, spark plasma sinter

(SPS), pressure less infiltration, laser deposition, and casting. The study emphasized that the disintegration of nanoparticles occurs to varying extents in each of the processes. The primary hindrance to the successful integration of nano reinforcements within the matrix was their density, which differed significantly in comparison to the matrix. This in turn led to the heterogeneous scattering of the nanomaterial in the matrix. Choi and Awaji [14] showed the importance of managing thermal stresses during nanocomposite production. The investigation showed higher residual stress generated in the matrix material due to differential thermal expansion between reinforcement and matrix during cooling after sintering. A focus area to explore in nanocomposites is Carbon nanotubes (CNTs). CNTs possess excellent electrical, mechanical, and thermal properties. Their novel properties motivated the researchers to develop new MMCs using CNTs as reinforcement [15, 16]. Tribological characteristics of composite depend on the characteristics of reinforcement and matrix constituents when two surfaces slide on each other. CNTs have higher surface areas and are harder than Al this improves the mechanical behaviour of the developed composite. Kim et al. [17] fabricated Al/CNT (1.0, 3.0 and 5.0 wt%) composites via SPS and hot pressing (HP) techniques. They reported minute amount of CNT addition was sufficient to pack the space among the aluminium particles and the higher addition of CNT caused the agglomeration of reinforcement molecules inside the matrix phase. Noguchi et al. [18] adopted a nanoscale dispersions technique to produce Al/MWCNT (13 nm) composites under the melting point temperature of Al. In SEM analysis, few voids were found in the fabricated composites. Liao et al. [19] developed Al/ CNTs (0.5 wt%) of 1.5 mm thickness material. The findings disclosed that the disconnection of CNT was mainly dependent on the particle blending and secondary process. The different grain shapes and sizes were noticed for sintered, hot extruded, and rolled composites. Bakshi et al. [20] reported mechanical ball milling as an efficient technique to obtain a homogeneous distribution of CNTs with reduced clustering. To enhance the homogeneity of nanomaterial dispersion in the matrix, sonication followed by compaction and sintering process was carried out to produce Al/CNTs composite. The microstructure and grain size of reinforcements depend on the ball milling duration. The incorporation of more than 1.0 wt% CNTs resulted in a reduction in strength, stiffness, and ductility because of the CNTs cluster. The reduction of MWCNT and Al powders diameter limits the agglomeration during mechanical alloying. In this method, the homogeneous dispersion occurred within the ductile matrix, due to collides and fractures among the powder particles. The implementation of ball milling and sonication processes before mechanical alloying limits agglomeration. Long CNTs dispersed uniformly in the aluminium matrix over the short CNTs were reported. Ali and Rubel [21] mentioned Al/MWCNT composite was produced successfully through PM, followed by a hot extrusion process. The alignment of MWCNTs in the Al matrix can be attained using the hot extrusion process. The shortening in the length of CNTs and formation of aluminium carbide was revealed in the matrix and grain boundary when Al/CNT composites were prepared via the stir casting route. The improved mechanical characteristics were noticed even when a small volume of MWCNT was reinforced into the aluminium matrix. An important aspect to be considered during the production of nanocomposites is controlling the dispersion and avoiding agglomerations which can be done using the PM technique.

The following sections present the production, characterization, materials used, experiments conducted, important results, and conclusions of research work on the assessment of wear resistance of MWCNT-reinforced Al nanocomposites.

2 Production and Characterization of Composites

2.1 Nanocomposite Production Procedure

PM technique employed to prepare MWCNT reinforced Al nanocomposites specimen for the present research work. The schematic diagram (Fig. 1) shows the procedure adopted to fabricate nanocomposite specimens.

2.1.1 Matrix and Reinforcement Materials

To synthesize the test samples Al with 200-mesh size was purchased from M/s Metal Powder Co., Chennai, Tamilnadu, India. MWCNT nanoparticles were procured from M/s Sigma Aldrich, Bangalore, Karnataka, India, and were produced using the CVD technique. The properties of as-received MWCNT from the suppliers are as shown in Table 1.

The obtained MWCNT was chemically treated by immersing it in concentrated nitric acid, filtered followed by washing with de-ionized water, and were dried at 120 °C to remove impurities.

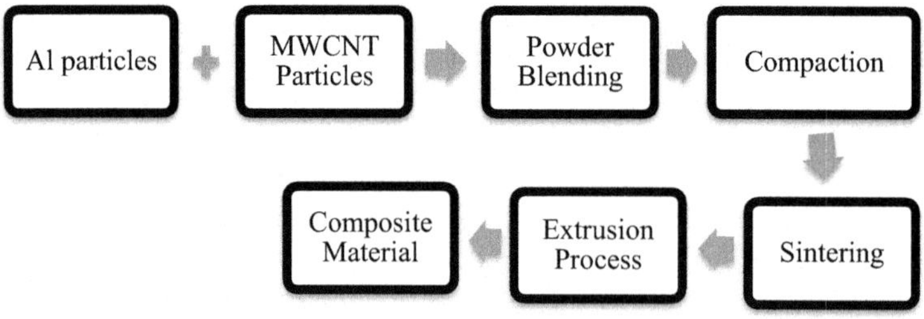

Fig. 1 Fabrication steps for preparing composites

Table 1 Properties of MWCNT (as tested and specified by the provider)

Sl. No.	Properties	Remarks
1	Purity	Carbon > 95% (trace metal basis)
2	OD × ID × L	7–15 nm × 3–6 nm × 0.5–200 μm
3	Total impurities	Amorphous carbon, none detected by TEM
4	Melting point	3652–3697 °C
5	Density	~ 2.1 g/mL at 25 °C

2.1.2 Blending

To obtain nanocomposite, 35 g of the metal powders was taken with the necessary quantity of MWCNTs nanomaterial in a ceramic mortar as shown in Fig. 2.

Using a pestle the blend was hand mixed for about 2 h for each batch. The powder mix thus obtained was added to ethanol, and the metal powder—MWCNT mixture was ultrasonicated in a shaker for 30 min to attain a homogeneous mixture. Finally, the mixed powders were dried at 120 °C under a vacuum (less than 10^{-2} Pa). The mixed composite materials were then introduced into a high energy ball mill to split the clusters produced in the powdered mixture stage. This was done using a high speed/energy planetary ball milling apparatus. This process of breaking agglomerates was accomplished for 10 min at 200 rpm to achieve uniform distribution of MWCNT in the matrix. A ratio of 1:10 in terms of the weight of the mix and the weight of the balls was maintained in the steel jar during ball milling. To explore the properties of developed composites, four samples were prepared with different wt% of MWCNTs in Al. Table 2 shows the proportion of MWCNT reinforcement for the production of nanocomposites.

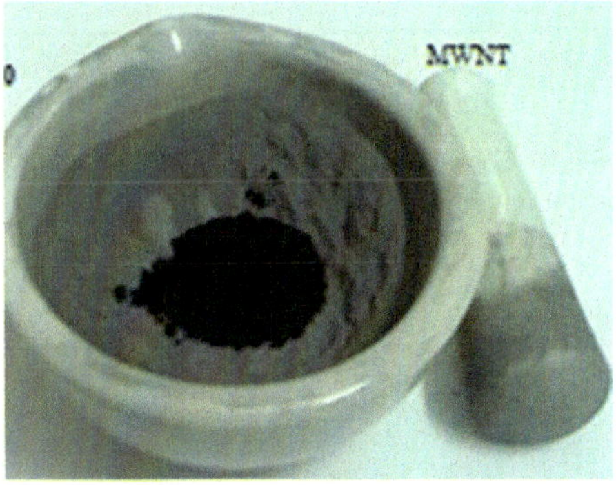

Fig. 2 Ceramic mortar and pestle with LM6 Alloy powder and MWCNT

Table 2 MWCNTs weight for different concentration percentages

Sl. No.	Percentage of MWCNT (%)	Weight of MWCNT (g)
1	0.25	0.0875
2	0.50	0.175
3	0.75	0.2625
4	1.0	0.35

2.1.3 Compaction

The compaction arrangement used consisted of a plunger, a compaction die, and a base plate that was made from EN31 steel. A cylindrical die having a hole of 25.4 mm diameter and 65 mm length was utilized to form circular cross-sectioned specimens as per ASTM B-92503 standard. Universal Testing Machine (UTM) was used for applying the compaction load. A photograph showing the compaction die setup and the compaction operation performed with the help of UTM is shown in Figs. 3 and 4 correspondingly.

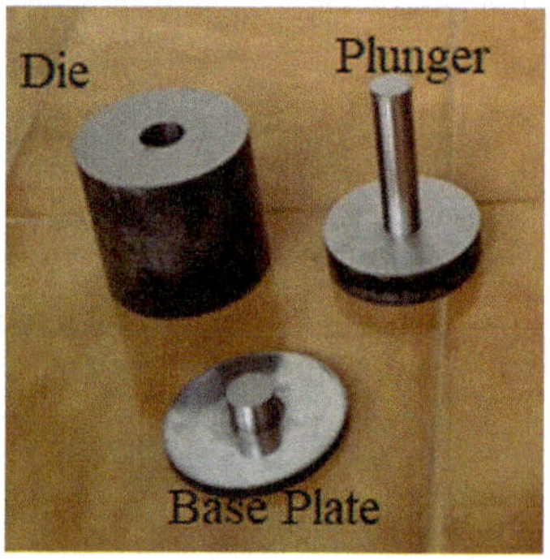

Fig. 3 Compaction die

Fig. 4 Compaction die setup kept on UTM setup

Fig. 5 Compacted specimen

Prior to compression, the plunger, compaction die and bottom plate were cleaned with cotton. To minimize the friction between the mating surfaces of the die, plunger, and bottom plate, acetone, and zinc stearate solution were applied. The matrix and reinforcement blended mixture was transferred into the die cavity which was placed on the base plate. After that, the plunger was inserted into the die cavity. Then the die setup was placed on the UTM table. Loads at the rate of 10 kN/min were applied using the UTM. The base plate was removed and the die setup was placed on the wooden base block having the hole to aid in removing the compacted billet from the die.

Pure Al powder size of 200 meshes was compacted in the die setup, the compacted specimen used in the density test is shown in Fig. 5.

Compaction of the composite was achieved using a universal tensile testing machine by compressing the Al/MWCNT composite according to the procedure detailed in ASTM B-925 03 standard. During compaction, the load was increased from 60 to 160 kN in steps of 20 kN. The density of the green compact was considered after every step of loading.

Figure 6 represents the deviation in density with applied load. A monotonous increase in density is observed with increasing compressive load. A higher rate of increase in the density of the composite was observed till 120 kN. A relatively negligible increase in the density of the prepared composite was witnessed beyond the load applied of 120 kN. Hence, 120 kN was considered as the optimum load for the fabrication of green compacts.

2.1.4 Sintering

Sintering of the compacted test samples were carried out at Bhatt Metals, Bangalore, Karnataka, India. The green compacts thus acquired were sintered at a particular temperature for 1 h. Under this process, bonding took place between the porous aggregate particles, and on cooling, the powder particles bound to form a solid billet.

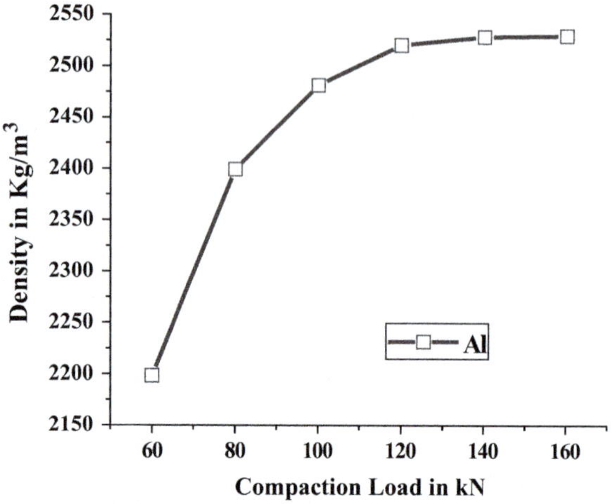

Fig. 6 Compactionload versus density for Al material

Fig. 7 Sintering process **a** samples kept inside the furnace and **b** outer view of the furnace

The sintering process was carried out under vacuum conditions to avoid oxidation. Figure 7 shows the sintering apparatus used with samples placed in the furnace.

A pressure of 10^{-2} bar was maintained inside the furnace while heating and cooling. The sintering temperature of 580 °C (0.8 times of melting temperature of the matrix) was used Al nanocomposite. The sintered specimens were allowed to chill to room temperature in a nitrogen atmosphere. No evidence of melting was noted in the sintered components. Minute microscopic cracks were present on the specimens containing higher wt% of MWCNT.

2.1.5 Extrusion Process

Extrusion is a secondary process in which component diameter is reduced and its length elongated. The extrusion procedure was done with the help of a hydraulic

Fig. 8 Hot extrusion setup

press. During extrusion, the metal was squeezed through a closed cavity known as a die. Figure 8 shows the extrusion setup, it consists of an extrusion die, plunger, 10 mm diametric circular eyepiece, dummy piece, and a heating coil.

At the onset of the extrusion process, the die and plunger were lubricated by applying a zinc stearate-acetone solution. The eyepiece was inserted at the bottom of the die to obtain the required size of the specimen. A sintered sample was placed in the die cavity. The dummy bar was placed between the plunger and the sample. The heating coil was arranged around the die to maintain the essential temperature required for smooth extrusion. The sample was heated up to a recrystallization temperature of around 600 °C. Hydraulic pressure of 21 N/mm^2 was applied through the plunger for extruding the specimen extrusion at an extrusion rate of 1.2 mm/s.

2.1.6 Characterization

To assess the chemical stability of the MWCNT during the sintering operation, an analysis of X-ray diffraction patterns was undertaken. Figure 9 shows the X-ray diffractometer (JEOL-JDX-8P) plots for the produced nanocomposite materials. The diffraction pattern is a graph of intensity versus diffraction angle. There is a marked similarity between the diffraction pattern obtained for the aluminium powder and those of the nanocomposites. This clearly shows that there was no chemical reaction between MWCNT and aluminium matrix after sintering and hot extrusion. Had there been a shift in the peaks with respect to the diffraction angle and or a change in the intensity of the peaks between the patterns for the composites it would indicate a change in the phase which would alter the properties significantly.

Fig. 9 XRD pattern for Al/ MWCNTnanocomposites

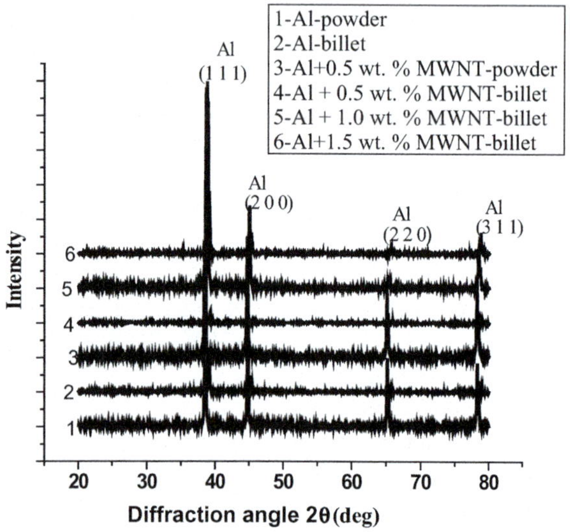

3 Result and Discussion

3.1 Density and Tensile Measurement

The density of aluminum is 2.70 g/cm^3. The density of multi-walled carbon nanotubes (MWCNTs) is about 1.35 g/cm^3. The density of MWCNT is half of that of Al. This implies that as the fraction of the MWCNT is increased in the nanocomposites the density of the nanocomposites will be decreased. This fact is also clear from the plot of density versus weight concentration of MWCNT (Fig. 10).

Fig. 10 Density versus Al/ MWCNT composite

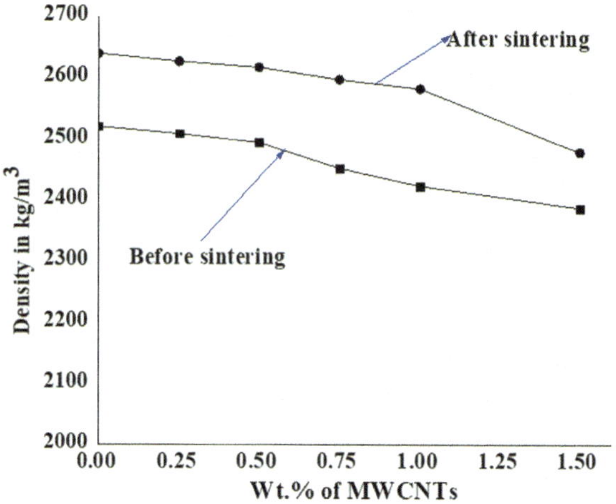

The density of the nanocomposite is further exaggerated by the degrees of compression that it has been subjected to different loads.

The density of the nanocomposite is not significantly influenced by the sintering temperature. The graph shown above was obtained for sintering at a temperature of 580 °C under an inert nitrogen atmosphere (10^{-2} bar pressure). Approximately a 5% higher density is recorded for the Al and its composites on sintering. This is less than the density changes observed when the compression load applied was raised from zero to 120 kN. However, the extent of bonding between reinforcements and the matrix is affected by the sintering temperature. In all the cases the plot clearly shows that the post-sintered density was found to be higher compared to the pre-sintered one.

Tensile strength is an important consideration in developing materials for engineering applications. A post-sintering step involving extrusion was implemented prior to testing the ultimate strength of the developed composites. The extrusion operation was conducted at a temperature of 600 °C. During the process of extrusion, a constant pressure of 21 N/mm^2 was maintained. To conduct the uniaxial tensile test the extruded composites were machined to standard dimensions in accordance with the ASTM E -8 standard. The modulus of elasticity of the Al and nanocomposites of Al was measured (Fig. 11). Young's modulus was found to increase slightly when the MWCNT concentration of 0.25 wt% was added to Al. Further improvement in the modulus (by approximately 3.5%) resulted when the concentration of the MWCNTs was increased by 50–0.5 wt%. At this concentration, the modulus of elasticity was at its highest (63.8 GPa). On increasing the concentration of the MWCNTs reinforcements to 0.75 and 1.0 wt% the modulus was observed to decrease implying deterioration of bond strength between the reinforcement (MWCNT) and the Al matrix phase. At these two concentrations, the modulus was lower than that obtained from the bare aluminium matrix material used. Such a behavior of the modulus with reinforcement concentration is unseen and is distinct from the behavior exhibited by conventional composite materials. A plausible reason for the distinct behaviour in composites could be due to the clustering and entanglement of the MWCNT during processing.

The SEM micrographs of the fractured surfaces are exhibited in Fig. 12. The dimple-like structures were fashioned on the Al material as shown in Fig. 12a indicating ductile failure. Al + 0.5 wt% MWCNTs composite deformed more when compared to the pure matrix material. The MWCNTs bonded strongly with the matrix and shared the applied force in the cases of composite. The MWCNTs act as obstacles to dislocation motion and delayed the early fracture in composite material. Figure 12b shows the MWCNTs present on the fractured surface. Clusters and pulled-out MWCNTs in the Al—1.0 wt% MWCNTare observed (Fig. 12c). Agglomerated MWCNTs and their non-uniform distribution were the reasons for the easy slip in the material.

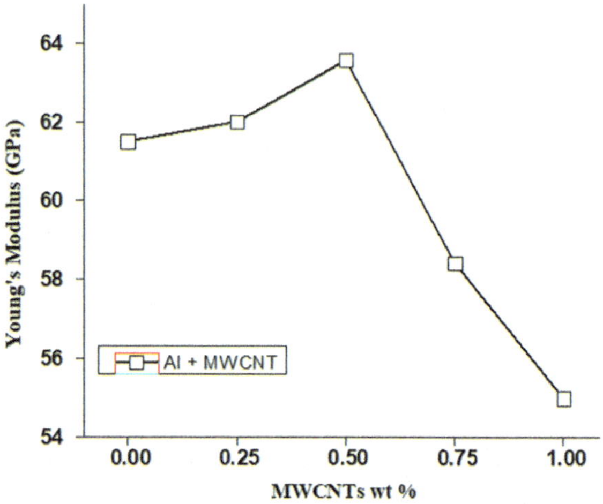

Fig. 11 Young's modulus versus Al/ MWCNT composite

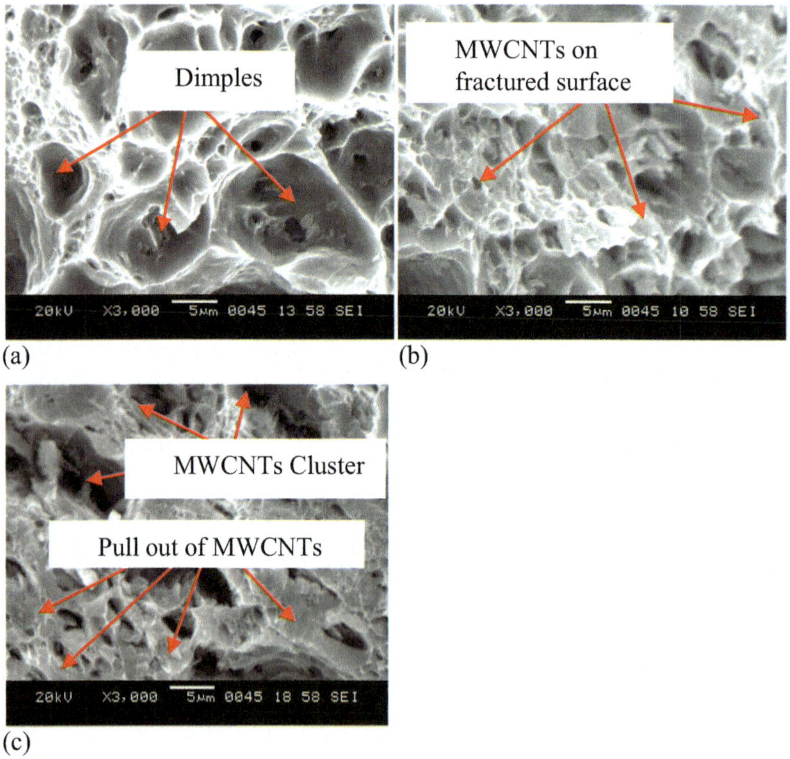

Fig. 12 SEM of tested sample surfaces of **a** Al, **b** Al + 0.5 wt%, and **c** Al + 1.0 wt% MWCNTs composites

3.2 Wear Investigation

Experiments to study the wear behaviour were conducted as per ASTM G 99-95a standard. Debris material from the surface was removed with acetone. Specimen weight was considered prior to and after each trial using an electronic weighing machine. The variables such as MWCNT composition, applied load, rotation speed, and distances were selected at 3 levels (Table 3). For 3 levels and 4 factors, L9 Taguchi's design was chosen for the dry sliding wear investigation.

The design of the experiment (DOE) and analysis of variance (ANOVA) was conducted using the Minitab software. The relationship between the independent factors and their effect on wear was discovered. The analysis findings were converted into a signal-to-noise (S/N) ratio. The lower the rate of the S/N ratio smaller would be its influence on wear. The S/ N ratio was determined using the following Eq. (1).

$$\text{S/N ratio} = -10\text{Log}10\left[\frac{1}{n}\sum y2\right] \quad (1)$$

where "y" refers to the response value and "n" is the number of observations.

Table 4 indicates Taguchi's design of the experiments (DOE). The wear investigation was investigated as per DOE and the rate of wear (mm^3/m) was calculated by considering volume loss. The S to N ratio was calculated using Eq. (1). The smaller S/N value is the better one for wear resistance.

Table 3 Process variables and their levels

Level	MWCNT composition (wt%)	Load (N)	Sliding speed (rpm)	Sliding distance (m)
1	0	10	250	500
2	0.5	15	500	1000
3	1.0	20	750	1500

Table 4 L9 Taguchi's design and wear rate

Sl. No.	Material	Load (N)	Speed (rpm)	Distance (m)	Wear rate × 10^{-3}
1.	0.0	10.0	250	500	1.443
2.	0.0	15.0	500	1000	2.660
3.	0.0	20.0	750	1500	150.220
4.	0.5	10.0	500	1500	1.320
5.	0.5	15.0	750	500	1.400
6.	0.5	20.0	250	1000	3.270
7.	1.0	10.0	750	1000	1.982
8.	1.0	15.0	250	1500	5.030
9.	1.0	20.0	500	500	9.729

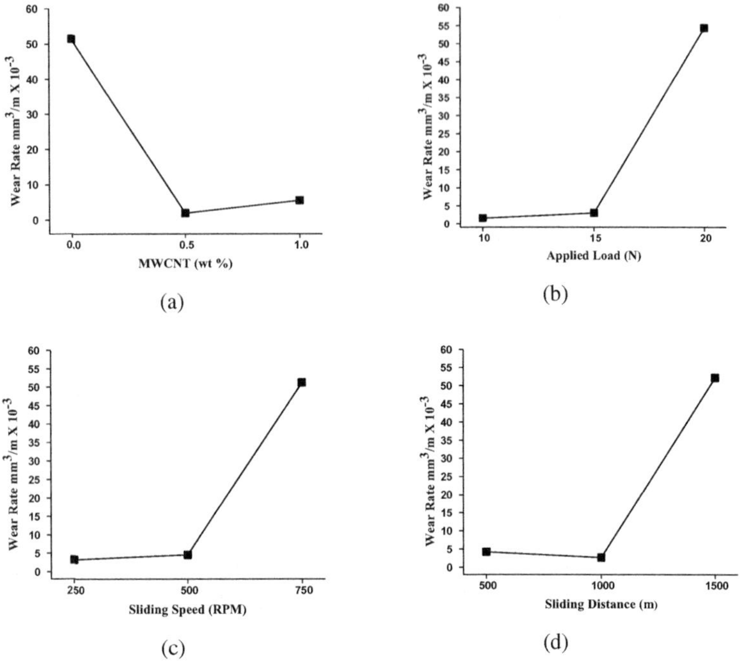

Fig. 13 Main effect plot for means of wear rate versus **a** MWCNT wt%, **b** applied loads, **c** sliding speeds, and **d** sliding distances

The main effect plots for the mean of the rate of wear are shown in Fig. 13. The higher rate of wear was recorded for the pure Al compared to MWCNT/Al composite in Fig. 13a. The maximum resistance to wear was noted at 0.5 wt% MWCNT/Al composite. A large difference concerning the rate of wear was noticed between Al and Al/MWCNT composite. The rate of wear was higher at increased load and speed (Fig. 13b, c) [2, 22].

Gradual wear was witnessed when the load increased from 10 to 15 N at speeds ranging from 250 to 500 rpm. Further, the highest wear occurred at 20 N and 750 rpm due to plastic deformation [23]. The wear behavior against sliding distance is shown in Fig. 13d. Wear was found to gradually reduce as the distance traversed increased from 500 to 1000 m. However, an increase in wear occurred when the sliding distance was increased to 1500 m. MWCNT in the Al matrix enhances material properties. MWCNT characteristics such as higher strength, higher thermal conductivity, and solid film self-lubrication provide resistance to wear. The lubrication nature of the MWCNT minimizes friction generation and dissipates the contact surface temperature [24, 25]. So, the wear is decreased in the uniformly dispersed MWCNT/Al composite.

The SN ratio plot (Fig. 14), indicates the independent factor's optimum values to enhance wear resistances. The 0.5 wt% MWCNTs reinforcement had the best S/N ratio for 10 N, 250 rpm, and 1000 m of sliding distance. Higher slopes were obtained on an increase in the reinforcement concentration from 0.0 to 0.5 wt%, between the applied load of 15–20 N, at speeds from 500 to 750 rpm, and distances ranging from 1000 to 1500 m [26]. The slope is indicative of the amount of impact on the material loss.

The outcome of the S/N ratio (the smaller the better) is shown in Table 5. The number of delta values for variables that influence the impact on the rate of wear and ranks was represented based on parameters causing more wear in ascending order. The load was an important parameter for wear followed by material, distance, and speed.

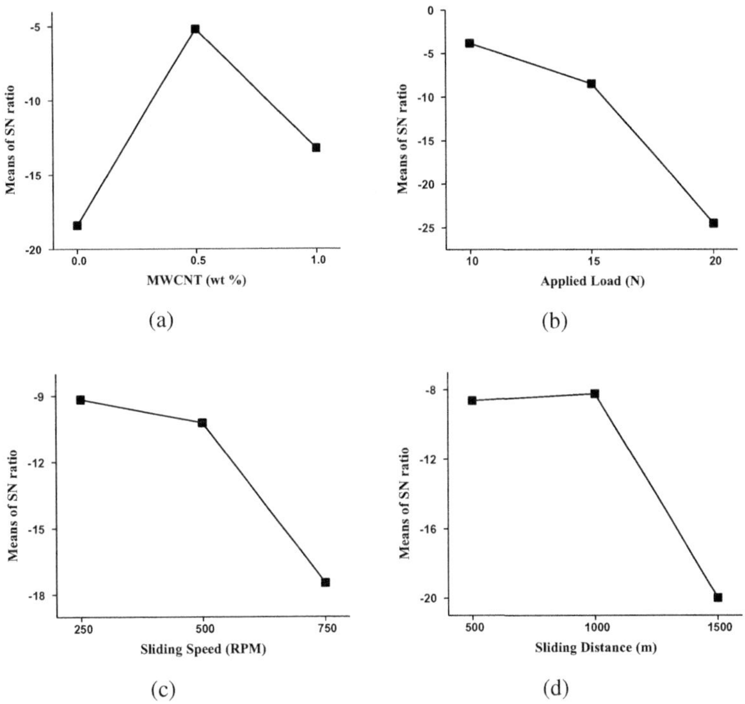

Fig. 14 Means of SN ratio versus **a** MWCNT wt%, **b** applied loads, **c** sliding speeds, and **d** sliding distances

Table 5 Response table for signal to noise ratio

Level	Material	Load (N)	Speed (rpm)	Distance (m)
1	− 18.406	− 3.846	− 9.169	− 8.623
2	− 5.208	− 8.484	− 10.223	− 8.244
3	− 13.245	− 24.529	− 17.467	− 19.992
Delta	13.197	20.683	8.298	11.749
Rank	2	1	4	3

4 Conclusion

The experimental analysis of Al/MWCNT nanocomposites leads to the following conclusions:

- The optimum compaction load was found to be 120 kN for the development of Al/MWCNT nanocomposites through the PM process.
- The density of the nanocomposite was lower compared to the aluminium material.
- XRD analysis showed the chemical stability of MWCNT in the composites.
- Al + 0.5 wt% MWCNT nanocomposite showed the highest tensile strength and wear performance owing to the excellent reinforcement behavior.
- DOE analysis revealed that wear depends strongly on the factors in the hierarchy of load, reinforcement, distance, and disk speed.

References

1. Khatkar SK, Verma R, Kant S, Suri NM (2020) Optimization of wear behavior of AZ91D-SiC-Gr hybrid composites using Taguchi experimental design. In: Soft computing: theories and applications. Springer, Singapore, pp 1151–1163
2. Radhika N (2017) Fabrication of LM25/SiO_2 metal matrix composite and optimization of wear process parameters using design of experiment. Tribol Ind 39(1)
3. Prasad KNP, Ramachandra M (2013) Evaluation of factors affecting sliding wear behaviour of Al-flyash metal matrix composites by using design of experiments. Int J Mod Eng Res 3(5)
4. Mishra AK, Sheokand R, Srivastava RK (2012) Tribological behaviour of Al-6061/SiC metal matrix composite by Taguchi's techniques. Int J Sci Res Publ 2:1–8
5. Singla M, Singh L, Chawla V (2009) Study of wear properties of Al-SiC composites. J Miner Mater Charact Eng 08(10):813–821. https://doi.org/10.4236/jmmce.2009.810070
6. Ghosh S, Sahoo P, Sutradhar G (2012) Wear behaviour of Al-SiCp metal matrix composites and optimization using Taguchi method and grey relational analysis. J Miner Mater Charact Eng 11(11):1085–1094. https://doi.org/10.4236/jmmce.2012.1111115
7. Radhika N, Subramanian R, Prasat SV (2011) Tribological behaviour of aluminium/alumina/graphite hybrid metal matrix composite using Taguchi's techniques. J Miner Mater Charact Eng 10(05):427–443. https://doi.org/10.4236/jmmce.2011.105032
8. Kundu S, Roy DB, Mishra AK (2013) Study of dry sliding wear behavior of aluminium/SiC/Al2O3/graphite hybrid metal matrix composite using Taguchi technique. Int J Sci Res Publ (3)
9. Malaki M, Xu W, Kasar A, Menezes P, Dieringa H, Varma R, Gupta M (2019) Advanced metal matrix nanocomposites. Metals 9(3):330. https://doi.org/10.3390/met9030330
10. Srivastava A (2017) Metal matrix nanocomposites (MMCs): a review of their physical and mechanical properties. Int J Nanotechnol Med Eng 2(8):152–154. https://doi.org/10.25141/2474-8811-2017-8.0152
11. Dadkhah M, Saboori A, Fino P (2019) An overview of the recent developments in metal matrix nanocomposites reinforced by graphene. Materials 12(17):2823. https://doi.org/10.3390/ma12172823
12. Veličković S, Garić S, Stojanović B, Vencl A (2016) Tribological properties of aluminium matrix nanocomposites. Appl Eng Lett 1(3):72–79
13. Muley AV, Aravindan S, Singh IP (2015) Nano and hybrid aluminum based metal matrix composites: an overview. Manuf Rev 2:15. https://doi.org/10.1051/mfreview/2015018

14. Choi SM, Awaji H (2005) Nanocomposites—a new material design concept. Sci Technol Adv Mater 6(1):2–10
15. Casati R, Vedani M (2014) Metal matrix composites reinforced by nano-particles—a review. Metals 4(1):65–83
16. Ramnath BV, Parswajinan C, Elanchezhian C, Pragadeesh SV, Ramkishore PR, Sabarish V (2014) A review on CNT reinforced aluminium and magnesium matrix composites. Appl Mech Mater 591:120–123
17. Kim I-Y, Lee J-H, Lee G-S, Baik S-H, Kim Y-J, Lee Y-Z (2009) Friction and wear characteristics of the carbon nanotube–aluminum composites with different manufacturing conditions. Wear Int J Sci Technol Friction Lubr Wear 267(1–4):593–598. https://doi.org/10.1016/j.wear.2008.12.096
18. Noguchi T, Magario A, Fukazawa S, Shimizu S, Beppu J, Seki M (2004) Carbon nanotube/aluminium composites with uniform dispersion. Mater Trans 45(2):602–604. https://doi.org/10.2320/matertrans.45.602
19. Liao JZ, Tan MJ, Ramanujan RV, Shukla S (2011) Carbon nanotube evolution in aluminum matrix during composite fabrication process. Mater Sci Forum 690:294–297. https://doi.org/10.4028/www.scientific.net/msf.690.294
20. Bakshi SR, Lahiri D, Agarwal A (2010) Carbon nanotube reinforced metal matrix composites—a review. Int Mater Rev 55(1):41–64
21. Ali MH, Rubel RI (2020) A comparative review of Mg/CNTs and Al/CNTs composite to explore the prospect of bimetallic Mg-Al/CNTs composites. AIMS Mater Sci 7(3):217–243
22. Arif S, Jamil B, Shaikh MBN, Aziz T, Ansari AH, Khan M (2020) Characterization of surface morphology, wear performance and modelling of graphite reinforced aluminium hybrid composites. Eng Sci Technol Int J 23(3):674–690
23. Deuis RL, Subramanian C, Yellup JM (1997) Dry sliding wear of aluminium composites—a review. Compos Sci Technol 57(4):415–435
24. Shivaramu HT, Nayak V, Umashankar KS (2020) Dry sliding wear characteristics of multi-walled carbon nanotubes reinforced Al-Si (LM6) alloy nanocomposites produced by powder metallurgy technique. Mater Res Express 7(4):045001
25. Zhang L, He XB, Qu XH, Duan BH, Lu X, Qin ML (2008) Dry sliding wear properties of high volume fraction SiCp/Cu composites produced by pressureless infiltration. Wear 265(11–12):1848–1856
26. Shivaramu HT, Nayak V, Umashankar KS (2022) Influence of process variables on wear testing of aluminium-silicon/multi-walled car-bon nanotubes nanocomposites. Eng Res Express 4(4)

Machining Challenges of Ceramic Matrix Composites

A Review on Conventional Machining Challenges of Ceramic Matrix Composites

Samuel Dayanand and Satish Babu Boppana

Abstract Ceramic matrix composites (CMC's) are presently an undeniably well-known selection of materials castoff to produce basic parts for an assortment of designing ventures. CMC's are presently a rising material decision for a few high worth parts, that has as of late started the need of understanding the impact of few machining methods. Because of the intricate idea of CMC's—for example heterogeneous construction, machining become very testing as the cycle can relent higher mechanical and thermal loads. The heterogeneous design of CMC's prompts us to think about the complex machining, that may eventually lead to extraordinary surface deformities. An overview pertaining to exploration carried out in the traditional and non-traditional machining of CMC's on basically assessing what various machining strategies mean for the machined surfaces is thoroughly discussed. Because of the upgraded properties, these materials have higher potential for use in novel and high-performance applications. This is accomplished by investigating the various material portrayal methods as of now, used to notice and measure the mechanical and subsurface defects.

Keywords CMC · Ceramic · Machining · Composites

1 Introduction

The ability of composite structures to improve the strength to weight ratio while being contrasted to non-strengthen materials has piqued the interest of several industrial approaches. Composites are often classified based on the material forming the

S. Dayanand (✉)
Department of Mechanical Engineering, Government Engineering College, Gangavathi 583227, India
e-mail: samueldayanand@gmail.com

S. B. Boppana
Department of Mechanical Engineering, School of Engineering, Presidency University, Bengaluru 560064, India
e-mail: satishbabu3@yahoo.co.in

S. B. Boppana et al. (eds.), *Structural Composite Materials*, Composites Science and Technology, https://doi.org/10.1007/978-981-99-5982-2_15

matrices [1]. Each of these composites is important in today's technology because of its super physical, thermal and mechanical properties. MMC's are usually flexible and provide excellent high flexibility, strength, dimensional stability and toughness, with aviation components being the most common application [2–4]. Ceramic Matrix Composites (CMC's) have come out as viable options out of to their magnificent physical, chemical and mechanical properties. CMC's are a new and favorable material for high-tech engineering ventures. Because of superior properties, for example low thermal conductivity, heat resistance, high strength, corrosion resistance, low density and wear resistance, they can perform under tough conditions. When collated to conventional and ceramic materials, these properties come up with a remarkable increase in service life [5, 6]. CMC's were heterogeneous materials in which the second phase was distributed throughout the ceramic base matrix. Because of their nature as a reinforcement material, CMC's have characteristics such as hardness, self-healing and functionality [7]. CMC's are classified based on the ceramic matrix's nature: oxide like SiO_2, Al_2O_3, or non-oxide like C, BN, SiC, AlB_2, TiB_2, and ZrB_2 [8–15]. Because of their mechanical robustness at higher temperatures, non-oxide CMC's are presently the most commonly used materials; particularly SiC reinforced based CMC's, which are a popular material option for high end accomplishment such as aeronautics engines [9, 10]. This review focuses on the most commonly on non-oxide CMC's, which are C/C, C/SiC, and SiC/ZrB_2 [11, 12]. Because of their enhanced oxidation and lowering resistance in the combustion surroundings, oxide-oxide CMCs are used and are covered in this review, despite the fact that there is a limited amount of literature available at the moment. Advanced ceramics, includes silicon nitride (Si_3N_4), have the unique properties of maintaining high hardness at elevated temperatures encountered in machining/grinding, and also good corrosion and wear resistance and chemical stability. These materials are useful in applications involving abrasion, high temperatures, and maintaining hardness at high temperatures [13, 14]. Biocompatible implants such as dental prostheses, refractory and wear parts such as valve seats and guides, precision bearings in the nuclear power industry, and automobile components such as insulators, sensors, pistons, and valve linings are examples of typical applications [15]. To defeat the issues while requesting necessities for materials ventures in the present years in different crucial ventures running between engineering, energy and transportation, there will be an impending need to make special chain and more unbending primary materials.

In any case, machining of ceramic production, keeping up with wanted surface completion is a moving errand because of their extremely high strength and low break sturdiness. Utilization of traditional coolants is likewise impractical because of their low thermal conductivity and shock made by use of low temperature coolant at a restricted region. Silicon carbide (SiC) fibers have been used broadly to build up the greater part of CMC's because of their higher flexible modulus and high strength. The expansion of stubbles with the shorter fiber CMC's works on its hardness and break opposition. The long fibers have high sturdiness that keeps load in any event, when the ceramic matrix breaks and decreases the crack. The short fibers and bristles are utilized to work on break opposition [16, 17]. The constant monofilament fibers in the scattered stage are produced utilizing synthetic fume with SiC in the underlayer which is made out of element tungsten in addition to carbon(C) fibers, conveying the most essential reinforcing. The fortifications work on the durability and, now and again, make the material thermally and electrically conductive. Because of the better sturdiness and related to conductivity, the impeding impact of coolant causing surface and subsurface breaks is hindered. Notwithstanding the benefits referenced above for the high-level and artistic lattice composites, there are constraints of these materials as well. The super regrettable part of these materials is that they are truly challenging to machine owing to their higher hardness and strength, which confines their modern double-dealing [9–11, 18–20]. The main choices for machining/grinding these materials are by a CBN (Cubic Boron Nitride) or a precious stone grating grinding wheel. Since these materials are utilized in high innovation applications, the parts made from them ordinarily require great layered control and surface completion. Thus, in practically all applications, some completing activity is required which is normally the grinding [18, 19]. The significant issues engaged with grinding of ceramic production and ceramic framework composites (CMC) are high extraneous and typical trimming powers, grinding temperature, surface unpleasantness, surface breaks and subsurface harms. A blend of ideal upsides of these reaction boundaries are looked for while keeping a decent removal rate of material. To keep away from high device wear, abrasives like CBN and precious stones are utilized [20, 21].

Regarding the actual machining of composites, a lot of exploration has been drained and recalled in a few surveys [20, 21]. Most of the ceramic materials are made by the sintering process. CMC's parts are processed to rigid resilience for aspects and nature of machined composite surfaces. In any case, in the wake of sintering, the parts go through extensive shrinkage, that makes ensuing machining undeniable. The actual hardness of the sintered processed CMC's essentially restricts material to be removed by various machining. The grinding operation with required instruments and ensuing it getting done while cleaning is the most often involved strategies for accomplishing layered exactness. The expense of machining interaction can arrive at 60–80% and once in a while even up to can reach 90% of the all-out creation cost [5–8, 22, 23]. It leads that in spite of the multitude of previously disclosed benefits, the utilization of CMC's is hindered by the significant expense of their respective machining. Increasingly newer CMC's with special possessions are being fostered consistently. A significant number of these newer materials are undeniably challenging to process with conventional innovations, and that implies working on the current cycle (or making new strategies) for handling is vital and critical. Subsequently, as the field of long fiber built up CMC's has not been exhaustively considered and due to the new development sought after, the material for higher temperature primary applications (for example aviation, car and atomic businesses) is in demand. It is accepted that the ongoing comprehension in machining CMC's is of vital logical importance [5–8, 10–17, 19, 22, 24, 25]. Moreover, the impact that various machining methods prompt to the surface respectability and the comparing assessments methods expected to portray these are not talked about either in the previously mentioned literature. Consequently, as the machined surface is most certainly an essential viewpoint in underlying parts, for example, the ones that CMC's are be intended for, this survey targets gathering and basically examining the various outcomes accomplished such a long way regarding the surface uprightness. A few literatures have concentrated on how the handling temperature and CTE influences the lingering stresses in the composite by utilizing Raman Spectroscopy equipment [5–8, 10–19, 22, 26]. Contingent upon CTE alongside working temperature, elevated degrees of decrease in residual stresses were noted in the composite's morphology which might influence and likewise get impacted by the way in which machining is done. CMC's are thought of as hard to be machined because of their exceptional nature. Thus, fully intent on giving an actual clarification of reason they are viewed as the hardest to machined composites, the primary material associated properties influence the machining system [6–17, 19, 22, 27, 28].

- Hardness (Hv): This tends to be connected with the nearby obstruction that an instrument finds the indent. Thus, harder material build have raised mechanical properties which bring about a hindering impact to the machined surface. In CMC's, extensively hard materials (contrasted with various composites) are usually utilized as a support (for example SiC strands, Al_2O_3) [5–9, 13, 19, 22, 29].
- Crack durability: Addresses how much energy is expected to break a material. Subsequently, lower upsides of crack sturdiness will be normal for materials tending to experience a weak expulsion system. Notwithstanding, to look at the meaning of these qualities in a trimming situation amid various materials, the hardness and break strength ought to be utilized together. Subsequently, the proportion Hv/K_{IC} for enormous qualities mean a more prominent propensity of happening fragile break. In this way, it tends to be seen that composites having reinforcements like SiC ought to be thought of as very challenging to get the materials machined that might yield high powers of metal removal (brought about by high hardness) and will be more inclined to experience weak break during the trimming system (brought about by higher upsides of Hv/K_{IC} proportion [19, 22, 30–32].

Consequently, it very well may be inferred that among every one of the composites, CMC's carrying SiC particulate or potentially aluminium oxide will show the most noteworthy machining powers and could often experience a fragile evacuation system. This mix is basic for the surface age, particularly while thinking about aviation and atomic applications, where little breaks can prompt a horrendous disappointment. In this process not just mechanical and customary machining cycles will be thought about yet in addition non-ordinary ones, for example, laser removal, grating waterjet and electrical release machining need to be considered [19, 22, 33–35].

2 Significance of the Surface Finish

CMC's are intended to oppose unforgiving circumstances like thermal loads and cyclic mechanical in extreme ventures (for example in aviation and automobile sector). Consequently, the surface and subsurface harms made amid the machining methods should be described to assess the part's life. Moreover, the progressive construction of nano oxide CMCs (particularly SiC based CMC's) is intended to likewise oppose in debasing conditions (for example O_2, H_2O fume, hydrocarbon or debasements like Na+) [13, 32, 36, 37]. In case safeguarding surface layers fizzle, the interior construction can experience a sped-up process prompting a weak and disastrous disappointment of the part. Subsequently, a cautious comprehension of the impact of the machining system into the outer layer of CMCs is of basic significance for the modern local area [19, 22, 23, 38].

2.1 *Surface Integrity*

The trustworthiness of the material surface subjected to machining depends on the proportion of the impact that has happened during the removing of material [18–20]. Surface uprightness has been a significant examination field while machining elite execution metals, for example, Titanium (Ti) and Nickel (Ni) based compounds. By and large, three fundamental viewpoints are considered while examining surface trustworthiness [19, 22, 23, 39].

2.2 *Surface Roughness*

The surface roughness incorporates the portrayal of surface structure like unpleasantness and waviness. Normal qualities for the whole surface (for example normal harshness, Ra) are utilized for statistical correlations as immediate correlation with the weakness execution have been accounted for [19, 22, 23, 40].

2.3 *Surface Metallurgy*

It incorporates the modifications on the surface alongside its underlayer surface. By and large, electron magnifying instrument procedures (for example TEM and SEM) combined with electron beam dispersed diffraction and XRD are utilized to notice the morphology subjected to metal removal while evaluating the strains initiated at underlayer surface. While considering materials involving harder metallic combinations, a machining-prompted coagulation surface is formed in such ceramic composites. The material density is usually utilized as a quality assurance boundary [19, 22, 23, 41, 42].

2.4 Mechanical Attributes

Changes in the hardness pertaining to material because of the machining actuated miss happening are measured commonly by XRD and SEM. Ordinarily, a profile inside and out of these two stuffs is given to grasp the subsurface mechanical modifications [19, 22, 23, 43].

Because of the unique and one of a kind sorts of CMCs, the material through which it is removed is totally unique to purely metallic materials and thus its surface trustworthiness. In addition, the reality of managing varying ceramic particles with more modest grain sizes, formless parts and a few connection points, obliges to investigate new techniques for portrayal, as the ordinary ones (for example XRD) probably won't be suitable for estimations [19, 22, 41–43].

3 Traditional Machining Process

Since CMC's are usually utilized for getting complex parts (for example air motor vanes), regular machining is tried to change the shape of the material. In this part, the various machining methods which depend on using a regular mechanical method of material removal process with a device need to be examined. To begin with, the basics are introduced by taking a gander at the work done through removal of material and scratch surface tests. Subsequently, the cycles of boring and grinding processing are made sense of in discrete segments [19, 22, 44, 45].

3.1 Material Removal in Orthogonal Cut and Scratch Surface Tests in Machining of CMC's

In machining (like processing, boring, turning) utilizing state of the art devices, ortho cutting is the most usually utilized essential trial to concentrate on the removal of material. To investigate the removal of material, authors [27, 44] led few tests on three normal fiber directions of carbide made composites. The material that was removed was altogether unique through various fiber directions and in various cutting profundities. While the cutting force is little, the cut is overwhelmed by the shear conduct of composites while the action leads through lengthy strands and the shear conduct of matrix during trimming through filaments and crossover strands, like the pliable expulsion. The slicing thickness increments to a specific worth, the cutting was overwhelmed by the inception and spread of the fiber matrix orientation broke during the machining through longitudinal strands and across filaments and the bowing incited crack of strands in the process of trimming through cross over strands. On this premise, a logical machining model in view of the inelastic way of behaving and crack mechanics of anisotropic and orthotropic, weak composites has been incorporated, which is useful to grasp about the surface roughness during machining operations [19, 22, 34, 35, 45–48].

Scratch tests are usually utilized as a worked-on situation fully intent on catching the essential peculiarities in the removal of material process happening in machining. Ortho based cutting is usually absorbed to machining processes with characterized forefronts (for example boring, processing), while scratch tests will generally be read up for the comprehension of the impact of a solitary molecule in a rough based cutting cycle. A few edges of the high-speed imaging framework got fully intent on understanding the various machining system that orthotropic and weak materials, like CMC's, can introduce. The material harm that shows up for huge whole chip thickness that was driven by a shaky break arrangement, was found to in line with the interface of fiber-matrix as it is the course with a lower crack strength. By the by, for little whole chip thickness, the translaminar break strength for cross over strands and the inelastic shear conduct for equal filaments were viewed as the ruling element in the machining system. Authors [49, 50] reported under dry grinding operation and scratch surface tests at various fiber directions and various angles from (0°– 30°– 45° and 90°) which it was observed that the scratched surface for 30° and 45° was more seriously harmed than for 0°. Owing to strands situated at various 30°, 45° and 90°, the filaments were not totally taken out because of the huge twisting during the development of the matrix. Regardless of whether the support of CMC's is mostly founded on lengthy strands, the investigation of the impact of ceramic matrixes built up with ceramics particulates is of higher importance as some higher temperature CMC's are molecule built up. Also, in a few long strands CMC's, the actual matrix can be a short particle ceramic [51]. Regardless of whether the primary goal of the work was not to grasp about the removal of the material but rather to fathom the effect opposition, likenesses in the outcomes can be gotten as tiny whole chip thickness are regularly utilized while machining hard CMCs. It was accounted for that microcracking is observed and microplasticity were ruling the scratch conduct prompting the end that both, pliable and weak peculiarities could seem while machining harder ones at little greater chip thickness. Generally, the particle nature prompted a grain break system in the silicon carbide molecule locales and slip planes in the particulate zirconium diboride regions. Moreover, the harshness of the machined surface was demonstrated to be profoundly subject to micro-cracking systems of the particulates. The impact in the leftover chips was assessed in a similar material by utilizing Raman Spectroscopy equipment in the silicon carbide particulates, showing an increment of pressure while rising the load during scratch tests [19, 22, 26, 33, 51, 52].

3.2 Drilling Operation

It tends to be gained from the machining that, while involving machining tasks for which the device turns through its hub (for example boring or processing), a mix of the various cutting systems brings about various impacts in the surface profile. Thus, it is vital to stress that normal or nonexclusive upsides of microstructural or mechanical hardness qualities probably won't be important while portraying the machined surface in the biaxial materials like CMC's [3, 19, 21, 23, 24, 26, 53].

The research in CMC's has mostly centered around the examination between regular boring and turning and ultrasonic machining utilizing rough devices; this isn't shocking exploration, since ordinary drills wear quickly in harder designs like CMC's [53].

Authors [54–57] likewise concentrated on what the harder heterogeneous nature of SiC based CMC's means for boring while machining little openings (somewhere in the range of 0.8 and 5 mm width) with standard drilling operation. It was accounted for that, due to this novel nature of particulate silicon carbide reinforced CMC's, untimely device breakage occurs while machining with tinier widths (for example 0, 8 mm width). This was logically and tentatively realized in view of a spiral power estimated during the boring preliminaries, not seeming while machining other hard materials, for example, solid SiC reinforced [3, 21, 23, 24, 26, 36, 54–57].

3.3 Grinding Process

3.3.1 Material Removal by Scratching Test

During machining like grinding operation, scratch tests are many times used to concentrate on the machining process. To concentrate on the grinding qualities of silicon carbide reinforced composites in various fiber directions, researchers [56–59] did a solitary rough scratch test through various things with a specific indenter on the grinding process and examined the grinding operation. Inside the surface scratch profundity range (10–50 μm), the weak expulsion joining the layered fragile break of carbon strands and the breaking of silicon carbide reinforced matrix was the process of material removal [19, 22, 60–65].

3.3.2 Material Removal Mechanisms

The material expulsion system in grinding can be extensively sorted in two kinds: (a) Indentation break mechanics (b) Machining approach. The main methodology depends on connections among rough and workpiece similar to distortions created by an indenter. In the subsequent methodology, shearing of the material in pliable mode is considered as is seen in machining tasks [19, 22, 46–49, 66, 67].

3.3.3 Problems in Grinding of Ceramics and Ceramic Matrix Composites

In sintered ceramic based materials, 90% expense of the item is caused in the phase of grinding. Thus, failure in grinding as lower material expulsion rate unfavorably influences the business feasibility of the thing.

Tool Wear

High hardness and high strength (ceramic matrix composite) materials are challenging to machine due to extreme wear. CBN wheels are utilized for grinding these materials [68].

Surface Roughness

Grinding with prevalent weak break of material causes surface and subsurface breaks and low quality of surface completion. Silicon Carbide (SiC) created by utilizing 3 μm size powder sintered at higher temperatures had a thickness of 3.17 gm/cm^3. Subsequent to grinding this material in various circumstances by changing various parameters, they reasoned that prevalently pliable material expulsion is vital for high part quality [19, 21, 22, 69].

Surface/Subsurface Damages

Grinding of ceramic matrix composites create surface and subsurface harms and twisted surface. Grinding with prevalently pliable disappointment will help in possibly lessening these deformities. It is feasible to further develop the reaction boundaries like grinding powers and surface harshness [19, 22, 46, 48, 49, 60–65].

Utilization of Coolants

Grinding liquids are utilized to cool the workpieces, carry out and moves the chips, grease up the grinding zone to lessen grating and clean the grinding wheel. In any case, involving coolant in grinding of nonconducting ceramic matrix composites reduces thermal shock because of cooling of the cutting region. Method of least amount oil (MQL) can be utilized to deal with the intensity created in grinding. The utilization of MQL worked on the nature of the sample workpiece and decreased the grinding process temperature [19, 22, 33, 64, 70, 71].

3.3.4 Variations in Input and Response Parameters

Test analysts shifted input conditions in grinding and noticed varieties. Input boundaries in grinding are cutting pace, table feed rate, profundity of cut, rough coarseness size or nonattendance of a trimming liquid. The reaction boundaries can be trimming powers, grinding temperature, surface harshness, surface and subsurface harms. Conditions of reaction boundaries on input boundaries were viewed as follows [34, 35, 72, 73]; expansion in fringe wheel speed caused decline in unrelated and typical grinding powers; surface harshness diminished with speed up; explicit grinding energy expanded with diminishing chip thickness which is called size impact. A speed up caused decrease of chip thickness which brought about decrease in grinding power and decrease in wheel wear. The surface completion was improved by utilization of coolant. The surface harshness diminished with table feed, coarseness size and coarseness thickness. The grinding power diminished with decline in table feed, coarseness size and coarseness thickness [19, 22, 34, 35, 56–65, 74].

3.4 Milling Process

Also, to what has been accounted for in boring, the majority of the examinations in processing of CMC's have been centered around the correlation among regular processing. Not many analysts dealt with force forecast models, while others assessed the surface wrap up by taking a gander at the harshness and the structure of the machined surface [75]. It was inferred that, for entrance profundities more prominent than 4 μm, the surface harshness unexpectedly expanded, introducing bigger number of scores and pits (because of a weak crack). Then again, for more modest upsides of entrance profundity, the Carbon strands would in general be more nonstop without huge broke regions. Comparable ends were drawn while processing carbon—carbon composites, where a somewhat better surface completion was gotten with regular processing. Proof of a pliable to-weak change during the machining was seen in the matrix, while a takeout system was ruling the fiber expulsion component [76].

To lessen the cycle incited powers and device wear and further improve the surface completion, examinations have been embraced to dissect the possibility of utilizing machining to relax the material in front of the device prior to cutting. In spite of the method, it has been demonstrated to be savvy, lessen the machining powers and device wear; little part of work has been tracked down in the writing regarding characterization of the surface profile subsequent to processing of CMC's [72, 75, 76].

4 Applications

CMC's are utilized in numerous high temperature processes. They have a high creep opposition, which empowers plans with huge mechanical loads. Hot gas valves are made from the same that are used to control the gas stream in gas terminated high temperature heaters. Contrasted with metallic valves, the life of the CMC's parts is significantly longer and over-repays their higher buying costs. Various uses of CMC's in high temperature processes are fire tubes, heat exchangers, defensive tiles, and various high temperature holders. The high wear opposition and the good rubbing properties of CMC's lead to applications like sliding contact course, brakes and grip plates [7, 10, 12, 15, 16, 19, 21, 35, 39, 77–100].

5 Conclusions

Because of the novel nature that these ceramic materials present (for example hard, diverse and orthotropic materials) they are viewed as truly challenging to-machine and thus a few regular and non-ordinary machining methods are considered. The uncommon characteristics of high hardness, strength and sturdiness (in the event of composites), great erosion opposition and low unambiguous weight make them the most impending material for future advances. In any case, the troubles experienced in machining these materials present serious difficulties in fostering these materials. The deformities created in grinding these materials, normally as the completing activity, are surface/subsurface breaks, plastic twisting of the surface. These deformities, in the event that not controlled at assembling stage, can deliver them ill-suited for the modern capabilities to be performed by them. The above challenges in grinding of ceramic matrix composites can be met by enhancement of grinding info boundaries, utilization of proper trimming liquids and strategies like MQL and cryogenic cooling. Unique methods can likewise be utilized, for example, pliable system grinding.

References

1. Kaw AK (1997) Mechanics of composite materials. CRC Press
2. Teti R (2002) Machining of composite materials. CIRP Ann Manuf Technol 51(2):611–634
3. Liu D, Tang Y, Cong WL (2012) A review of mechanical drilling for composite laminates. Compos Struct 94(4):1265–1279
4. Dandekar CR, Shin YC (2012) Modeling of machining of composite materials: a review. Elsevier
5. Rayat MS, Gill SS, Singh R, Sharma L (2017) Fabrication and machining of ceramic composites—a review on current scenario. Mater Manuf Process 32(13):1451–1474
6. Di Carlo J (2014) Advances in SiC/SiC composites for aero-propulsion. In: Ceramic matrix composites: materials, modelling and technology. Wiley, Hobooken, pp 217–235
7. Krenkel W, Georges Thébault J (2014) Ceramic matrix composites for friction applications. In: Ceramic matrix composites. Wiley, Hoboken, pp 647–671

8. Sauder C (2014) Ceramic matrix composites: nuclear applications. In: Ceramic matrix composites. Wiley, Hoboken, pp 609–646
9. Evans AG, Zok FW (1994) The physics and mechanics of fibre-reinforced brittle matrix composites. J Mater Sci 29(15):3857–3896
10. Heidenreich B (2014) C/SiC and C/C-SiC Composites in ceramic matrix composites. Wiley, Hoboken, pp 147–216
11. Spriet P (2014) CMC applications to gas turbines. Ceramic matrix composites. Wiley, Hoboken, pp 591–608
12. Hatta H, Weiss R, David P (2014) Carbon/carbons and their industrial applications in ceramic matrix composites: materials, modelling and technology. Wiley, Hobooken, pp 87–146
13. Keller K, Jefferson G, Kerans R (2014) Oxide-oxide composites in ceramic matrix composites: materials, modelling and technology. Wiley, Hobooken, pp 236–272
14. Kagawa Y, Guo S (2014) Ultrahigh temperature ceramic-based composites in ceramic matrix composites: materials, modelling and technology. Wiley, Hobooken, pp 273–292
15. Savino R, Criscuolo L, Di Martino GD, Mungiguerra S (2018) Aero-thermochemical characterization of ultra-high-temperature ceramics for aerospace applications. J Eur Ceram Soc 38(8):2937–2953
16. Naslain RR (2016) SiC-matrix composites: tough ceramics for thermo structural application in various fields. Eng Ceram Curr Status Futur Prospect 84:142–159
17. Morscher GN (2010) Stress-environmental effects on fiber-reinforced SiC-based composites in ceramic matrix composites: materials, modelling and technology. Bansal and Wiley
18. Jannotti P, Subhash G, Zheng J, Halls V (2017) Measurement of microscale residual stresses in multi-phase ceramic composites using Raman spectroscopy. Acta Mater 129:482–491
19. Liu Q, Huang S, He A (2019) Application requirements and challenges of CMC-SiC composites on aero-engine. J Mater Eng 47(2):1–10
20. Qinglong AN, Jie C, Weiwei M, Ming C (2021) Machining of SiC ceramic matrix composites: a review. Chin J Aeronaut 34(4):540–556
21. Liu, Q, Huang, G, Xu, X, Fang, C, Cui C (2018) Influence of grinding fiber angles on grinding of the 2D C/C—SiC composites. Ceram Int (Feb, 0–1)
22. Diaz OG, Luna GL, Liao Z (2019) The new challenges of machining ceramic matrix composites (CMCs): review of surface integrity. Int J Mach Tools Manuf 139:24–36
23. Thakur A, Gangopadhyay S (2016) State-of-the-art in surface integrity in machining of nickel-based super alloys. Int J Mach Tools Manuf 100:25–54
24. Ding K, Fu Y, Su H, Chen Y, Yu X, Ding G (2014) Experimental studies on drilling tool load and machining quality of C/SiC composites in rotary ultrasonic machining. J Mater Process Tech 214(12):2900–2907
25. Wang J, Feng P, Zheng J, Zhang J (2016) Improving hole exit quality in rotary ultrasonic machining of ceramic matrix composites using a compound step-taper drill. Ceram Int 42(12):13387–13394
26. Feng P, Wang J, Zhang J, Zheng J (2017) Drilling induced tearing defects in rotary ultrasonic machining of C/SiC composites. Ceram Int 43(1):30, 791–799
27. Diaz OG, Axinte D, Butler-Smith P, Novovic D (2018) On understanding the microstructure of SiC/SiC ceramic matrix composites (CMCs) after a material removal process. Mater Sci Eng A 713(26)
28. Tawakoli T, Azarhoushang B (2011) Intermittent grinding of ceramic matrix composites (CMCs) utilizing a developed segmented wheel. Int J Mach Tools Manuf 51(2):112–119
29. Ghosh D, Subhash G, Orlovskaya N (2008) Measurement of scratch-induced residual stress within SiC grains in ZrB_2-SiC composite using micro-Raman spectroscopy. Acta Mater 56(18):5345–5354
30. Li ZC, Jiao Y, Deines TW, Pei ZJ, Treadwell C (2005) Rotary ultrasonic machining of ceramic matrix composites: feasibility study and designed experiments. Int J Mach Tools Manuf 45(12–13):1402–1411
31. Du J (2018) New observations of the fiber orientations effect on machinability in grinding of C/SiC ceramic matrix composite. Ceram Int 44:31, 13916–13928

32. Marshall DB, Davis JB (2001) Ceramics for future power generation technology: fiber reinforced oxide composites. Curr Opin Solid State Mater Sci 5:29, 283–289
33. Azarhoushang B (2014) Wear of non-segmented and segmented diamond wheels in high-speed deep grinding of carbon fibre-reinforced ceramics. Int J Adv Manuf Technol 74(9–12):1293–1302
34. Diaz OG, Axinte D (2017) Towards understanding the trimming and fracture mechanism in ceramic matrix composites. Int J Mach Tools Manuf 118–119:12–25
35. Diaz OG, Axinte DA, Novovic D (2018) Probabilistic modelling of tool unbalance during trimming of hard-heterogeneous materials: a case study in ceramic matrix composites (CMCs), Compos Part B 148:217–226
36. Hocheng H, Tai NH, Liu CS (2000) Assessment of ultrasonic drilling of C/SiC composite material. Mater Sci 31:133–142
37. Wang J, Zhang J, Feng P (2017) Effects of tool vibration on fiber fracture in rotary ultrasonic machining of C/SiC ceramic matrix composites. Compos Part B Eng 129:233–242
38. Bertsche E, Ehmann K, Malukhin K (2013) Ultrasonic slot machining of a silicon carbide matrix composite. Int J Adv Manuf Technol 66(5–8):1119–1134
39. Zhang L, Ren C, Ji C, Wang Z, Chen G (2016) Effect of fiber orientations on surface grinding process of unidirectional C/SiC composites. Appl Surf Sci 366:424–431
40. Liu Q, Huang G, Xu X, Fang C, Cui C (2017) A study on the surface grinding of 2D C/SiC composites. Int J Adv Manuf Technol 1–9
41. Frank E, Steudle LM, Ingildeev D, Spörl JM, Buchmeiser MR (2014) Carbon fibers: precursor systems, processing, structure, and properties. Angew Chemie Int Ed 53(21):5262–5298
42. Cao X, Lin B, Wang Y, Wang S (2014) Influence of diamond wheel grinding process on surface micro-topography and properties of SiO2/SiO2 composite. Appl Surf Sci 292:181–189
43. Cao X, Lin B, Zhang X (2013) A study on grinding surface waviness of woven ceramic matrix composites. Appl Surf Sci 270:503–512
44. Oguntuyi SD, Johnson OT, Shongwe MB (2021) Spark plasma sintering of ceramic matrix composite of TiC: microstructure, densification, and mechanical properties: a review. Int J Adv Manuf Technol 116(1–2):69–82
45. Enya K, Naagawa T, Kaneda H, Onaka T, Ozaki T, Kume M (2007) Microscopic surface structure of C/SiC composite mirrors for space cryogenic telescopes. Appl Opt 46(11):2049–2056
46. Nagaraja AM, Gururaja S, Udayakumar A (2022) Tensile behavior of ceramic matrix mini composites with engineered inter phases fabricated by chemical vapor infiltration. J Eur Ceram Soc 42(6):2659–2671
47. Ghosh D, Subhash G, Radhakrishnan R, Sudarshan TS (2008) Scratch-induced microplasticity and microcracking in zirconium diboride-silicon carbide composite. Acta Mater 56(13):3011–3022
48. Ghosh D, Subhash G, Bourne GR (2009) Inelastic deformation under indentation and scratch loads in a ZrB2-SiC composite. J Eur Ceram Soc 29(14):3053–3061
49. Agarwal S, Rao PV (2008) Experimental investigation of surface/subsurface damage formation and material removal mechanisms in SiC grinding. Int J Mach Tools Manuf 48:698–718
50. Zhang B, Howes TD (1994) Material-removal mechanisms in grinding ceramics. Ann CIRP 43(1):305–308
51. Agarwal S, Rao PV (2010) Grinding characteristics, material removal and damage formation mechanisms in high removal rate grinding of silicon carbide. Int J Mach Tools Manuf 50:1077–1087
52. Li Y, Ge X, Wang H (2019) Study of material removal mechanisms in grinding of C/SiC composites via single-abrasive scratch tests. Ceram Int 45(4):4729–4738
53. Chen J, An Q, Chen M (2020) Transformation of fracture mechanism and damage behavior of ceramic-matrix composites during nano-scratching. Compos A Appl Sci Manuf 130:105756
54. Wing BL, Halloran JW (2017) Micro stress in the matrix of a melt-infiltrated SiC/SiC ceramic matrix composite. J Am Ceram Soc 38(1):42–49

55. Ayrikyan A (2018) Investigation of residual stress in lead-free BNT-based ceramic/ceramic composites. Acta Mater
56. Xing Y, Deng J, Zhang G, Wu Z, Wu F (2017) Assessment in drilling of C/C-SiC composites using brazed diamond drills. J Manuf Process 26:31–43
57. Zou F, Chen J, An Q (2020) Influences of clearance angle and point angle on drilling performance of 2D Cf-SiC composites using polycrystalline diamond tools. Ceram Int 46(4):4371–4380
58. Cao X, Lin B, Zhang X (2015) Investigations on grinding process of woven ceramic matrix composite based on reinforced fiber orientations. Compos Part B Eng 71:184–192
59. Azarhoushang B, Tawakoli T (2011) Development of a novel ultrasonic unit for grinding of ceramic matrix composites. Int J Adv Manuf Technol 57(9–12):945–955
60. Ding K, Fu Y, Su H, Cui F, Li Q, Lei W (2017) Study on surface/subsurface breakage in ultrasonic assisted grinding of C/SiC composites. Int J Adv Manuf Technol 3095–3105
61. Tuersley IP, Jawaid A, Pashby IR (1994) Review: various methods of machining advanced ceramic materials. J Mater Process Technol 42:377–390
62. Venugopal A, Rao PV (2002) Modeling of grinding of silicon carbide with diamond wheels. Miner Process Extr Metall Rev 23:51–63
63. Malkin S, Hwang TW (1996) Grinding mechanisms for ceramics. Ann CIRP 45(2):569–580
64. Mao C, Tang X, Zou H, Zhou Z and Yin W (2011) Experimental investigation of surface quality for minimum quantity oil–water lubrication grinding. Int J Adv Manuf Technol. https://doi.org/10.1007/s00170-011-3491-3
65. Kopac J, Krajnik P (2006) High-performance grinding-a review. J Mater Process Technol 175:278–284
66. Chen J, Shen J, Huang H, Xu X (2010) Grinding characteristics in high speed grinding of engineering ceramics with brazed diamond wheels. J Mater Process Technol 210:899–906
67. Huang H, Yin L, Zhou L (2003) High speed grinding of silicon nitride with resin bond diamond wheels. J Mater Process Technol 141:329–336
68. Agarwal S, Rao PV (2011) Improvement in productivity in SiC grinding. In: Proceedings of the institution of mechanical engineers. Part B J Eng Manuf 225:811–830
69. Diaz OG, Axinte DA, Butler-Smith P (2019) On understanding the microstructure of SiC/SiC ceramic matrix composites (CMCs) after a material removal process. Mater Sci Eng A 743:1–11
70. Qu S, Gong Y, Yang Y (2019) Grinding characteristics and removal mechanisms of unidirectional carbon fibre reinforced silicon carbide ceramic matrix composites. Ceram Int 45(3):3059–3071
71. Qu S, Gong Y, Yang Y (2018) Surface topography and roughness of silicon carbide ceramic matrix composites. Ceram Int 44(12):14742–14753
72. Li B, Li C, Zhang Y (2016) Grinding temperature and energy ratio coefficient in MQL grinding of high-temperature nickel-base alloy by using various vegetable oils as base oil. Chin J Aeronaut 29(4):1084–1095
73. Adibi H, Esmaeili H, Rezaei SM (2018) Study on minimum quantity lubrication (MQL) in grinding of carbon fiber-reinforced SiC matrix composites (CMCs). Int J Adv Manuf Technol 95(9–12):3753–3767
74. Jia D, Li C, Zhang Y (2019) Experimental evaluation of surface topographies of NMQL grinding ZrO_2 ceramics combining multiangle ultrasonic vibration. Int J Adv Manuf Technol 100(2019):457–473
75. Hu M, Ming W, An Q (2019) Experimental study on milling performance of 2D C/SiC composites using polycrystalline diamond tools. Ceram Int 45(8):10581–10588
76. Zhang X, Yu T, Li M (2020) Effect of machining parameters on the milling process of 2.5D C/SiC ceramic matrix composites. Mach Sci Technol 24(2):227–244
77. Zhang X, Yu T, Zhao J (2020) An analytical approach on stochastic model for trimming force prediction in milling ceramic matrix composites. Int J Mech Sci 168:105314
78. Diaz OG, Axinte DA (2017) Towards understanding the trimming and fracture mechanism in ceramic matrix composites. Int J Mach Tool Manuf 118:12–25

79. Wang X, Gao X, Zhang Z, Cheng L, Ma H, Yang W (2021) Advances in modifications and high-temperature applications of silicon carbide ceramic matrix composites in aero-space: a focused review. J Eur Ceram Soc 41(9):4671–4688
80. Zivic F, Busarac N, Milenkovic S, Grujovi´c N (2021) General overview and applications of ceramic matrix composites (CMCs). Encycl Mater Compos 2:3–19
81. McDanels DL, Serafini TT, DiCarlo JA (1986) Polymer, metal, and ceramic matrix composites for advanced aircraft engine applications. J Mater Energy Syst 8(1):80–91
82. Descamps P, Tirlocq J, Cambier F (1991) Ceramic matrix composites: properties and applications. In: Riley FL (ed) 3rd European symposium on engineering ceramics, pp 109–125
83. Schmidt S, Beyer S, Knabe H (2004) Advanced ceramic matrix composite materials for current and future propulsion technology applications. Acta Astronaut 55(3):409–420
84. Liu G, Qiu J, Chen M (2019) Research progress of composites nozzle extension for liquid rocket engine. J Rocket Propul 04:1–8
85. Christin F (2002) Design, fabrication, and application of thermos structural composites (TSC) like C/C, C/SiC, and SiC/SiC composites. Adv Eng Mater 4(12):903–912
86. Wang P, Liu F, Wang H, Li H, Gou Y (2019) A review of third generation SiC fibers and SiCf-SiC composites. J Mater Sci Technol 35(12):2743–2750
87. Gao T, Hong Z, Yang J (2014) Application and prospect of ceramic matrix composite components for commercial aircraft engine. Aeron Manuf Technol 450(6):14–21
88. Tontisakis A, Simpson W, Lincoln J, Dhawan R, Opliger M (2021) Evaluation of surface finish technology in the manufacture of oxide-oxide ceramic matrix composites. Ceram Int 47(4):5347–5363
89. DiCarlo JA, van Roode M (2006) Ceramic composite development for gas turbine engine hot section components. In: Turbo expo: power for land sea and air, ASME conference proceedings, vol GT2006–90151, pp 221–231
90. Arai Y, Inoue R, Goto K (2019) Carbon fiber reinforced ultra-high temperature ceramic matrix composites: a review. Ceram Int 45(12):14481–14489
91. Padture NP (2016) Advanced structural ceramics in aerospace propulsion. Nat Mater 15(8):804–809
92. Liu X, Shen X, Gong L, Li P (2018) Multiscale thermodynamic analysis method for 2D SiC/SiC composite turbine guide vanes. Chin J Aeronaut 31(1):117–125
93. Liu Q, Huang G, Xu X (2018) Influence of grinding fiber angles on grinding of the 2D–Cf/C–SiC composites. Ceram Int 44(11):12774–12782
94. Zhang L, Wang S, Li Z (2019) Influence factors on grinding force in surface grinding of unidirectional C/SiC composites. Appl Compos Mater 26(3):1073–1085
95. Liu Q, Huang G, Xu X (2017) A study on the surface grinding of 2D C/SiC composites. Int J Adv Manuf Technol 93(5–8):1595–1603
96. Du J, Ming W, Ma J (2018) New observations of the fiber orientations effect on machinability in grinding of C/SiC ceramic matrix composite. Ceram Int 44(12):13916–13928
97. Yang M, Li C, Zhang Y (2019) Predictive model for minimum chip thickness and size effect in single diamond grain grinding of zirconia ceramics under various lubricating conditions. Ceram Int 45(12):14908–14920
98. Qu S, Gong Y, Yang Y (2020) Investigating minimum quantity lubrication in unidirectional Cf-SiC composite grinding. Ceram Int 46(3):3582–3591
99. Liu H, Yang J, Jiao J (2017) Fabrication technique and application prospect of continuous SiC fiber reinforced ceramic matrix composites for aeroengine. Aeron Manuf Technol 60(16):90–95
100. Naslain R (2004) Design, preparation and properties of non-oxide CMCs for application in engines and nuclear reactors: an overview. Compos Sci Technol 64(2):155–170

Polymer Matrix Composites: Machining Challenges

Machining Challenges of Polymer Matrix Composites

K. S. Lokesh, C. G. Ramachandra, G. Ravindra Babu, and D. Shrinivasa Mayya

Abstract Numerous sectors benefit from the usage of fiber-reinforced composite materials for structural purposes. Delamination is regarded as the main issue in the production of the components. In the production process, drilling is frequently utilized in materials have an impact on chip deformation and drilling behavior. By carefully choosing the tool, method, and operating circumstances, both conventional and unconventional drilling procedures are capable of producing small holes for composite materials. This article reviews the approach to drilling composite material without delamination. The main scenes are shown, along with the analytical approach's elements and the application of unique drill bits in practice. The current review is to maximize the process variables, particularly the drilling process's cutting parameters and milling of various types of polymers reinforced with different fibres. The acquired experimental data were examined using analysis of variance (ANOVA).

Keywords Machining challenges · FRP composites

K. S. Lokesh (✉)
Department of Aeronautical Engineering, Srinivas Institute of Technology, Mangalore 574143, India
e-mail: lokeshvijay@sitmng.ac.in

C. G. Ramachandra
Department of Mechanical Engineering, Presidency University, Bangalore 560064, India

G. Ravindra Babu
Department of Aeronautical Engineering, Karavali Institute of Technology, Mangalore 575029, India

D. Shrinivasa Mayya
Department of Mechanical Engineering, Srinivas Institute of Technology, Manglore 574143, India

S. B. Boppana et al. (eds.), *Structural Composite Materials*, Composites Science and Technology, https://doi.org/10.1007/978-981-99-5982-2_16

1 Introduction

Combining two or more different materials results in a composite material. They are mixed together in a way that gives the resulting composite materials better qualities. It cannot be produced using a substance with only one component [1]. The best attributes of its constituent parts are bone [2]. Glass fibre reinforced plastics (GFRP), which are created from plastic and glass fibre, are the most popular type of synthetic composite material [3]. The separate parts are completely distinct from the GFRP composite material in terms of their characteristics. Plastics may be easily molded into any complicated shape and are lightweight, strong, and have great corrosion resistance. However, due of insufficient strength, stiffness, and dimensional stability, they are unsuitable for applications requiring weight bearing. As opposed to that, glass fiber is suitably rigid and strong, with a very high strength. It is similarly unsuitable Because of their brittleness and fibrous nature; plastics are not suitable for load-bearing applications. However, when these two are joined in the right ratios and with a certain arrangement of glass fibres, we are left with GFRP, which has superior mechanical capabilities and other characteristics suited for load bearing applications [4]. Composite materials are taking the place of many conventional engineering materials in a number of different application sectors. A variety of items including domestic and among other industries, use a family of materials known as glass fiber reinforced composite materials. Since the days of the Bible, when brick and chopped, straw was combined to create construction materials, composite materials have been used. In the early 1900s, thermoset phenolic was reinforced with fibre or particle material, the era of modern composite materials officially began. In the USA, glass fibre was first produced for commercial use in 1937. Since then, new matrix materials, reinforcing materials, and production techniques have all seen significant development glass fibre reinforced fibre reinforced plastics (GFRP). As can be observed [4], plastics are able to meet the strict technological and economic criteria of numerous industries. Figure 1 Factors affecting machining characteristics in turning of PMC's.

2 Drilling of Composite Materials

Fiber-reinforced plastics present a number of issues during the drilling process due to their geometry and other characteristic defects [5]. Because of their distinctive qualities, including Fiber-reinforced composite materials have experienced a substantial growth in application in the automotive and aerospace industries recently due to their high specific stiffness and strength, great damping, superior corrosion resistance, and low thermal expansion. For these applications, hole-making techniques are mostly used. For instance, a tiny single-engine aircraft has over 100,000 holes while a huge transport aircraft has millions of holes, the majority of which are for fasteners like rivets, bolts, and nuts [6].

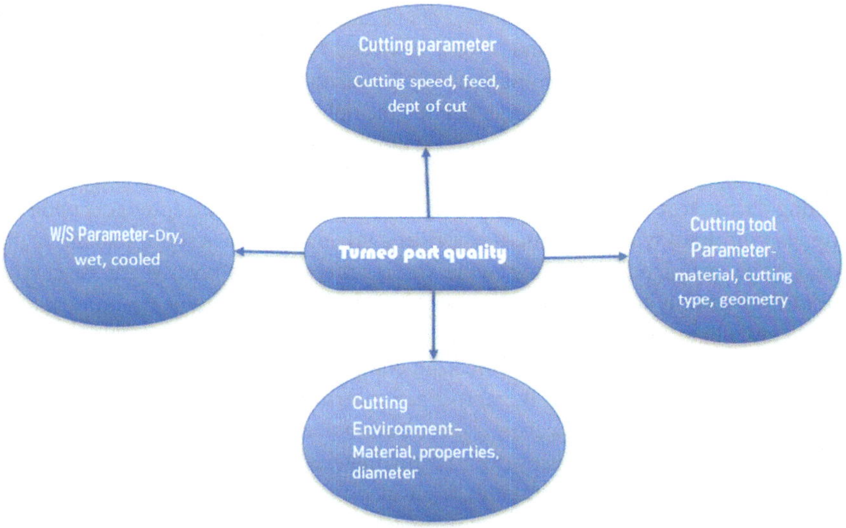

Fig. 1 Factors affecting machining characteristics in turning of PMC's

Aircraft assembly accounts for over half of the total cost of producing an airframe. The quality of the drilled holes may have a substantial impact on the life of the riveted joints for which they are used. The axial straightness, roundness, and waviness of the hole's cross-section may place a lot of strain on Revit and cause it to malfunction [7, 8]. For repairs and/or assembly work, traditional drilling techniques are frequently utilized in the automotive and aerospace sectors. One of the major issues with drilling traditional or composite materials is drill wear, which results in changes to the properties of the created holes and cutting tool failure. Linking process variables such stresses [9, 10], surface polish [11–13], delamination [14], vibration [15], and acoustic emission [16, 17] in order to understand how tools wear, numerous experimental methodologies have been used. A lot of effort has been spent in recent years to developing due to the challenges associated with the dependability, calibration, and expense of using these experimental detection approaches. Numerous works have been carried out in adding filler to thermoset bonded fibres with enhanced mechanical features [18–26]. In many cases, delamination is what prevents composite materials from being used in structural applications. In the aviation sector, delamination-related drilling was the cause of 60% of all parts that were rejected during final assembly of an aircraft structure. Delamination analysis in composite material drilling is a serious issue that needs to be kept to a minimum. Many studies have studied the drilling of GFRP composite material. Natural fibres play greater role in enhancing stiffness and UTS which in turns directs the utility of NF's to consider for machinability aspects. When drilling GFRP composites, the effects of various drill types, such as, candlestick on thrust force and delamination are discussed [27]. They contrasted the twist drill with the significant thrust force just before delamination began [28]. The signal to noise ratio delamination analysis was also evaluated when drilling GFRP composites. [29]. The results proved that a very low feed rate and a quick speed of cutting favor the least amount of delamination [30]. Experiments show that the delamination size reduces as the feed is decreased, but that the delamination size is not significantly affected by the cutting speed. Mechanical studies reveal the potentiality of bio-fillers in enhancing the cutting parameters of bio-based materials [31]. Results show that drilling thrust force, which depends on feed, speed, and cutting-edge condition, directly correlates with delamination. Drilling infringement is a significant issue that needs to be examined, according to the studies mentioned above. When discussing the drilling of polymer matrix composites using ordinary equipment, many authors have shown that the drilling parameters have a substantial impact on the quality of the cut surfaces. For usage in load-bearing components, glass fibre reinforced polymers (GFRP), particularly in the aerospace industry, have gained more and more attention in recent years, however, if these parameters are not chosen properly [32]. Numerous outstanding qualities exist in this material, including exceptional corrosion resistance, high specific strength, high specific elastic modulus, and low weight [33]. As the domains of application expand, turning, drilling, milling, and cutting-off processes are being employed increasingly frequently in GFRP production. However, the glass fibre component frequently makes machining GFRP challenging. In order to obtain accuracy and efficiency while machining composite materials, one must have a deeper understanding of cutting processes [34]. Drilling is a particularly stochastic process because polymeric composites are anisotropic and inhomogeneous. It is difficult for traditional drilling technologies to increase the caliber of holes drilled in composites. Therefore, finding new drilling techniques is crucial. Vibration assisted cutting includes vibratory drilling as a subset, this differs from conventional drilling fundamentally. The innovative vibratory drilling method has attracted a lot of attention recently [35]. Multiple potentially appealing features exist for matrix composites with several forms of fibre reinforcement [30]. The bulk composite's tribological and mechanical characteristics can be significantly enhanced by reinforcing polymer with fibres [31, 32]. By including the mica particles, the compressive strength and hardness both the tribological properties of UD fibre composite [36] were enhanced. Researchers looked into the surface damage properties of polyamide group reinforced with cross orientation of fibrous structures [37, 38] showed how increasing the amount of glass fibre tends to increase the hardness of the bulk composite. while very slightly increasing the friction coefficient. The friction coefficient is decreased as the applied loads are increased. Numerous researchers looked into how the orientation of the fibres affected the various composites' varying wear rates and coefficients of friction.

3 Wearability Factors Considered for the Tool

Cutting cannot go on forever using a tool. It has a set lifespan. If a cutting tool is to have a long lifespan, its face must be as smooth as possible. The amount of time a tool may be used productively before getting worn out is known as its tool life. A dull tool increases cutting forces and power consumption, leads to poor surface smoothness, chatter during milling, and overheats the tool. Tool wear is the gradual deterioration of the cutting edge's sharpness with use. There are five different forms of tool wear: Wear from adhesion (a) This method of wear involves welding the tool material to the workpiece. High cutting forces, high temperatures, and softer tool material are all factors in this form of tool wear. (b) Abrasion wear: The frictional force causes this form of wear created when chips move across the tool's face. (c) Diffusion wear: Tool wear is caused by the solid-state diffusion phenomenon. This is determined by the temperature and surface area at which the tool and work are in contact. As temperature increases, the rate of diffusion exponentially increases.

4 Experimental Results and Discussions

Several publications on creating and evaluating methods for foreseeing the Studies on the machinability of similar work groups were discussed. Experiments reported the work have the primary goal of is to look at how some factors may affect the thrust force, torque and other operations. According to the findings, the torque cycle's start point starts sometime after the push force (depending on the feed value).

The following results were drawn from the experimental work: during full drill engagement, the once the cutting lips and the chisel edge had both left the laminate, the thrust force was gradually lowered until it was zero. In contrast, the torque increased when the drill emerged from the laminate and continued to climb until the cycle was complete before abruptly increasing to a number that was around 10 times the peak value. The cutting speed and feed barely make a difference to how harsh the epoxy glue is on the surface. On the other hand, raising the cutting speed and fibre volume fraction for (GFRP) improved the surface roughness.

Several parameters for drilling and milling operations conisidered by several researchers on recent studies for different polymers are discussed. Two types of majorly used adhesives such as Epoxy and polyester based polymer composites are detailed in Table 1.

The machinability factors were thrust force, torque, peel-up and push-out delamination, and surface roughness of drilled holes. They ultimately came to the conclusion that drill pre-wear had a significant impact on the behavior of driving force during the drilling process. High feed and cutting rates intensify this effect, which also raises surface roughness and delamination. Increased drill pre-wear destroys microcracking at the ply interfaces due to the matrix, degrading the surface's texture. Low transition temperatures and a low coefficient of heat conduction for (GFRP) composites

Table 1 Parameters of drilling and milling operations [39]

1. For epoxy based composites

Parameters	Drilling operation				Milling operation		
Fibre type	Sisal and E-glass fiber, flyash + graphene	Screw pine fiber	Rice husk	Camellia sinensis and Ananas comosus, glass fiber	Jute	Kenaf fibre	Flax fibre
Geometry of the drill/tool material	HSS twist drill bit of diameter of 12 mm	3, 4, 5 mm dia	HSS drill bit	Drill bit of 6 mm	TiAlN-coated cemented carbide (WC) end mill	HSS	A high-speed steel (HSS) end mill
Speed of the spindle	700, 900, 1200 rev/min	600, 852, 1260 rev/min	500, 800, 1250 rev/min	300, 600, and 900 rev/min	25, 50, 75 m/min	509, 1019, 764 rev/min	1000, 2000, 3000 rev/min
Feed rate	–	0.1 mm/rev	0.02, 0.04, 0.06 m/min	0.02, 0.04, 0.06 m/min	0.04, 0.08, 0.12 mm/rev	204, 713.5, 1223 mm/min	0.1, 0.2, 0.3 mm/rev
Depth of cut (milling)	–	–	–	–	2	2	-

2. For polyester based composites

Parameters	Drilling operation			Milling operation	
Fibre type	Sisal fiber	Hemp fiber	Jute fiber	Hemp, jute, banana, glass	Woven natural jute
Geometry of the drill/tool material	HSS M2 -drills of 8 mm Ø with three different ϴ-90°, 104° and 118°	4 mm drill dia	Brad and spur drill tool dia 5, 7, and 10 mm	Carbide end mill	7 mm, 4-fluted end mill made of high-speed steel
Speed of the spindle	500, 100, 1500 rev/min	800, 2800, 4800 rev/min	355, 710, and 1400 rev/min	16, 24, 32 rev/min	210, 660, 1750 rev/min
Feed rate	10, 15, 20 mm/min	0.07, 0.17, 0.27 mm/rev	50, 108, 190 mm/min	0.1, 0.2, 0.3 mm/rev	0.04, 0.06, 0.15 mm/rev
Depth of cut (milling)	–	–			1, 1.5, 2

help to reduce the generated temperature, which in turn causes significant surface roughness is produced during high-speed drilling and drill pre-wear. Major findings noted the growing use of composite materials made of glass fiber-reinforced plastic (GFRP) in structural, automotive, and aeronautical applications. Due to its propensity to delaminate, drilling is a complicated operation used to attach composite structures.

Within the factors and their constraints that are explored, the created GFRP composites' delamination during drilling can be predicted using a model. As a result, while the impact of the fibre orientation angle was less noticeable, the amount of the delamination grew along with the feed rate and drill diameter. The spindle speed when drilling GFRP composites has very little bearing on delamination. Owing to a larger drill's increased drive force diameter, drilling composite materials has a higher delamination factor. Similar smaller holes could be utilized in place of larger ones [39]. Glass-polyester composites were subjected to drilling experiments using conventional HSS tools; drilling was halted at the current depth to examine harm growth when drilling. The samples, which had undergone metallographic polishing method, were checked for damage using optical microscopy. The data are helpful in characterizing the history of damage and in designing drill designs intended specifically for composite machining. The following are the main findings from the earlier analysis. The type of drilling-related damage to composite materials is significantly influenced by the feed rate. Failures are primarily caused by delamination that starts close to where the hole's cylindrical surface meets the conical surface made by the primary cutting edges if it is set to suitably low values. Most delamination's at low feed rates are produced when the tool exit edge is near. The average length of the delamination closest to the sample's back face is greater, which suggests a stronger inclination to grow. Damage determined as indicative of the observed defect condition from micro analysis. In fact, attaching fasteners to laminate composites during assembly drilling is one of the most popular manufacturing procedures. The material anisotropy brought on by fibre reinforcement has a significant impact on how machinable a material is. Therefore, accurate machining is required to assure dimensional stability and increase component productivity. Within the bounds of the experiments, the following can be deduced from the study of drilling GFRP composite plate findings utilizing conceptual S/N ratio method, response surfaces, and ANOVA. (1) The Taguchi approach provides a strategy for designing process parameter optimization that is systematic and efficient and yields the least, as demonstrated in this study. (2) The delamination impact is seen to be primarily influenced by the feeding rate, cutting rate, and thickness of the material The least amount of delamination at the entry and exit of the drilling favors fast cutting speed and low feed, which enhances surface polish and tool life. (3) Using conceptual S/N ratio and ANOVA techniques to data analysis, similar outcomes are discovered [40]. The findings majorly claim that because of better mechanical qualities, lightweight materials are utilized in a variety of industries. In order to place fasteners for the construction of laminates, drilling is a highly common machining activity.

Compressive residual strength, corrected delamination factor, and thrust force of unit-directional glass/epoxy resin were all examined in relation to drilling parameters' impact. To increase the drilled laminates' compressive residual strength, the results emphasize the significance of feed rate [41]. Investigation into how the drilling process's chip generation and thrust force are impacted by the anisotropy of fiber-reinforced polymers (FRP) was done by S. Arul et al. in 2005. One of the main reasons for damage to fibre reinforced plastics during drilling is delamination, it potentially leads to long-term performance decline in addition to compromising the structural integrity. By using a traditional drilling technique, it is challenging to generate holes of excellent quality and efficiency. This study on drilling polymeric composites attempts to develop a technique that will guarantee fewer flaws and longer tool life. Waste glass and rubber derived thermosets proved to be higher wear resistance hence difficult to machinable [42, 43]. Composite material drilling is a delicate and crucial process where materials delaminate when subjected to the forces associated with drilling operations. While the trials were being conducted. The exit hole of the drilled area was scanned to determine the amount of delamination factor. The acquired experimental data were examined through means of analysis of variance (ANOVA). The results demonstrated that the delamination factor increased at both low and high levels within the parameters' selected experimental range. The ideal parameter values for minimizing delamination were ultimately established. The recent investigation depicted that the diameter affects the thrust force and the feed rate affects the torque. Nonlaminated composites are frequently employed in ballistic applications because they have better mechanical qualities than laminated composites. An inquiry was conducted to evaluate the Using coated tungsten carbide twist drills because there is little information on the machinability of nonlaminated composites. The pultrusion process with a high fibre weight percentage was used to create the GFRP composite rods. Using a coordinate measuring machine, the drilled holes' ovality (hole diameter inaccuracy) was evaluated. ANOVA method is used to investigate how process variables like feed and spindle speed affected the ovality of the drilled holes. In order to drill pultruded GFRP composite rods defect-free, the optimal level of process parameters towards lowest ovality was attained. Statistical software MINITAB 16 was used to determine a correlation between ovality and process parameters. On ovality, speed had little of an impact. The feed had a big impact on how novel the holes were that were bored [44]. Major findings of recent works reported on optimizing drilling of GFRP composites, namely cutting speed, feed, point angle, and chisel edge width. A L9 orthogonal array was used for testing in this work to assess the effects of various combinations of process parameters on hole quality. These studies followed the Taguchi experimental strategy. An analysis of variance (ANOVA) test was conducted to determine the significance of each process parameter on drilling. The results show that speed, chisel edge width, and point angle are the other important variables that have a substantial impact on the thrust force; This research is helpful in determining the best values for kinds of process factors that would lower the delimitation, improve the drilling hole's quality [45].

5 Conclusion

In order to satisfy the need of classifying the issues under consideration, successful review completed in producing with neat validation of various factors which are enumerated in different papers. The cutting speed and feed barely make a difference to how harsh the epoxy glue is on the surface. GFRP, on the other hand, raising the fibre volume fraction and cutting speed reduced surface roughness. Increased drill pre-wear causes an increase in thrust force, which destroys the matrix and damages the surface finish by causing microcracking at the ply interfaces. Because of a rise in thrust force caused by a larger drill diameter, drilling composite materials has a higher delamination factor. A superior surface polish, longer tool life, and a high cutting speed and low feed result in an increase in the delamination within the experimental known range of the parameters. The ideal parameter values for minimizing delamination were ultimately established. Additionally, it has been noted that in all sandwich composites, as feed rates and speed are reduced, it is discovered that the drill profile rupture is escalating. The ovality of the drilled holes is affected by process variables including feed and spindle speed. In order to drill pultruded GFRP composite rods defect-free, the optimal level parameters near to lowest ovality was attained. In addition to the acquired experimental data. The present review also concludes the optimized process parameters, particularly the drilling and milling parameters for several types of polymers reinforced with various fibres, including the feed, cutting speed, and drill bit diameter.

References

1. Harris B (1999) Engineering composite materials. The Institute of Materials, London
2. Dhingra AK (1980) New fibres and their composites [Watt W, Harris B, Ham AC (eds)]. Royal Society of London (Phil Trans Roy SocLond, A294, 411–417 Fitzer E and Heine 1988. Composite Materials Series, volume 2, editor AR Elsevier, Amsterdam, 73–148)
3. Aveston J, Cooper GA, Kelly A (1971) Proceedings of NPL conference on the properties of fibre composites. IPC Science and Technology Press Ltd., Guildford, pp 15–26 (Cahn RW, Harris B, 1969, Nature, 221, 132–141)
4. Harris B, Habit FA, Cooke RG (1992) Proc Roy Soc (Lond), A437, pp 109–131 [Parratt NJ and Potter KD, 1980, Advances in Composite Materials (Proc ICCM3) vol 1 (editors AR Bunsell, C Bathias, A Martrenchar]
5. Khashaba UA (2009) Delamination in drilling polymeric composites: a review. In: PauloDavim J (ed) Drilling of composite materials. Nova Science Publishers Inc.
6. Sudhakaran S (1988) Mapping the cutting force to the hole quality using neural networks. In: Artificial neural engineering, vol 8. ASME Press, pp 773–778
7. Radhakrishnan T, Wu SM (1981) On-line hole quality evaluation for drilling composite material using dynamic data. J Eng Ind ASME 103:119–25
8. Eriksen E (2000) The influence of surface roughness on the mechanical strength properties of machined short-fibre-reinforced thermoplastics. Compos Sci Technol 60:107–113
9. Gindy NNZ (1988) Selection of drilling conditions for glass fiber reinforced plastics. Int J Prod Res 26(8):1317–1327
10. Rawat S, Attia H (2009) Wear mechanisms and tool life management of WC–Co drills during dry high-speed drilling of woven carbon fiber composites. Wear 267:1022

11. Velayudham A, Krishnamurthy R, Soundarapandian T (2005) Evaluation of drilling characteristics of high-volume fraction fibre glass reinforced polymeric composite. Int J Mach Tool Manuf 45:399–406
12. Arul S, Vijayaraghavan L, Malhotra SK, Krishnamurthy R (2006) The effect of vibratory drilling on hole quality in polymeric composites. Int J Mach Tool Manuf 46:252–259
13. Pai R, Bangarappa L (2022) Hybridization effect on water absorption and flexural properties of E-glass/banana fibre/epoxy composites. Mater Today Proc 52(Part 3):1841–1845. ISSN 2214-7853. https://doi.org/10.1016/j.matpr.2021.11.491
14. Tsao CC (2008) Experimental study of drilling composite materials with step-core drill. Mater Des 29:1740–1744 [Tsao CC, Hocheng H (2008) Evaluation of thrust force and surface roughness in drilling composite material using Taguchi analysis and neural network. J Mater Process Technol 203:342–348]
15. Campos Rubio J, Abrao AM, Faria PE, EstevesCorreia A, Paulo DJ (2008) Effects of high speed in the drilling of glass fibre reinforced plastic: evaluation of the delamination factor. Int J Mach Tool Manuf 48:715–720
16. Hocheng H, Tsao CC (2003) Comprehensive analysis of delamination in drilling of composite materials with various drill bits. J Mater Proc Technol 140:335–339
17. Chen W-C, Tsao CC (1999) Cutting performance of different coated twist drills. J Mater Proc Technol 88:203–207
18. Pinto T (2018) Experimental study on effect of metallic oxides mixture on tribological behaviour of aluminium based metal matrix composites. Int Res J Eng Technol (IRJET). e-ISSN: 2395-0056 05(06), p-ISSN: 2395-007
19. Konur O, Mathews FL (1989) Effects of the properties of the constituents on the fatigue performance of composites: a review. Composites 20(4):317–328. http://ijsrmme.com/IJSRMME18 2210
20. Ummaji B, Gururaj P, Rayappa K, Yashavantha J (2018) Effect of red mud particles on scratch resistance of aluminum based metal matrix composites. Am J Aerosp Eng 5(1):24–29. https://doi.org/10.11648/j.ajae.20180501.14
21. Karthik BY, Avinash CG, Ganesh Kamath SL (2018) Experimental study on tribological characteristics of E-waste filled fibre reinforced plastics. Int J Sci Res Mech Mater Eng (IJSRMME). ISSN: 2457–0435, 2(1):11–16. http://ijsrmme.com/IJSRMME182203 (Volume 2|Issue 2|March-April 2019|) www.ijsrmme.com
22. Gowspeer M. Effect of E-waste rubber on wear behaviour of glass fibre reinforced with epoxy composites, pp 20–25. https://doi.org/10.7324/IJASRE.2018.32624
23. Lokesh KS, Pinto T, Ramachandra CG (2017) Effect of tool wear; machinability studies on polymer composites; a review. Int J Eng Inform Syst 1(5):71–77 ⟨hal-01571294⟩
24. Pinto T, Shrinivasa Mayya D et al (2022) Effect of Wollastonite filler on the experimental and microstructural analysis of epoxy composite reinforced with E-glass fibre. J Inst Eng India Ser D. https://doi.org/10.1007/s40033-022-00347-9
25. Pinto T (2019) Preparation and mechanical testing of E-waste glass filled polymer composites. Int J Sci Res Mech Mater Eng (IJSRMME). ISSN: 2457–0435 3(1):18–23. http://ijsrmme.com/IJSRMME19312
26. Pranav S, Pratibha, Kumar A, Chetan G (2017) Comparative study on tensile behaviour of e-glass (woven & chopped) fibre reinforced with epoxy composites. Int J Eng Sci Res Technol 06(03)
27. Arul S, Samuel Raj D, Vijayaraghavan L, Malhotra SK, Krishnamurthy R (2006) Modeling and optimization of process parameters for defect tolerance drilling of GFRP composites. Mater Manuf Process 21:357
28. Koenig W, Grass P, Wulf C, Willerscheid H (1985) Machining of fiber reinforced plastics. Ann CIRP 34(2):537–548
29. Komanduri R, Zhang B, Vissa CM (1991) Machining of fiber reinforced composites. Process Manuf Compos Mater PED-Voi 49/MD-Vol 27:1–36
30. Koenig W, Grab P (1989) Quality definitions and assessment in drilling of fibre reinforced thermosets. Ann CIRP 38(1):119–124

31. Tagliaferri V, Caprino G, Diterlizzi A (1989) Effect of drilling parameters on the finish and mechanical properties of GFRP composites. Int J Mach Tool Manuf 30(1):77–84
32. Lu Z, Friedrich K (1993) Wear and friction of a unidirectional carbon fiber-glass matrix composite against various counterparts. Wear 162–164:1103–1113
33. Naga SAR (1992) Effect of glass fiber content on the frictional behavior of glass fiber reinforced polyamide-66. In: 8th Internationals Kolloquium, Tribologie 2000, Technische Akademie Esslingen, Ostjildren, Germany, pp 14–16, paper 25.7
34. Naga SAR (1992) Wear behavior of glassfiber reinforced polyamide-66. In: Proceedings 5th Conj: on applied mechanical engineering (AME), vol 5–7. Military Technical College (MTG), Cairo, pp 157–162
35. Sung NH, Suh N (1979) Effect of fiber orientation on friction and wear of fiber reinforced polymeric composite. Wear 53:129
36. Mainford E, Prewo K (1985) Friction and wear of graphite fiberreinforced glass matrix composites. Wear 102:253
37. Rajamurugan TV, Shanmugam K, Rajakumar S, Palanikumar K (2012) Modelling and analysis of thrust force in drilling of GFRP composites using response surface methodology (RSM). Procedia Eng 38:3757–3768. ISSN 1877-7058. https://doi.org/10.1016/j.proeng.2012.06.431
38. Sharath Chandra HS, Ravindra Babu G, Shrinivasa Mayya D, Naveen Kumar JR (2023) Impact of fibre orientation on mechanical properties of GFRP composites. Mater Today Proc. ISSN 2214-7853. https://doi.org/10.1016/j.matpr.2023.03.754
39. Raj SSR, Dhas JER, Jesuthanam C (2021) Challenges on machining characteristics of natural fiber-reinforced composites—a review. J Reinf Plast Compos 40(1–2):41–69. https://doi.org/10.1177/0731684420940773
40. Gokulkumar S, Thyla PR, Arun Ramnath R et al (2020) Acoustical analysis and drilling process optimization of Camellia sinensis/Ananas comosus/GFRP/epoxy composites by TOPSIS for indoor applications. J Nat Fibers 1–18. https://doi.org/10.1080/15440478.2020.1726240
41. Bhat R, Mohan N, Sharma S, Pai D, Kulkarni S (2020) Multiple response optimisation of process parameters during drilling of GFRP composite with a solid carbide twist drill. Mater Today Proc 28(Part 4):2039–2046. ISSN 2214-7853. https://doi.org/10.1016/j.matpr.2020.02.384
42. Pinto T, Lokesh KS, Bharath Kumar S, Ramachandra C (2016) G3, Effect of E-waste filler on tensile and flexural behavior of E-glass/Epoxy composites. Int J Eng Trends Technol 02(05)
43. Pai KR, Naveen Kumar JR, Hebbale AM (2021) Experimental study on preparation and mechanical characteristics of jute/silk/coco-peat reinforced with epoxy polymers. Mater Today Proc 46(Part 7):2764–2769. ISSN-2214-7853. https://doi.org/10.1016/j.matpr.2021.02.511
44. Malakar C, Trishul MA, Shreyas (2019) Investigation of drilling characteristics of screwpine fibers reinforced composite. Mater Today Proc. https://doi.org/10.1016/j.matpr.2019.09.038
45. Jayaprakash V, Sivasaravanan S, Bupesh Raja VK et al (2020) Optimization of drilling parameters of epoxy/rice husk composite material. Mater Today Proc 21:104–107

Processing of Composites: Challenges

Challenges Faced in Processing of Composites

K. S. Lokesh, C. G. Ramachandra, and D. Shrinivasa Mayya

Abstract Composites made of fiber-reinforced resin are being used more and more in the aerospace and auto industries. However, owing to the complex physical and chemical characteristics of the constituent materials moulding techniques, making composites presents highly difficult hurdles. As a result, while assessing the quality of composites, knowledge on how to spot production-related issues is crucial. The matrix of composites' residual stress development, Vacuum flaws and resin-rich flaws are first summarised. This chapter describes many resin-related processes, such as the curing of heat responsive resins. Resin penetration during hot pressing, RTM and 3D printing, and resin-rich flaws during the moulding process. Second, the method by which fibre reinforcement flaws such fibre waviness and wrinkle occur in composites is introduced, and the impact of such flaws on the creation of the composite structure is underlined. Supporting structure modulus, strength, and stability may be significantly reduced by fibre misalignment defects, according to several research reports. Finally, difficulties brought on by interfacial defects, like layer peel ups and unbinding are elaborated at the interface between reinforcements and matrix. By combining the different difficult aspects that cause manufacturing flaws in laminated and additively made structures so that the inculcated information may provide a prognosis for composite manufacturing.

Keywords Challenges in PMC processing · Defects in manufacturing

---

K. S. Lokesh (✉)
Department of Aeronautical Engineering, Srinivas Institute of Technology, Mangalore 574143, India
e-mail: lokeshvijay@sitmng.ac.in

C. G. Ramachandra
Department of Mechanical Engineering, Presidency University, Bangalore 560064, India

D. S. Mayya
Department of Mechanical Engineering, Srinivas Institute of Technology, Manglore 574143, India

S. B. Boppana et al. (eds.), *Structural Composite Materials*, Composites Science and Technology, https://doi.org/10.1007/978-981-99-5982-2_17

1 Introduction

It is not well known how characteristics of the manufacturing process have an impact on structural strength, durability, and damage tolerance. The fatigue-critical, flight-critical performance and service life of thick carbon/epoxy and glass/epoxy composite fatigue-critical components are particularly affected by the consequences of insufficient design methods and manufacturing processes, which appear as flaws like wrinkles and porosity. Such flaws impair matrix-dominated characteristics, resulting in weaker structural behaviour and fatigue resistance. In order to determine the state of the product and prevent making assumptions about the worst-case scenario, accurate measurements to quantify manufacturing flaws are crucial [1]. Accurate 3-D measuring skills are needed since the wrinkling shape and porous dispersion are three-dimensional. To demonstrate computed section representation capacity for precise 3-D scaling, characterisation of composite structural flaws, it's clear that it is possible to measure porosity and wrinkling in dense structural details. Additionally, the requirements for advancements to transform the defect profile information into FE models for evaluating the consequences of the faults are covered. Though the parts are naturally more sensitive to fluctuations in production over metals, composite designs now employ use the metal design philosophy and the same factor of safety to calculate the ultimate design load from the limit load. Differences in the part fabrication process, such as operator skill, tooling setup, humidity fluctuation, equipment control, and so on, are common causes of part quality variation, in addition to material variations in resin content, bulk factor, and fibre alignment [2]. Production yields are reduced as a result of the increasing susceptibility of composite component quality to material and process variables. Composite manufacturing communities must comprehend and manage their actions in order to maximise production yields. Majority of the portion remain non-trackable in defining the accuracy of the yield.

2 Various Challenges in Processing of Composites

2.1 Matrix Defects of Composites

The Compressive strength, interlaminar shear performance, and impact resistance are mechanical qualities dominated by resin matrix in composites would be reduced by the flaws of resin matrix at many sizes, including meso scale to macro scale, which are challenging to regulate. Additionally, residual stresses and curing deformation caused by the crystallisation of the matrix will influence the precision and security of using engineering composite structures, which will result in the early failure of composites. As a result, the three topics of stress due to residual loads with voids and excess resin defects in materials are explored in depth in this section's research of resin matrix. The common challenging factors which arise during processing of composite structures are represented in Fig. 1.

Fig. 1 Common challenges from the constituents of composites

Numerous studies have documented the most significant and frequent flaws that occur when processing metal matrix and polymer matrix composites with a variety of natural and mineral fillers and both synthetic and natural fibre. Consolidated report of the same has been highlighted in Table 1,

Table 1 Challenges faced during the processing of FRPs with various filler combination

Type of reinforcement	Type of production method	Fillers used	Challenges faced
Fibre reinforced-woven E-glass and chopped fibres	Hand-lay-up method	Casio3, waste rubber, E-waste glass, mineral filler	Voids and bubble formation occurs if layup is not done properly [1–4]
Particulate reinforced-micro and nano particulate with Al-alloys	Stir casting method	Red mud particles, graphite particles, Sic	Requirement of reinforcement particles counts more than calculated due to vaporization during the preparation [5–9]
Fibre reinforced-fibre orientations fibres, fibre mat	Hand lay-up method	Mineral fillers-waste and scrap glass	Non-uniform dispersion of fillers in matrix leads to agglomeration due to fine grains [9–14]
Fibre reinforced-jute/silk/ CSM fibre	Hand lay-up method	Fish scale, Cocopeat powder, waste rubber	Improper binding of fillers with binder due to random sizing of fillers which overcome by sizing them to maintain uniformity in blending with resin [15]

2.2 *Induced Residual Stress*

Resin undergoes physical and chemical changes throughout the solidification or crystallisation process, which results in repairing deformity and residual strains in adhesively bonded composites. Dry-out distortion refers to a degree of discrepancy between the structure's final shape and the absolute geometry specified in the architecture. The engineering application of composites may suffer greatly from these dimensional variances [16, 17]. Induced distortion that remain in the composites for post dry-out are referred to as curing residual stresses. Their presence will impact the fibre and resin interface, decreasing the structural stability and durability of composites. Residual stresses and deformation brought on crystallisation, therefore significant basic issues for the production of these Light structures. The application through composites in engineering would be hampered by a lack of research.

2.2.1 For Thermoset Based Matrix

There are two types residual stresses at the macro- and microscales in thermosetting resin matrix composites [18]. Hygroscopicity, chemical resin shrinkage during polymerization, and an imbalance in the thermal expansion of the resin and the fibre are the main causes of micro-scale residual strains. Chemical shrinkage during polymerization, consolidation, variations in crystallinity or degree of cure throughout thickness, distribution of the volume fraction of the fibres, and component interaction, non-thermoelastic residual stresses also have an irreversible creation mechanism [19]. Recent research results have examined the issue of thermosetting resin curing in great detail. Three factors, to accurately mimic the resin curing process, one needs have a complete understanding of the reaction kinetics, heat conduction, and thermal behaviour.

2.2.2 For Thermoplastic Based Matrix

Different processes of residual stress production are brought about by the manufacturing of thermoplastic composites [20], for instance, shrinkage caused by crystallisation in semi-crystalline thermoplastics as contrasted to shrinkage caused by curing in thermoset matrices. This could lead to more demands being placed on the experimental techniques, such as their applicability for non-transparent matrix composites. In this case, the overall crystallinity is typically utilised to describe internal structure macroscopically.

2.3 *Existence of Voids*

As seen in Fig. 2, void flaws are a common occurrence in composite materials. The flow rate of resin, the temperature at which it cures, and the consolidation pressure are examples of manufacturing process variables that can be used to regulate void formation [21]. The next sections of this section cover the mechanisms by which voids develop in composite produced via advanced manufacturing methods.

2.3.1 Existing Voids for Specific Manufacturing Methods

The laying up and curing processes are when void flaws in advanced manufacturing methods are primarily created [22]. During impregnation [23] or laying up [24], air can become trapped results in flaws that cause entomb and in between laminar defects is one of the main causes of voids. Some parameters, including hot-pressing process of composites, surface roughness, laying up environment, laminate thickness, ply orientation, internal ply, drop-offs, and tooling all contributed to void formation [25]. Based on the results of thermogravimetry and mechanical spectrometry experiments, it is possible to determine the best curing pressure settings to reduce void content and the parameters in the curing cycle that affected the size, shape, and distribution of voids. The surface tension was taken into consideration in the resin voids in composites diffusion model, which was able to predict the resin's entrapped voids' collapse [26]. In autoclave moulding and hot pressing of composites, pressure plays a critical in preventing voids. Additionally, pressurized compression void formation will be impacted by curing factors including residual humid nature, and residual solvent content. To reduce the number of void defects, all of the aforementioned elements should be taken into account.

2.3.2 Challenges in RTM Method

Composites with intricate interior structures can frequently be produced using resin transfer moulding (RTM). Here preformed layer must be soaked with liquid resin.

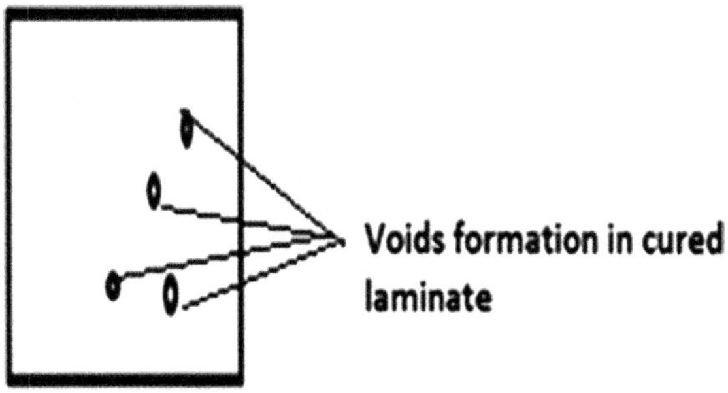

Fig. 2 Void formation in produced FRP laminate

During RTM, resin bubbles that have grown because to the non-uniform flow will deform, migrate, and combine. The dominating features of materials, which are directed by the resin and include resistance to axial compression, toughness and shear response are impacted by void defects. As a result, work on the causes of vacuum defects in matrix transfer mold is a scientific issue that must be resolved in the field of structural development. The void-content is determined by the air entrapment. The void defects can be categorised into three groups based on the shape and location of the stuck air. Dry macro voids develop if fibre bunch is not un-filled. Such gaps will be created by both unequal penetration and early resin gelation. Mesoscopic and microscopic voids are the two main categories of voids. The micro gaps are cylindrical while the mesoscopic spaces are spherical. Therefore, as matrix accelerates ahead and the pressure alters, the void defects that were created at matrix discharge will distort. Investigation of the void distribution in 3D braided composites led to the development of the probability density formula [27]. The meso-/micro-scale voids in RTM were statistically predicted [28, 29] established the correlation between the gaps and capillary number. Internal resin of the structure is cured, causing the internal components of composites to crystallise and changing the void pressure. According to the simulation's findings, a composite material's void flaws would diffuse more readily as temperature rose, and the dry-out flow would accelerate the model's load and cause void defects to shrink in size.

2.3.3 Challenges in Additive Manufacturing

The 3D printing of composites is gaining popularity because to its benefits, including adaptability and compatibility for creating complicated structures. Due to the layer-by-layer coating manufacturing process, there exists developing flaws in Additive Structures. Furthermore, the interlayer shear strength of composites produced through 3D printing will be directly impacted by the interfacial contact. The inner surface characteristics of additive structures made out of carbon with PLA were thoroughly examined with an emphasis on how thermal affect the adherence of beads [30]. When the nozzle was heated to between 200 and 230 °C, it was discovered to achieve the mixing of specific matrix with fibre bunches and also guarantee the binding between laminates. The performance of interlayer bonding in 3D printed composites was examined, and it was discovered that interlayer performance would deteriorate as interlayer thickness increased [31]. Researchers also discovered from the results of experiments that better strength for UD-composites and fewer inside flaws than other corresponding categories [32]. Theoretical frameworks and numerical simulation techniques have been offered by several academics for the analysis of 3D printing void flaws. The investigation also revealed that valve dimension is the primary feature influencing the dispersion of defects in Additively manufactured structures [33]. The discussion is meant for to know how voids occur in composite materials produced by advanced techniques. The interface and matrix of the composite will become weaker as a result of such flaws. Additionally, when the structure is experienced with outside load, the void defects will cause a concentration of stress, causing the surface to exceed the internal barrier to regulate earlier than expected and compromising the composite material's overall strength. Table 2 depicts Nature of damage incurred while processing of Composites with types of various Processing Tasks.

Table 2 Nature of damage incurred while processing of composites with types of various processing tasks

Type of defect	Nature of defect	Nature of effect
Residual stresses	Residual stresses in the matrix of thermosetting resin	The inner space availability for the adhesive reduces drastically as the random link of molecules are regenerated Microscale residual stresses will result in matrix cracks, but Robust scale cured transformation results as macroscale left out stresses because of its prime importance [33]
	Residual stresses in the matrix of thermoplastic resin	The clubbed molten structure of thermoplastic likely changes to an arranged neat non-amorphous segment as it cools from the molten state volatiles produced by resin as it cures
Void defects	Formation of void defects in hot pressing and autoclave	Void flaws can also result from moisture that has dissolved in the resin. The void development during curing, which asserted that thermal expansion and diffusion could cause void sizes to fluctuate. Moisture-based voids' dissolution and development during resin curing [34] At low curing pressure, a sizable no of big blows created in the single layer interfaces, and as the pressure was raised, the size and number of these voids considerably decreased [35]
	Formation of void defects in RTM	Due to the uneven resin flow during this procedure, air stuck with blows produced near the flow front [36] Multiple causes, including uneven dry-out can result in void defects in RTM [37] Mesoscopic voids will form between fibre bundles if the discharge rate of matrix is less than that of fibre bunches [38] On the other hand, micro-voids form b/w each segment in the bunch of fibres [39] The resin flow velocity and pressure will also be impacted by the creation of voids in composites [29]
	Formation of void defects in 3D printing	The interlayer bonding performance of composites produced by additive production method was examined, and it was discovered that interlayer performance would deteriorate as interlayer thickness increased. Void builds due to lean binding with matrix interface [31] Additionally, the void defects will cause a concentration of stress, results in compromising the composite material's overall strength [33]

2.4 Excess Resin Consumption

Resin-rich zones, which are a common occurrence in composites are defined as the regions where fibres are locally deficient [40, 41]. Localized fibre splitting and fibre bunches can result in excess matrix regions when composites are being made. Additionally, excess matrix flaws are challenging to regulate while structure moulding. The attributes of flaws that are excess matrix have been researched by several academics. RTM U-beam performance and production were investigated [42]. The reinforcement had a tendency to draw tightly around corners when the mould closed, removing these areas from the resin-rich area. In order to determine the compositional link in correlation with layer segments, the examination of excess matrix region of 0 unidirectional and cross-ply composites [43]. In braided or woven composites in particular, faults rich in resin are more common. There are micro-scale and minute scale resin-rich defects in these types of structures as per the reports [44, 45] on the distribution of excess matrix flaws in woven fiber-reinforced composites. When the fibre count binder was low, the resin-rich flaws were reduced. Mechanical carbon fiber/epoxy composites will be affected by the existence of resin-rich flaws, and it was discovered that these defects would lessen the strain at which a composite would fail. The demonstration of shearing and plasticity of composites were not significantly affected by micro-sized resin-rich flaws [46]. Large sized excess matrix flaws, would result in localized stress leads to early failure of a material [47] as a result, capability and operation cycle of structure will both be significantly impacted by excess matrix, the primary fault of composites. Resin rich defect in cured composite laminate is shown in Fig. 3.

3 Fibre Wrinkle and Local Mis-alignment Defects in Cured Composites

Considering the structural applications of composite member, the key element is fibre, which functions as a type of reinforcement. It serves as core member and hold up the load transfer structure. As a result, the scaffold of the entire composite material will become unstable due to manufacturing flaws in the fibre, which will

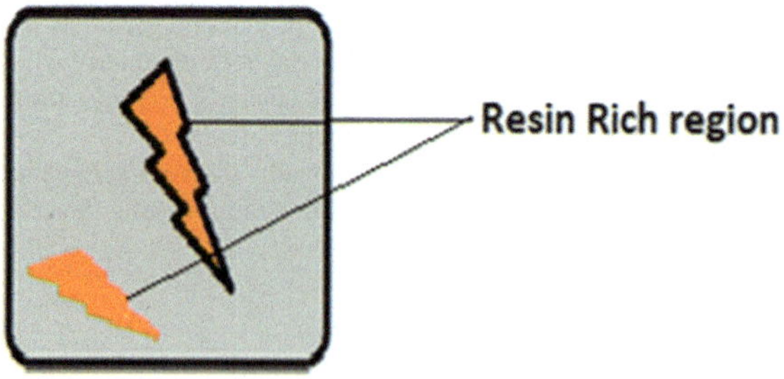

Fig. 3 Resin rich defect in composite laminates

Fig. 4 Wrinkles in fibre

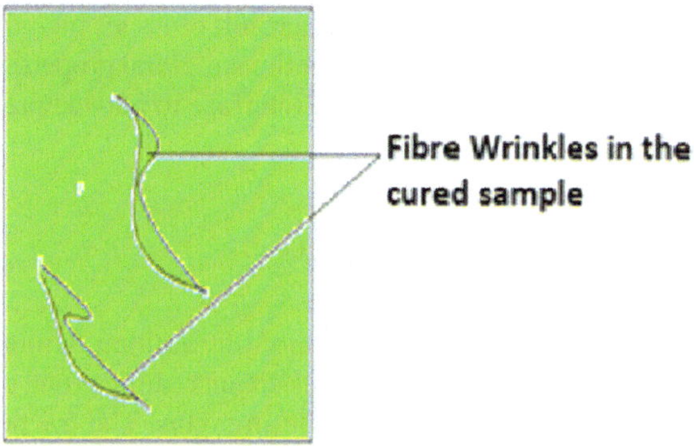

significantly diminish the material's strength, modulus, stability, and lifetime [48]. It is simple to manufacture fibre curling, which results in localized distortion in fibre alignments with the issues denoted as wrinkles in fibres, because the fibre shrinks when the binders cure during the moulding of material structures. The main cause of fibre wrinkle and waviness faults has been shown to be buckling of fibres, tows, and plies [49]. At various stages of production, the creation of fibre misalignment given the wide variety of manufacturing techniques and composite structures. For instance, in RTM, dry fabric deposition could result in fibre misalignment. Additionally, resin penetration during the injection stage may exacerbate this misalignment [50]. As a result, dry textile materials' drape and shear locking angles have an impact on fibre wrinkling [51]. The impact of the layers of materials with barrier inside results in the interlayer friction coefficient, which directly affects wrinkle formation [52], was considerably influenced by the interaction of numerous parameters, including the kind of thermoplastic modifier, fibre volume fraction, and moulding temperature. Fibre Wrinkles in the cured sample is represented in Fig. 4.

The influence of stacking sequence of layers during development of composite structure was revealed and demonstrated that co-stacking essential layers could boost layer buckling resistance and prevent wrinkle formation [53]. Generation of local misalignment of fibres has been shown to be influenced by various plies stacked internally within the system which supports the importance of manufacturing parameters includes several curing factors which aims to reduce production flaws majorly [54]. Due to their effect on the structural features and dimensions of the composite, fibre misalignment faults are particularly significant. This frequently causes assembly-stage disruptions, resulting in expensive surface shaping steps [55]. The mechanical properties of composite constructions are significantly harmed by wrinkle faults, which frequently develop in numerous neighbouring layers and cause the structure to break early [56]. These fibre misalignments present an additional risk and are challenging to identify [57], which has a direct influence on quality, safety, and economic efficiency. Interface flaws in composite materials. Different types of interfaces in composite structures, including those in between laminates and those in between fibre and matrix. Interfaces are a distinctive structural form of composites.

Interfacial flaws may exist as a result of issues including unequal wetting and cures. These flaws will weaken the entire structural properties of materials by exposing two components that make up the interface to outside pressures independently.

4 Delamination Defects

Delamination is one of the major life-limiting failure types in composite laminates. Advanced composite laminates are incredibly prone to delamination because of their poor inter-ply shear and tensile strengths. The fracture toughness, is the amount of external energy needed to cause a fracture and is connected to the onset of delamination damage in composite laminates. The structural properties of composites will be significantly impacted by delamination damage, leading to an early failure under operating conditions [58]. The phenomena of delamination may occur due to flaws production or the influence of outside elements throughout operational span of composites. The insufficient curing techniques create uneven pressure on the various locations, which results in delamination faults. Matrix cracking typically begins before delamination and grows as a result [59]. This failure is also influenced by the high interlaminar stresses, which are typically linked to the lowest through-thickness strength. This happens because the composite must rely on its relatively weak matrix to carry loads in that direction because the laminate plane's fibres cannot support the thickness [60]. The use of these materials could result in substandard functionality due to this form of failure. Inadequate curing methods cause irregular pressure in different spots, which causes delamination there. These delaminated patches have the potential to significantly reduce the compressive strength of composite materials. This happens as a result of the buckled laminated structure.

5 Other Challenges

The risk of making too many crucial design decisions during production before taking the manufacturing process into account. It gets harder to execute design changes without harming component performance or cost as design maturity rises. Reliance on geometrically constrained isotropic material-based digital design technologies that restrict the design optimization domain to material thickness and geometry. This method ignores the main benefits and drawbacks of various production processes, which has material/defects probability and shape limits because of continuous fibre engaged in the process. Briefly, the black metal design methodology. Engineers use the black metal design method for a variety of reasons. One benefit of the defined rules is that the design process is simple. The difficulty lies in the fact that determining a composite structure's strength is more difficult than determining the strength of a metal structure. This is due to the fact that the layered composites are constructed from a number of plies, each having unique spatial extents and angles. Designers currently use techniques and technologies that are either focused on metal manufacturing processes or inadequately take into account composite manufacturing, making them a deceptive solution. As a result, present tools either have an approach that is too general or too component-specific. Basically, the black metal design philosophy. Engineers use the black metal design method for a variety of reasons. One benefit of the defined rules is that the design process is simple. The difficulty lies in the fact that determining a composite structure's strength is more difficult than determining the strength of a metal structure. This is due to the fact that the layered composites are constructed from a number of plies, each having unique spatial extents and angles.

5.1 Design for Manufacturing; New Challenge

Two issues for composites result from the DFM method' need that manufacturing restrictions be known at the time of design generation. Creating for innovative manufacturing methods: if the production restrictions are not well understood and the design uses a novel manufacturing material or technique. Best practises when using digital tools: if the design was produced using digital tools intended for use with a different method or material. In the framework of DFM, a strategy for using digital design tools is necessary. Understanding the aforementioned difficulties gives the current issue a solid foundation. We will delve deeper into the most cutting-edge response to such difficulties in the next blogs.

6 Conclusion

By combining the different difficult aspects that cause manufacturing flaws in laminated, woven, braided, and additively made composites, this chapter provides a prognosis for composite manufacturing. The occurrence of residual stress, the development of voids in fibres reinforced with various categories of polymers, and their impact on cured samples are discussed as manufacturing flaws. The risk of making too many crucial design decisions during production before taking the manufacturing process into account. It gets harder to execute design changes without harming component performance or cost as design maturity rises. Reliance on geometrically constrained isotropic material-based digital design technologies that restrict the design optimization domain to material thickness and geometry. This method ignores the main benefits and drawbacks of various production processes, which has material/defects probability and shape limits because of continuous fibre engaged in the process. In the study that develops the alternative model to assess performance based on density compensation while comparing the modified fibres with neat reinforcements and also with defects associated with 3-D printing of composites, resin rich content in the manufacturing of composite structures also encountered. The main problems are the possibility of delamination, the extremely shortens the tool lifespans, and the extraction of emerging dust particles. The entire composites manufacturing process chain must be addressed to find a solution to these issues, starting with the machine concept, process parameters, component quality, and tools involved.

References

1. Lokesh KS (2018) Evaluation of toughness on varied thickness of chopped strand Mat/PU-foam sandwich structures. Int Res J Eng Technol (IRJET) 05(09). e-ISSN: 2395–0056, p-ISSN: 2395–0072
2. Pranav S, Pratibha, Kumar A, Chetan (2017) Comparative study on tensile behaviour of e-glass (woven & chopped) fibre reinforced with epoxy composites. Int J Eng Sci Res Technol 06(03)
3. Ummaji B, Gururaj P, Rayappa K, Yashavantha J (2018) Effect of red mud particles on scratch resistance of aluminum based metal matrix composites. Am J Aerosp Eng 5(1):24–29. https://doi.org/10.11648/j.ajae.20180501.14
4. Pinto T, Bharath Kumar S, Ramachandra CG (2016/5) Effect of E-waste filler on tensile and flexural behavior of E-glass/epoxy composites. Int J Eng Trends Technol 02(05)
5. Naveen Kumar JR, Prasad P, Hebbar N, Savitha MB, Navaneeth Gowda N (2021) A potential application of polyethyleneimine-reduced graphene oxide nanocomposite sensing film coated on interdigitated electrode prepared from copper-clad for carbon dioxide detection. Mater Res Innov 25:6, 363, 371. https://doi.org/10.1080/14328917.2020.1826674
6. Naveen Kumar JR, Prasad P, Hebbar N, Savitha MB, Navaneeth Gowda N (2021) A potential application of polyethyleneimine-reduced graphene oxide nanocomposite sensing film coated on interdigitated electrode prepared from copper-clad for carbon dioxide detection. Mater Res Innov 25:6, 363, 371. https://doi.org/10.1080/14328917.2020.1826674
7. Pai R, Bangarappa L (2022) Hybridization effect on water absorption and flexural properties of E-glass/banana fibre/epoxy composites. Mater Today Proc 52(Part 3):1841–1845. ISSN: 2214–7853. https://doi.org/10.1016/j.matpr.2021.11.491
8. Sriramamurthy LK, Hunasikatti S, Ramegowda NK, Kannantha V, Pai R (2019) Effect of E-waste rubber on mechanical behavior of glass a fiber reinforced with epoxy composites. AIP Conf Proc 2080:020003. https://doi.org/10.1063/1.5092886
9. Chethan IC, Kumar N (2018) Determination of compressive strength of graphene reinforced with aluminium-7075 metal matrix composites. Int J Mech Eng Technol 9(1):327–335
10. Naveen Kumar JR, Sampreeth U (2020) Experimental evaluation of substrate and annealing conditions on ZnO thin films prepared by sol-gel method. Mater Today Proc 24(Part 2):201–208. ISSN: 2214–7853. https://doi.org/10.1016/j.matpr.2020.04.268
11. Kumar N, Pai R, Hebbale AM (2021) Development and experimental investigation of mechanical properties of graphene-based aluminum 6061 alloys. Mater Today Proc 46(Part 7):2421–2424. ISSN: 2214–7853. https://doi.org/10.1016/j.matpr.2021.01.303
12. Lokesh KS, Shanmugam BK, Panduranga BP, Naveen Kumar JR, Hanumanthappa H (2021) Experimentation and prediction analysis on the mechanical performance of fish scale and coconut shell powder based composites. J Nat Fibers. https://doi.org/10.1080/15440478.2021.1958410
13. Pinto T, Ramachandra CG (2017) Effect of tool wear ; machinability studies on polymer composites; a review. Int J Eng Inform Syst 1(5):71–77 <hal-01571294>
14. Pinto T (2018) Comparative study on axial loading conditions and effect of mineral filler on CSM and WF fibres. Int Res J Eng Technol (IRJET). e-ISSN: 2395-0056, 05(10), p-ISSN: 2395-0072
15. Ummaji B, Gururaj P, Rayappa K, Yashavantha J (2018) Effect of red mud particles on scratch resistance of aluminum based metal matrix composites. Am J Aerosp Eng 5(1):24–29. https://doi.org/10.11648/j.ajae.20180501.14

16. Chetan IC (2017) Influence of graphene on hardness number of aluminium-7075 based metal matrix composites. Int J Innov Sci Res Technol 2
17. Li J, Ya X, Liu Y, Cen Z, Kou Z, Dai D (2009) A study of the integrated composite material structures under different fabrication processing. Compos A Appl Sci Manuf 40:455–462
18. Baran I, Cinar K, Ersoy N, Akkerman R, Hattel JH (2017) A review on the mechanical modeling of composite manufacturing processes. Arch Computat Methods Eng 24:365–395
19. Wisnom MR, Gigliotti M, Ersoy N, Campbell M, Potter KD (2006) Mechanisms generating residual stresses and distortion during manufacture of polymer-matrix composite structures. Compos Part A 37:522–529
20. Twigg G, Poursartip A, Fernlund G (2004) Tool–part interaction in composites processing. Part I: experimental investigation and analytical model. Composites Part A Appl Sci Manuf 35(1):121–133. ISSN: 1359–835X. https://doi.org/10.1016/S1359-835X(03)00131-3
21. Hamidi YK, Aktas L, Altan MC (2004) Formation of microscopic voids in resin transfer molded composites. Trans ASME 126:420–426
22. Mehdikhani M, Gorbatikh L, Verpoest I, Lomov SV (2019) Voids in fiber-reinforced polymer composites: a review on their formation, characteristics, and effects on mechanical performance. J Compos Mater 53(12):1579–1669
23. Judd N, Wright W (1978) Voids and their effects on mechanical-properties of composites-appraisal. SAMPE J 14:10–14
24. Huang H, Talreja R (2005) Effects of void geometry on elastic properties of unidirectional fiber reinforced composites. Compos Sci Technol 65:1964–1981
25. Campbell FC, Mallow AR, Browning CE (1995) Porosity in carbon-fiber composites—an overview of causes. J Adv Mater 26(4):18–33
26. Loos AC, Springer GS (1983) Curing of epoxy matrix composites. J Compos Mater 17:135–169
27. Gao X, Yuan L, Fu Y, Yao X, Yang H (2020) Prediction of mechanical properties on 3D braided composites with void defects. Compos B 197:108164
28. Schell JSU, Deleglise M, Binetruy C, Krawczak P, Ermanni P (2007) Numerical prediction and experimental characterization of meso-scale-voids in liquid composite moulding. Compos A Appl Sci Manuf 38:2460–2470
29. Tan H, Pillai KM (2012) Multiscale modeling of unsaturated flow in dual-scale fiber preforms of liquid composite molding I: isothermal flows. Compos A Appl Sci Manuf 43:1–13
30. Tian X, Liu T, Yang C, Wang Q, Li D (2016) Interface and performance of 3D printed continuous carbon fiber reinforced PLA composites. Compos A Appl Sci Manuf 88:198–205
31. Caminero MA, Chacon JM, García-Moreno I, Reverte JM (2018) Interlaminar bonding performance of 3D printed continuous fibre reinforced thermoplastic composites using fused deposition modelling. Polym Test 68:415–423
32. Justo J, Tavara L, García-Guzmán L, París F (2018) Characterization of 3D printed long fibre reinforced composites. Compos Struct 185:537–548
33. Fu Y, Kan Y, Fan X, Xuan S, Yao X (2022) Novel designable strategy and multi-scale analysis of 3D printed thermoplastic fabric composites. Compos Sci Technol 222:109388
34. Kardos JL, Dudukovic MP, Dave R (1986) Void growth and resin transport during processing of thermosetting-matrix composites. Adv Polym Sci 80:102–123
35. Li SJ, Zhan LH, Chenl R, Peng WF, Zhang YA, Zhou YQ, Zeng LR (2014) The influence of cure pressure on microstructure, temperature field and mechanical properties of advanced polymer-matrix composite laminates. Fibers Polym 15(11):2404–2409
36. Liu L, Zhang B-M, Wang D-F, Wu Z-J (2006) Effects of cure cycles on void content and mechanical properties of composite laminates. Compos Struct 73:303–309
37. Park CH, Lee WI (2011) Modeling void formation and unsaturated flow in liquid composite molding processes: a survey and review. J Reinf Plast Compos 30:957–977
38. Lundstrom TS, Frishfelds V, Jakovics A (2010) Bubble formation and motion in noncrimp fabrics with perturbed bundle geometry. Compos A Appl Sci Manuf 41:83–92
39. Lawrence JM, Neacsu V, Advani SG (2009) Modeling the impact of capillary pressure and air entrapment on fiber tow saturation during resin infusion in LCM. Compos A Appl Sci Manuf 40:1053–1064

40. Pai KR, Naveen Kumar JR, Hebbale AM (2021) Experimental study on preparation and mechanical characteristics of jute/silk/coco-peat reinforced with epoxy polymers. Mater Today Proc 46(Part 7):2764–2769. ISSN: 2214–7853. https://doi.org/10.1016/j.matpr.2021.02.511
41. Li X, Shonkwiler S, McMains S (2021) Detection of resin-rich areas for statistical analysis of fiber-reinforced polymer composites. Compos B 225:109252
42. Holmberg JA, Berglund LA (1997) Manufacturing and performance of RTM U-beams. Compos A 513–521:28A
43. Dong C (2011) Model development for the formation of resin-rich zones in composites processing. Composites A 42(2011):419–424
44. Idrees M, Ibrahim AMH, Tekerek E, Kontsos A, Palmese GR, Alvarez NJ (2021) The effect of resin-rich layers on mechanical properties of 3D printed woven fiberreinforced composites. Composites A 144 (2021):106339
45. Mahmood AS, Summerscales J, James MN (2022) Resin-rich volumes (RRV) and the performance of fibre-reinforced composites: a review. J Compos Sci 6:53
46. Ahmadian H, Yang M, Soghrati S (2020) Effect of resin-rich zones on the failure response of carbon fiber reinforced polymers. Int J Solids Struct (188–189):74–87
47. Al-Shawk A, Tanabi H, Sabuncuoglu B (2018) Investigation of stress distributions in the resin rich region and failure behavior in glass fiber composites with microvascular channels under tensile loading. Compos Struct 192:101–114
48. Nartey M, Zhang T, Gong B, Wang J, Peng S, Wang H, Peng H (2020) Understanding the impact of fibre wrinkle architectures on composite laminates through tailored gaps and overlaps. Compos B 196:108097
49. Farnand K, Zobeiry N, Poursartip A, Fernlund G (2017) Micro-level mechanisms of fiber waviness and wrinkling during hot drape forming of unidirectional prepreg composites. Composites A 103:168–177
50. Jochum C., Grandidier JC, Smaali M (2008) Proposal for a long-fibre microbuckling scenario during the cure of a thermosetting matrix. Composites A 39:19–28
51. Boisse P, Hamila N, Vidal-Sall´e E, Dumont F (2011) Simulation of wrinkling during textile composite reinforcement forming. Influence of tensile, in-plane shear and bending stiffnesses. Compos Sci Technol 71:683–692
52. Larberg YR, Åkermo M (2011) On the interply friction of different generations of carbon/epoxy prepreg systems. Composites A 42:1067–1074
53. Hallander P, Sjolander J, Åkermo M (2015) Forming induced wrinkling of composite laminates with mixed ply material properties; an experimental study. Composites A 78:234–245
54. Erland S, Dodwell TJ, Butler R (2015) Characterisation of inter-ply shear in uncured carbon fibre prepreg. Composites A 77:210–218
55. Piggott MR (1995) The effect of fibre waviness on the mechanical properties of unidirectional fibre composites: a review. Compos Sci Technol 53:201–205
56. Wilhelmsson D, Gutkin R, Edgren F, Asp LE (2018) An experimental study of fibre waviness and its effects on compressive properties of unidirectional NCF composites. Compos A 107:665–674
57. Suriani MJ, Rapi HZ, Ilyas RA, Petru M, Sapuan SM (2021) Delamination and manufacturing defects in natural fiber-reinforced hybrid composite: a review. Polymers 13:1323
58. Ravi TP (2017) Evaluation of mechanical properties and wear characterization of polymer composites under varying temperature conditions: a review. Int J Eng Inform Syst (IJEAIS) 1(4):64–68. https://doi.org/10.5281/zenodo.821168
59. Carraro PA, Novello E, Quaresimin M, Zappalorto M (2017) Delamination onset in symmetric cross-ply laminates under static loads: theory, numerics and experiments. Compos Struct 176:420–432
60. Wisnom MR (2012) The role of delamination in failure of fibre-reinforced composites. Philos Trans R Soc A Math Phys Eng Sci 370:1850–1870. https://doi.org/10.1098/rsta.2011.0441

SPRINGER NATURE

GPSR Compliance

The European Union's (EU) General Product Safety Regulation (GPSR) is a set of rules that requires consumer products to be safe and our obligations to ensure this.

If you have any concerns about our products, you can contact us on ProductSafety@springernature.com

In case Publisher is established outside the EU, the EU authorized representative is:

Springer Nature Customer Service Center GmbH
Europaplatz 3
69115 Heidelberg, Germany

The manufacturer's authorised representative in the EU is Springer Nature Customer Service Centre GmbH, Europaplatz 3, 69115 Heidelberg, Germany. If you have any concerns regarding our products, please contact ProductSafety@springernature.com

Printed and bound by CPI Group (UK) Ltd, Croydon, CR0 4YY
01/10/2026
02225900-0009